Quantum Dots
Properties and Applications

Edited by

Inamuddin[1], Tauseef Ahmad Rangreez[2], Mohammad Faraz Ahmer[3] and Rajender Boddula[4]

[1]Department of Applied Chemistry, Zakir Husain College of Engineering and Technology, Faculty of Engineering and Technology, Aligarh Muslim University, Aligarh-202002, India

[2]Department of Chemistry, National Institute of Technology, Srinagar, Jammu and Kashmir 190006, India

[3]Department of Electrical Engineering, Mewat College of Engineering and Technology, Mewat-122103, India

[4]CAS Key Laboratory of Nanosystem and Hierarchical Fabrication, National Center for Nanoscience and Technology, Beijing 100190, PR China

Published by **Materials Research Forum LLC**
Millersville, PA 17551, USA

Published as part of the book series
Materials Research Foundations
Volume 96 (2021)
ISSN 2471-8890 (Print)
ISSN 2471-8904 (Online)

Print ISBN 978-1-64490-124-3
eBook ISBN 978-1-64490-125-0

This book contains information obtained from authentic and highly regarded sources. Reasonable efforts have been made to publish reliable data and information, but the author and publisher cannot assume responsibility for the validity of all materials or the consequences of their use. The authors and publishers have attempted to trace the copyright holders of all material reproduced in this publication and apologize to copyright holders if permission to publish in this form has not been obtained. If any copyright material has not been acknowledged please write and let us know so we may rectify this in any future reprints.

Distributed worldwide by

Materials Research Forum LLC
105 Springdale Lane
Millersville, PA 17551
USA
https://www.mrforum.com

Manufactured in the United States of America
10 9 8 7 6 5 4 3 2 1

Table of Contents

Preface

Quantum dots are man-made semiconductor nanocrystals with diameters in the range of 2-10 nm. Since the discovery of cadmium-based quantum dots in the 1980s, a variety of quantum dots have been explored. Due to their small size, quantum dots have rendered quantum confinement effects, along with optical and electrical properties intermediate between those of bulk semiconductors and discrete molecules. Quantum dots are new generation nanomaterials with multifunctional applications in various attractive fields such as photovoltaics, light-emitting diodes, field-effect transistors, lasers, photodetectors, solar cells, biomedical diagnostics, quantum computing, among many others. The unique quantum dots properties and applications have triggered researchers and industries to consider using them in various disciplines. This book provides a survey of current research in quantum dots synthesis, properties, and applications. Chapters on fundamentals; optical; chemical; electronic and physical properties; various synthetic methodologies and theoretical insights along with their versatile applications are included. This book is an outstanding reference guide for professionals learning/working in nanotechnology, biomedicine, chemistry, physics, and electronics.

Key features:

1. Detailed overview of quantum dots and their applications
2. Explorations of the most recent advances and developments
3. Cutting-edge nanotechnologies for quantum dots
4. Focuses on the viewpoints of widespread presentations of quantum dots

Summary

Chapter 1 describes the most recent green techniques that have been used to synthesis quantum dots, including biogenic methods (e.g. plant-mediated, microorganisms-mediated methods), wet-chemical, and solid-state methods. From the point of view of precursors and by-products, green methods define as environment-friendly biocompatible techniques with economical approaches. Ultimately, the selected methods were compared.

Chapter 2 elaborates an important and detailed discussion of various conventional techniques used for the fabrication of quantum dots. Both the conventional top-down and bottom-up approaches for QDs synthesis have been briefly explained. This chapter highlights the recent advances so that the current development in this field can be grasped.

Chapter 3 examines the synthesis of carbon dots from dried mint leaves without using any chemicals and the characterization of investigated carbon dots. Additionally, metal ions sensing results revealed high sensitivity of mint derived carbon dots to Ag ions.

Chapter 4 discusses the recent research progress of graphene quantum dots, focusing on their synthesis, typical physicochemical properties, and antibacterial actions. The future development of effective graphene quantum dots is also addressed to highlight current limitations and motivate more research on this promising area.

Chapter 5 provides a comprehensive analysis of molecular and quantum mechanics computational approaches used to study the properties of quantum dots.

Chapter 6 details the application of quantum dots in sensors. It classifies the quantum dots based sensors into chemo- and bio-sensors; highlights the outstanding quantum dots properties explored for sensing, their sensing signals, and amplification strategies.

Chapter 7 discusses the five representative types of quantum dots and their role as an excellent interface to stimulate an enhanced interaction between the electrode and electrolyte resulting in superior charge storage properties of the supercapacitors.

Chapter 8 highlights the application of quantum dots (QDs) in the biomedical field especially in drug delivery and summarizes the current research interests of fine engineered QDs for biomedical applications. The potentially toxic effects of QDs have also been described for the development of QDs formulations for further studies.

Chapter 9 discusses the current advancements in quantum dots, including a recent overview of various methods to synthesize quantum dots and their futuristic possibilities. This chapter also elaborates the quantum dots based hybrid composites and carbonic materials for advanced supercapacitors applications.

Chapter 10 describes an outline of the work related to the applications of QDs in various separation techniques viz., quantum dots (QDs) in separation membranes and their utilization in various fields like heavy metal remediation, magnetic quantum dots (MagDots) for cellular/molecular separation and chromatographic separation column has been provided.

Chapter 11 summarizes the quantum dots, their classifications, and properties. Various synthesis approaches have been discussed with their pros and cons. Also, attempts have been made to discuss the application of carbon-based quantum dots in the fields of water and wastewater treatment.

Chapter 12 discusses the preparation, characterization, properties, and applications of semiconductor quantum dots (QDs). Because of their very small size and special electronic properties, QDs are expected to be building blocks of many electronic and optoelectronic devices. These particles possess tunable quantum efficiency, continuous absorption spectra, narrow emission, and long-term photostability.

Chapter 13 addresses the optical properties that characterize quantum dots, the manufacturing methods, and some types of these dots, also, the mechanism of luminescence of quantum carbon dots. Finally, we shed light on some chemical and biological applications of quantum nanoparticles.

Quantum Dots – Properties and Applications   Materials Research Forum LLC
Materials Research Foundations **96** (2021) 1-52   https://doi.org/10.21741/9781644901250-1

Chapter 1

Eco-Friendly Techniques to Synthesize Quantum Dots

Zeinab Fereshteh[1,2*]

[1]Department of Biomedical Engineering, University of Delaware, Newark, DE, USA

[2]Memorial Sloan Kettering Cancer Center, New York, New York, USA

[*] fereshte@udel.edu

Abstract

Quantum dot defines as a nanoparticle with particle size smaller than its exciton Bohr radius. Due to the remarkable quantum effects such as optical and electronic properties, they have attracted a great deal of attention by researchers and industries. Therefore, quantum dots have become a major topic in nano-technology. Here, we describe the most recent eco-friendly techniques that have been used to synthesize quantum dots, including biogenic methods, such as plant-mediated, microorganisms-mediated methods, wet chemical and solid-state methods.

Keywords

Quantum Dots, Synthesis Method, Environment-Friendly, Green Method, Biogenic Methods, Wet Chemical Methods

Contents

1. Introduction

Quantum dots (QDs) are defined as nanoparticles with particle size smaller than its exciton Bohr radius which is approximately <20 nm [1-3]. Due to very small particle size of QDs, they show significant quantum effects and their optical and electronic properties are followed by quantum mechanics rules instead of classical physics [4, 5]. Therefore, they have exceptional properties that cannot be found in larger particles e.g. high quantum efficiency, tunable photoluminescence, broad luminescence excitation spectra, broad absorption, narrow symmetrical emission spectra with large Stokes shifts, and excellent resistance toward photo-bleaching. As a result, they are considered promising candidates with a variety of different applications in displays, solar cells, light emitting diodes (LED), photovoltaic cells, photo-catalysis, sensors, fluorescence probes, and bio-imaging[3, 6, 7]. Due to these unique and remarkable features, quantum dots are a major topic in nano-technology.

On the other hand, the properties of QDs can be manipulated by varying compositions and particle size. Also, the particularity of QDs directly depends on their morphology, particle size and size distribution [3, 5] so that the synthesis methods of QDs have been considered by many researchers. In order to produce QDs with reliable and consistent properties, it is necessary to develop a technique to synthesize a reproducible product with small particle size and narrow size distribution of nanoparticles [1, 3, 5, 7]. On the other hand, the potential of using QDs in various applications increases the need to synthesis them on an industrial scale with environmentally friendly and green processes.

There are lots of proven advantages of the green methods such a being environmentally friendly, cost-effective, and easy to scale up in many cases, the lack of complex chemicals as well as toxic contaminants, and using low-priced precursors [4, 5, 7].

In this chapter, the most recent green techniques that have been used to synthesis quantum dots, including biogenic methods, such as plant-mediated, microorganisms-mediated methods, wet chemical and solid-state methods are described. To choose a green method, it must be environment-friendly, biocompatible, and safe, from the point of view of precursors and by-products. Also, from the perspective of scalability and industrial-friendly, they should have a cost-effective and economical methodology. As a final point, a brief discussion is provided on the comparison between selected methods.

2.   Green synthesis of quantum dots via biogenic methods

There are numerous types of phototrophic eukaryotes, such as plants, microbes, algae, fungi, and prokaryotes, e.g. bacteria, which have been used as efficient and eco-friendly green nano-bio factories to synthesize quantum dots. Due to being highly stabilized and capability of reducing the metal ions to metal /metal oxide nanoparticles, many researchers have introduced various synthesis methods to produce quantum dots. The biogenic techniques have a facile, economical, and environment friendly approach. In the following sections, each method is explained in detail.

2.1   Green synthesis of quantum dots by plant

Synthesizing nanoparticles by using different parts of plants has started since ~ 2 decades ago [8, 9] which has been taken a proper attention from researchers all over the world by now. So that, nowadays, producing QDs by employing plants or a part of them such as leaves, roots, fruits, flowers, or their extracts has become a distinguished method [10, 11]. Using plants to synthesis QDs there is no need for high pressure or high temperature, therefore this method is energy efficient and feasible to customize for many kinds of research or industrial laboratories [11-13].

The plant extracts include many chemical components such as phytochemicals, proteins, enzymes, etc. [11, 14]. Redox mediators such as polyphenols, sugars, and ascorbic acids are the important ingredients for synthesizing nanoparticles which acts as an oxidizing or reducing agent. Along with facilitating the formation of QDs, using plant extract can stabilize them and play as surfactants to prevent them from oxidation. And as a result nanoparticles synthesized by green methods can be stored for a longer time. On the other hand, their morphology and particles size can be controlled by plant's capping agents [11, 15, 16]. Compared to other green methods, this method needs a shorter time which depends on the quality and quantities of phytochemicals in the plants. Therefore, using a combination of plant species is suggested [10, 16, 17].

In general, the bio-reduction occurs between metal ions and phytochemicals of plant extracts in aqueous solution. In order to produce metal solution, nitrates, sulfates, chlorates, or any aqueous solvable metal salt can be used. In this way, the metal ions spontaneously react with different extract biomolecules. Afterwards, they reduce to metal or metal oxide/sulfide nanoparticles. Fig.1 [11] briefly shows a proposed mechanism of NPs synthesis by using plant extracts. In this reaction few non-hazardous byproducts are produced that can be solved in water and washed subsequently. Due to the fact that synthesizing nanoparticles by using plant extract is a single, low temperature reaction that occurs in aqueous solution, this approach is known as a cost-effective and environment-friendly method. In addition, in many cases, there is no requirement to have any surfactant or capping agent to stabilize the nanoparticles produced by this technique. As mentioned above, it is possible to synthesis metal or metal oxide/ sulfide NPs by this method. Gold, silver, copper, and platinum as metal NPs and zinc oxide, iron oxides, and titanium oxide, cadmium sulfide, silver sulfide, and copper sulfide as oxide/sulfide QDs are the top nanoparticles that produced by this method.

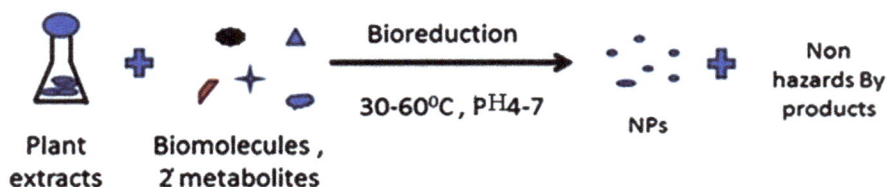

Fig. 1. Proposed mechanism of nanoparticle synthesis by applying plant extracts. (Reproduced with permission from Ref. [11]).

Table 1 [12, 18-78] represents a list of plant extracts used for synthesizing different quantum dots. However, the majority of the bioactive ingredients accumulate in the leaves, there are also several studies where other parts of plant have been used, for instance seeds, flowers, fruits, stems, peels, etc. It is possible to use fresh or dried plant for preparing the plant extract. In case of using dried plant, after washing thoroughly, they must be dried in a dark at very low temperature, e.g.in an oven set at 40-60 °C or lyophilize them. Then, the dried or fresh plant should be smashed, and distilled water added, ~5-10 wt.%. Also, ethanol, acetone, or liquid CO_2 can be used as a solvent. The mixture is heated/boiled for 10-30 minutes, followed by filtration through filter paper,

Whatmann No. 1, and stored at 4 °C for further experiments. The longer incubation time can extract higher amount of biochemical molecules. Because the ingredients of the extract would vary by changing the plant weight percentage, temperature, or time of soaking in the solvent, all steps should be kept consistent. Otherwise, the results would not be reproducible. In the next stage, the metal solution is added to the plant extract in different ratios. The mixtures are kept at various temperatures at multiple periods. In several studies, using a reductant or playing with pH helped to produce QDs more rapidly and monodispersed. Following incubation, precipitate was collected by centrifuge.

Table 1 Plant mediated synthesis of quantum dots and some practical information about synthesized[12, 18-78]

Plant	Metal/Metal Oxide	Parts of plant	Application	Particle size (nm)	Shape	Ref.
Cistus incanus	Au	Dried and powdered leaf	Multiphoton-excited luminescence properties	40-500	Popcorn-shaped nanostars	[18]
Silybin	Au	-	Antimicrobial chemotherapy	58	S	[19]
Sansevieriaroxburghiana	Au	Leaf extract	Organic pollutants	5-400	R	[20]
Simaroubagluca	Au	Leaf extract	Anti-microbial properties	<10	R	[21]
Honey	Au	-	In vitro cytotoxicity studies in mice	9-18	R	[22]
honey assisted combustion method	Ag substituted CoFe$_2$O$_4$	-	Magnetic and antibacterial properties	24–41	R	[23]
Ficusretusa	Ag	Leaf	Sensor	10-70	S	[24]
Halymeniadilatata	Hd-AuNPs Composite	Whole plant	Antioxidant, anti-cancer and antibacterial activities	16	S	[25]
Solanumnigrum	Dispersed porous carbon (G5/c and G10/C) composites Composite	Leaf	Electrochemical energy storage	13-30	S	[26]
Carica papaya	Au	Seed	Drug delivery	16	S	[27]
Camellia sinensis (black and green tea)	Ag	Leaf	Antimicrobial	10	S	[28]
puerarialobata	Au	Whole plant		18	S	[29]

Solanumnigrum Ricinuscommunis Morusnigra	CMC/Au nanocomposites	Leaf	CMC/Au nanocomposites to effect on the optical, thermal and electrical properties	15 8 38	S	[30]
Palm oil	Au	Leaf	Pharmaceutical application	92-112	R	[31]
Lonicera japonica	Au	Flower	Anticancer activity	10-40	S	[32]
Coleus aromaticus	Au	Leaf	Antibacterial activity UV blocking property anticancer activity	75	S, rod, and Tri	[33]
Curcumin, Turmeric, Quercetin Paclitaxel	Au	Root and fruit	Anti-cancer agents chemotherapy drug	5–25, 3–20, 15–60, 15–45	S Tri Pyramid	[34]
Anacardiumoccidentale	Au	Leaf	Antimicrobial and anticancer properties	40	S	[35]
Camellia japonica L.	Au	Leaf	Antimicrobial activity	20	S	[36]
Starch	Starch-stabilized gold nanoparticles (ss-AuNPs)		Sensor	22	S	[37]
Alcearosea	Au	Leaf	Catalysis	4-95	S	[38]
eucalyptus globulus honey ginger	Au	Root	Catalysis	3.3-1.9	S	[39]
B. ciliate, B. stracheyi, R. dantatus, and R. hastatus	Ag	Root	Antimicrobial properties	25-73	S	[40]
Selaginellabryopteris	Ag	Leaf	Antimicrobial properties	5-10	S	[41]
Hibiscus sabdariffa	Au	Leaf	Electrooxidation of nitrite	7	S	[42]
Withaniacoagulans	Ag		Antibacterial, Antioxidant and cytotoxic	14	S	[43]
Morindacitrifolia (noni)	Ag and Au	Bark	Sensor and Iin vitro anticancer studies	20 25		[44]
Cissusquadrangularis	Au/C, Ag/C, Pd/C	Stem	Antibacterial activity/ Electrochemical properties	11 11 20	S	[45]

Materials Research Forum LLC
https://doi.org/10.21741/9781644901250-1

Eryngiumcampestre	Cu/Cr/Ni	Leaf	eEvironmental remediation	<100	R	[46]
thirty chinses plants	Ag	Leaf Flower Fruit Seed Root Stem Aerial part	Wound healing activity	8-35	Mostly S	[47]
Ocimum sanctum (basil)	Ag	Root Stem	-	5-10	S	[48]
xerophyte (Bryophyllum sp.),Mesophyte (Cyprus sp.), Hydrophyte (Hydrilla sp.)	Ag	Whole	-	2-5	S	[12]
Jacaranda mimosifolia	ZnO	Flower	Antibacterial activity	2-4	S	[49]
Trifoliumpratense	ZnO	Flower	Antibacterial activity	100-190	R	[50]
Nepheliumlappaceum L	ZnO	Peel	Antibacterial applications	50	Needle like	[51]
Sedum alfredii	ZnO	Shoot	Photodegradation properties	15.8	Columnar	[52]
Carissa edulis	ZnO	Fruit	Photocatalytic property	50-55	Flower shaped	[53]
MoringaOliefera	ZnO	Leaf	Electrochemical property	13–28 and 32–61	Rod and longitudinal shape	[54]
AspalathusLinearis	NiO, Pd/PdO	Whole	Electrochemical activity	32, 23	quasi S, S and cubic	[55]
Taraxacumofficinale	Cobalt oxide	Leaf	Catalytic reduction	50-100	S	[56]
Ruelliatuberosa	CuO	Leaf	Antimicrobial and photocatalytic activities	83	Rod	[57]
Sugarcane	CuO	Juice	Antibacterial activity	23-30	Irregular	[58]
Cissusarnotiana	Cu	-	Antibacterial and antioxidant potential	60-90	S	[59]
Phyllanthusacidus	Ag	Leaf and twig	Biocatalytic activity	48	S	[60]
Cacao	Ag	Bean	-	11-15	S	[61]
Grape	Ag	Seed	-	13-33	S	[61]
Lyciumchinense	Ag and Au	Fruit	Anticancer activity	50-200 and 20-100	mostly S	[62]

Ruelliatuberosa	FeO	Leaf	**Antimicrobial properties and photocatalytic degradation**	**53**	Rod	**[63]**
Ficuscarica	FeO	Dried fruit	-	9	S	[64]
Moringaoleifera	Fe	Leaf and seed	Removal of nitrate from water and antibacterial activity	3.4-7.4	S	[65]
Punicagranatum	ZnO	Fruit peels	Cytotoxicity and antibacterial activities	33	Hex and S	[66]
Ricinuscommunis	ZnO	Seed	Tioxidant, antifungal and anticancer activity	12-14	S	[67]
Azadirachtaindica	Pt	Neem leaf	-	5-50	S	[68]
Swertiachirayaita	MgO	Whole	Antibacterial activity	<20	S	[69]
Costuspictus	MgO	Leaf	Antimicrobial and anticancer activity	30-70	Hex	[70]
Omani mango	Iron oxide	Leaf	Heavy oil viscosity treatment	3×15	Rod	[71]
Gooseberry	Iron oxide	Leaf	Glucose biosensor	10	S	[72]
Lantana camara	Iron oxide	Leaf	Biological activities	10-20	Rod	[73]
Juglansregia	Iron oxide	Green husk	Magnetic properties	6-13	S	[74]
Juglansregia	Core-shell Fe_3O_4/Au	Green husk	Anticancer	6	S	[75]
Banana	Core-shell $Pt_3Co@Pt$	Peel	-	15	S	[76]
Banana	CdS	Peel	-	1.48	S	[77]
Lycopersiconesculentum (tomato), Citrus sinensis (orange), Citrus paradisi (grapefruit), Citrus aurantifolia (lemon)	ZnO	Peel	Photocatalytic activity	9.7	S	[78]

S: Spherical, Oct: Octahedral, R: Round, Hex: Hexagonal, Tri: Triangular

2.1.1 Possible formation mechanism

Preparation of QDs from a solution progresses in three steps, nucleation, growth and stabilization [42, 79, 80]. Nucleation step includes producing clusters of several atoms which normally occurs rapidly, e.g. less than 15 minutes. After nucleation, the growth phase begins which means aggregating several nanoclusters or developing a former cluster [42, 78, 79]. This step can take place in several to 24 hours [48, 81], or 72 hours [82]. If the aging stage continues, the particles become bigger and they cannot show their quantum effects. Therefore, for the second step timing is a vital key to have uniform and monodispersed QDs. For this reason, surfactants, stabilizers or capping agents have been applied to stop growing nanoparticles. Using plant extracts is a privilege to have all steps in one pot. Plant extract not only plays as a reductant, but also acts as a capping agent. Some molecules such as phytochemicals reduce metal ions, and some components such as polyphenols, due to having hydroxyl functional groups, are adsorbed on the surface and prevent them from agglomeration [79, 83].

In order to understand the potential mechanism of nanoparticle formation by plant extract, a basic knowledge of bio-molecular chemistry is necessary. Organic chemistry can help to explain the role of each biochemical molecule in the plant extract and how they can control the particle size, morphology, and crystallinity of nanoparticles and afterwards stabilize them.

It is worth mentioning that the composition of extract prepared from one part of a plant would vary from the other parts or even sometimes can vary with the region of the cultivate. Therefore, researchers have extracted physicochemicals from all parts of a plant, for example, roots, seeds, barks, stem, flowers, fruits, and leaves. However, leaf extract has been consumed in the majority of studies to synthesize nanoparticles by plant-mediation, Table 1. It might be because of the fact that physicochemicals primarily accumulate in leaves. Also the leaves are more abundant than other parts of a plant/tree. In addition, there are numerous varieties of physicochemicals only in one type of extract. Therefore, characterization of the extract assists to predict the best mechanism for each case, requiring lots of time and expenses. As a matter of fact, the studies which provide the actual mechanism behind synthesizing NPs are in a minority. Whereas many researches have been carried out to introduce a new plant extract to synthesize nanoparticles.

Developing a rational methodology is required for illumination of the factual biochemical pathways. Therefore, it is necessary to elaborate the fundamental molecular concepts of active biochemicals in the extract to understand and then, control NPs formation. Several hypothetical mechanisms have been suggested for plant-mediated nanoparticles synthesis. However, the biochemical pathway and mechanism is unknown and additional studies are needed to attain more details.

One of the most important compounds in plant extracts is phytochemicals which stands for chemicals of plant origin ("phyto" means "plant" in Greek) [84]. They are naturally bioactive and responsible for growth or defense against competitors, diseases, pathogens, or predators. Some of the phytochemicals have been used as traditional medicines in Persian, Egyptian, Chinese, and Japanese cultures which are well-known ancient cultures. Based on the chemical structure, phytochemicals are categorized into three main group, phenolic acids, flavonoids, and stilbenes/lignans [84-87]. Additional details can be found in Fig. 2 [85]. In addition, Fig. 3 [42, 54] shows the schematic structures of biological active compounds in Moringa Oleifera [54] and Hibiscus sabdariffa leaf extract [42] which are common in many plant extracts.

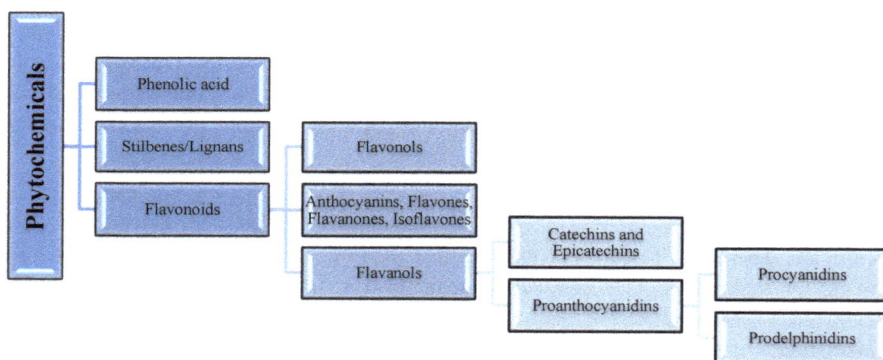

Fig. 2. Categories of phytochemicals.(Reproduced with permission from Ref.[85]).

Individual components in a plant can be served as reduction agents for nanoparticles synthesis, specifically polyhydroxyl groups of the secondary plant metabolites. They scavenge against free radicals and act as antioxidants, and finally perform a crucial role to reduce metallic ions into metal nanoparticles [88]. As mentioned above, the structures and compositions of phytochemicals in plant extracts are complicated and unclear in many cases. Besides, purification and isolation of the selected phyto-constituents are presently impossible. Therefore, an anticipation of the exact molecular pathway and the mechanism of plant-mediated NPs synthesis is challenging [89, 90].

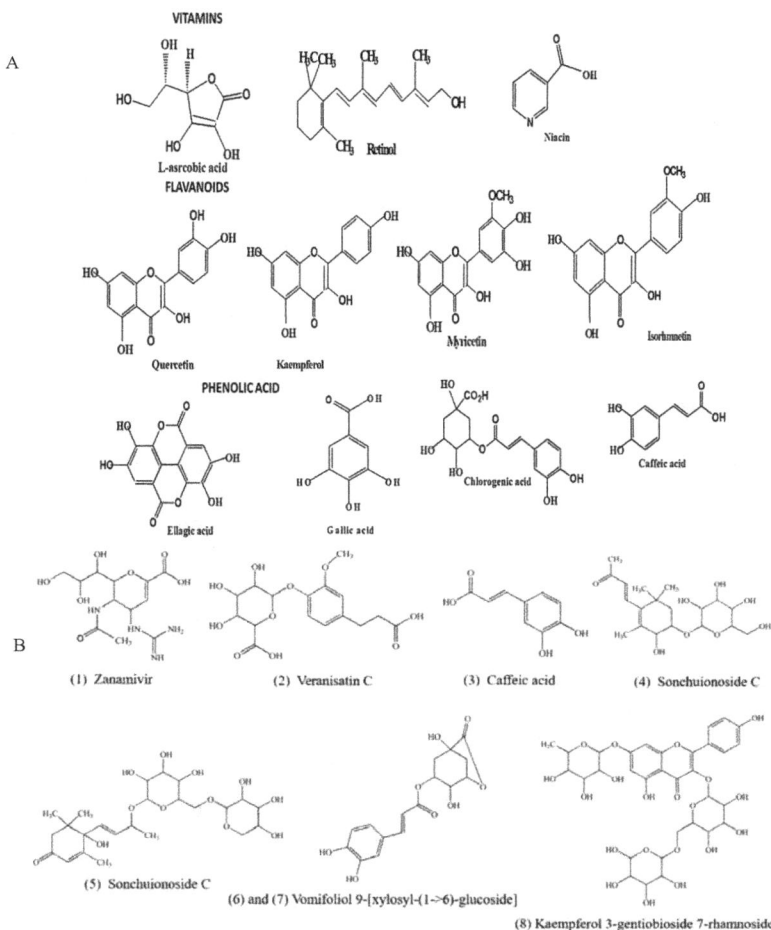

*Fig. 3. Schematic diagrams of biological active compounds in Moringa Oleifera.
(Reproduced with permission from Ref.[54]), and Hibiscus sabdariffa leaf extracts.
(Reproduced with permission from Ref.[42]).*

Despite these difficulties several mechanisms have been suggested during the past decade. According to Zhou et al. [29], Puerarialobata extract contains a number of flavonoids such as daidzin, puerarin, daidzein, and puerarin-7-xyloside which have multiple aldehyde groups, hydroxyl alcohols, phenolic hydroxyl groups. These functional groups have at least one oxygen atom with unshared a pair of electrons and negatively charged, and it can attack to the metal ions. As a results, Au^{+3}can be reduced to Au^o in an appropriate condition, 5.0 mL of puerarialobata extract (PLE), pH=11.2, 30 °C. Then, AuNPs are stabilized by covering organic components. It is confirmed via a blue-shift of the carbonyl bond absorption peak in FTIR which indicates oxidation of the hydroxyl group to carbonyl group and its adsorption on the Au NPs surface. Due to present of multiple hydroxyl groups and carbonyl groups, hydrogen bonds can be created between hydroxyl groups in puerarin molecule and the Au NP surface, which in turn leads to stabilize nanoparticles [29]. On the other hand, by measuring the amount of aldose and flavonoids in PLE before and after the reaction, investigators suggested aldoses were involved in a biosynthesis process, and flavonoid [29] and protein [76, 77] concentrations were approximately consistent after NPs fabrication. Nava et al. [78] claimed that the aromatic hydroxyl groups of flavonoids, limonoids, and carotenoids molecules, extracted from tomato, orange, grapefruit, and lemon peel, act as reducing agents. The nucleation process begins after the biomolecules ligate with metal ions which creates a reverse micelle. In this way, the morphology and particles size of NPs are controlled [78]. The finding is consistent with the findings of Mohd Taib et al. [42], where they synthesized Au NPs by Hibiscus sabdariffa leaf extract. They suggested that the orthodihydroxyl groups of the caffeic acid forms into a five membered chelate ring with Au^{3+} ions. Then, the dihydroxyl groups in the chelated ring are transformed to their oxidized forms, quinones, and which has two free-electrons to reduce gold ions. While the Au NPs are synthesized, they also become stable by quinones and polyphenolic molecules [42].

2.1.2 Effective parameters on plant-mediated nanocomposite synthesis

In order to produce monodispersed nanoparticles with desired properties, a reproducible protocol is required. To achieve that, it is necessary to investigate all factors that have an influence on the morphology or particle size of NPs synthesized by plant extracts. It is worth mentioning that studying all of the impact parameters is challenging, especially the plant extracts have a complex composite. However, there are several parameters that play crucial roles to control size and shape in the phytosynthesis of NPs, such as concentration of plant extract, pH, temperature, and incubation time.

Owing to having active ingredients e.g. reducing and capping agents, using different volume of plant extract or even the initial amount of dried/fresh plant can influence on the results. Increasing the percentage of Calotropisgigantea extract caused to form smaller Ni NPs with higher stability [91]. However, there is always an optimal ratio of metal ion to plant extract. The high amount of plant extract leads to polymerize some of active complexes, for instance polyphenols convert into high molecular compounds which in turn results in instability and agglomeration of the NPs [61]. Zhou et al. [34] suggested a three-step agglomeration mechanism of NPs in puerarialobata extract (PLE), Fig. 4 [29]. Step A accrues if the PLE< 3 mL, where the reaction is very slow and the particles grow, due to the lack of the puerarin and aldose groups in the solution. in the range of 5.00 to 9.00 mL of PLE; step B, the maximum steric hindrance and minimum particle size is achieved for the sake of enough amount of reduction and capping agents. However, the PLE volume < 9 mL facilitated self-agglomeration of the functional groups and it caused to sediment the NPs on the bottom of the flask [29]. After optimizing $AgNO_3$ concentration and reaction time, Saikia et al. [92] synthesized silver-carbon nanocomposites utilizing Ipomoea carnea stem at RT.

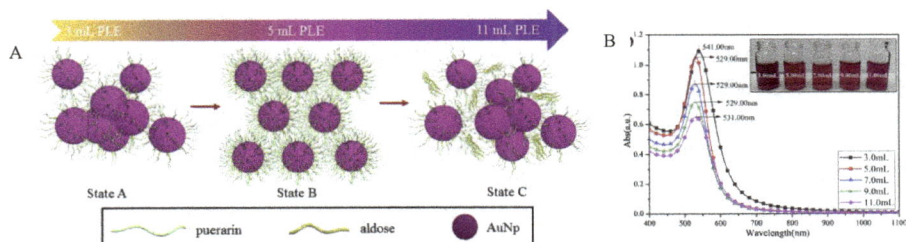

Fig. 4. A) Schematic diagram of soft agglomeration of AuNPs by increasing puerarialobata extract (PLE), B) Visual observations and UV–vis absorption spectra of reaction mixtures: varying contents of PLE at 30 °C with invariability HAuCl4, pH = 11. (Reproduced with permission from Ref.[29]).

Because the majority of the NPs plant-mediated synthesis contains metal or metal oxide, pH becomes a significant factor in the majority of cases. Due to the efficacy of pH on the nanoparticles properties, many researchers have devoted efforts to find out how to control pH [93]. Besides, changing pH is feasible and straightforward that might be worth to look after, on account of having monodispersed NPs. Sathish kumar et al. [82] investigated the effect of pH solution ranging from 1 to 9. They reported that the particle size of Pt synthesized by Cinnamomzeylanicum bark extract in pH>5 is bigger than pH<5, while the morphology of NPs has insignificantly changed. According to their study, the results at pH 5 or above would not be consistent [82]. A similar effect was reported by Desai et al. [94]. The findings show that increasing pH caused a red shift in the absorption maxima which in turn means the larger particle size of Au NPs. Therefore, the optimal pHs were at 4 and 7 according to Desai et al. [94] and Singh et al. [95] findings, respectively. However, the optimal pH strongly depends on the type of metal and plant extract. Din et al. [91] displayed the opposite results while the modified pH, ranging from 2-12. The reaction did not start until the pH was increased up to 10. By changing the pH to 12, they could synthesize NiO NPs instead of Ni NPs at pH = 10. They suggested that the reducing agents in Calotropis gigantean extract are deactivated at acidic pH, and only activated 10 <pH. It might be because of de-protonation of reductant or capping agents at higher basic conditions [91].

The effect of incubation time on the quantum dots properties has been studied in many earlier works. The reaction time starts from when the plant extract is added to the metal solution. M. Taib et al. [42] reported a blue shift and an increase in the intensity of the maximum absorption peak of Au NPs synthesized by H. sabdariffa L. extract, at 537 nm, by increasing the incubation time from 10 to 120 minutes, Fig. 5a [20]. Similarly, the color of reaction mix has changed from light orange to purple over time, Fig. 5b [20]. These findings show Au NPs synthesis needs 120 minutes to complete the reaction with Hibiscus sabdariffa extract, with 69.4% of yield [42]. This is in agreement with the results from Kumar's lab findings [43]. The Au NPs formation started after adding Sansevieriaroxburghiana leaf extract and completed 30 minutes later [20]. While some investigators suggested the optimal reaction time is less than 2 h, some studies showed the reaction needed longer time to get accomplished e.g. 11 [48], 24 [96], and 72 hours [97]. It is suggested that in each individual plant extracts, specific bioactive molecules are responsible for the reaction speed. If the plant extract is rich of phytochemicals, the reaction time would be shorter. Regarding the amount of bioactive chemicals in the Ocimum sanctum extracted from dried root and stem, Ahmad et al. [48] reported the Ag NPs produced 1 hours after mixing plant extract to silver ions solution and it needed 11 hours to complete the reaction. After 15 hours, Ag NPs slightly agglomerated and the surface plasmon resonance (SPR) of the NPs shifted toward a longer wavelength, as a results of increasing the size distribution [48].

The structure and particle size of QDs strongly depends on the reaction and annealing temperature [98, 99]. Via higher temperature, there is a possibility to synthesize QDs with homogenous structure or even single crystal. However, the chance of generating bigger NPs is unavoidable [55, 100, 101]. Ismail et al. [100] annealed Pd NPs in different temperature from 100 to 600 °C. They demonstrated that by increasing temperature from 100 to 600 °C the crystallography transited from fcc Pd to tetragonal PdO [100]. In another study, when the temperature of mixture was raised from room temperature (RT) up to 50 °C, the particle size also increased from 19 to 21 nm [102].

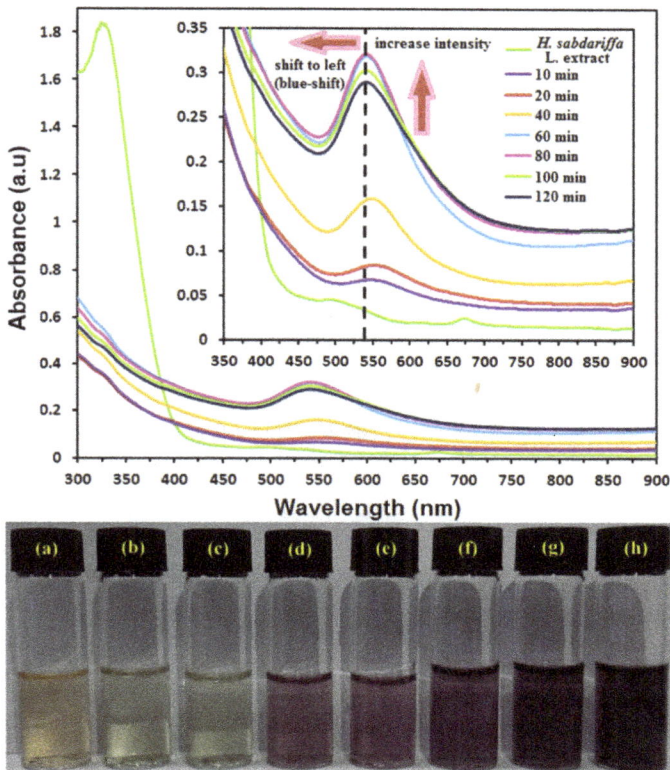

Fig. 5. A) UV–vis spectrum for Hibiscus sabdariffa extract and biosynthesized Au-NPs at different time periods. B) Hibiscus sabdariffa extract (a), and the synthesized Au-NPs with time period from 10, 20, 40, 60, 80, 100 and 120 minutes, as b to h, respectively. (Reproduced with permission from Ref.[42]).

2.2   Green synthesis of quantum dots by microorganisms

Microorganisms can be applied for synthesizing QDs by two methods including 1) extracting and purifying of the specific bioactive molecules from microbes, 2) manipulating them to synthesize QDs, through intracellular or extracellular mechanisms [103]. However, the exact mechanism of synthesize nanoparticles by microorganisms is not revealed yet. There are two major categories for NPs synthesis mechanism by using microbes including intra and extracellular approaches with different biological agents. Additionally, some researchers have used the spent media of microorganisms. The supernatant contains reducing as well as capping agents [45, 46]. The intercellular approach needs an ion transportation from solution into the microorganism cell. As it is known, the cell membrane is negatively charged, therefore, metal ions can easily attract to the cell membrane. The enzymes in the cell membrane reduce the metal ions to NPs and then NPs get diffused off via the cell membrane [46]. Bacteria have a specific metal defense system to detoxify heavy metals. When the concentration of metal ions increases in cells, they are assembled into foci and then secreted outside of cells [47]. It is suggested that synthesizing nanoparticles by microbes includes three steps: trapping, reducing, and capping. In the initial step, metal ions electrostatically make contact with the cell membrane. Afterwards, they are reduced into NPs by the enzymes present in the cell membrane or secreted by the cell [48-50]. However, in some cases, ion reduction occurs in the first step. After reduction, nanoclusters are transmitted to the bacterial cell membrane [51]. In order to extracellularly synthesize NPs, metal ions bind on the surface of the cell membrane and interact with reductant secreted from cell. There are various types of anionic functional groups on the cell surface such as sulfhydryl, sulfonate, hydroxyl, carboxyl, amine and amide groups. Similarly, extracellular polymers, e.g. proteins, polysaccharides, and humic substances, can trap metal ions [52, 53]. In some cases, bacteria can use both methods or a combination of them to synthesize QDs [104, 105].

2.2.1 Green quantum dot synthesis by bacteria

Synthesizing QDs by bacteria has many benefits due to their nature. For instance, they are found in large diversities and quantities, therefore, it is possible to find a type of bacteria which matches with the final products, from the point of view of composition, metal ions, or size and shape of NPs. On the other hand, they are able to adapt to extreme conditions which in turn leads to easily control reaction parameters such as temperature, pH, incubation time etc. Additionally, they can be cultured inexpensively, and grow fast [106]. As a result, using bacteria recently became an interesting method for synthesis quantum dots. Table 2 [104, 106-125] shows a list of researches that have used various bacteria to synthesize different nanocomposites.

Table 2 Green synthesized quantum dots by using various bacteria[104, 106-125].

Biogenic source	Metal/Metal Oxide	Cellular location	Application	Particle size (nm)	Shape	References
Rhodopseudomonascapsulata	Au	Extracellular	ND	10-20	S	[106]
Sulfate-reducing bacterial	Au	Cell envelope	ND	<10	sub-Oct and Oct	[107]
Bacterium *Shewanella algae*, ATCC 51181	Au	Extracellularly	ND	15-200	variety	[108]
Streptomyces SP MBRC-91	Ag	ND	Antibacterial activity	10-60	S	[109]
Streptomyces sp. ERI-3	Ag	Extracellular	ND	10-100	S, flower-like	[110]
Leuconostoclactis	Ag	Bacterial exopolysaccharide	Degradation of azo-dyes	35	S	[111]
Magnetotactic bacterium MV-1	Fe_3O_4	Intercellular	Parallelepiped	40	S	[112]
Escherichia coli	CdS	Intracellular	Nanoelectronics	2-5	Elliptical	[113]
Shewanellaoneidensis	CdS	ND	Cancer therapy	<100	Irregular	[114]
Shewanellaoneidensis MR-1	ZnS	Extracellular	Catalytic activity	5	S	[115]
Shewanellaoneidensis MR-1	CuS	Extracellular	Removing Cr(VI)	Thickness= 36.7	Hollow/ microshell	[116]
Shewanellaoneidensis MR-1	Pd/C	Extra/intracellular	Mass catalytic activity	13	S	[117]
Rhodopseudomonasacidophila	PVDF membrane	extracellular polymeric substances	Membrane filtration	ND	ND	[118]
Shewanellaoneidensis MR-1	CuS	Extracellular	Photothermal agent	5	S	[119]
Shewanellaoneidensis MR-1	FeS	Extra/interacellularly	ND	30	S	[104]
Escherichia coli	CdS	Intracellular	Antibiotic drug sensitivity	10	S	[120]
Shewanellaoneidensis	Ag2S	Extracellular	Biocompatibility	9	S	[121]
Shewanellaputrefaciens CN32	FeS	Intracellular	Dechlorination activity	100	S	[122]

Thermoanaerobacterethanolicus	Co, Cr, or Ni-substituted magnetite	**Extracellular**	**Magnetite**	<100	Oct	[123]
Shewanella oneidensisMR-1	Ferrihydrite/ graphene	Extracellular	Reduced Graphene oxide materials	ND	ND	[124]
Fusariumoxysporum	Ag	Culture supernatants	Antibacterial activity	22-30	S	[125]

S: Spherical, Oct: Octahedral.

Xiao et al. [116] synthesized hollow CuS nano/micro shell by Shewanellaoneidensis MR-1 which is also used as a sulfur supplier. The cupper ions were trapped on the bacterial surface and then nanorod CuS precipitated extracellularly. The CuS nanorods aggregated in the external bacteria membrane and eventually a hallow CuS micro shell prepared on the surface of cell, Fig. 6 [116]. In another study, Xiao et al. [104] shows that FeS nanoparticles can be synthesized intra and extracellularly, according to TEM images, see Fig. 7 [53]. FeS NPs were found on the surface of the cell, periplasmic space, and cytoplasm, which is in agreement with [105]. However, metal sulfide QDs are mostly produced only extracellularly [115, 121, 126]. This might be because of the fact that *S. oneidensis* MR-1 actively absorbs iron ions through siderophore molecules. In this way, the inside of the cell is rich of iron which can be the source iron for FeS NPs [104]. The analysis of subcellular localization by using TEM after 6 days shows that ZnS NPs were loosely precipitated on the cell surface and also could be found in the medium [115]. Unlike Marshall et al.'s findings [127], Thiosulfate reduced in the periplasm of *S. oneidensis* MR-1 and formerly S^{2-} extracellularly reacted with zinc ions and formed ZnS NPs. In this way, the formation of ZnS is occurred without association with an extracellular polymeric substance [115]. Xiong et al. [117] synthesized Pd NPs with carbon support by *Shewanella* cells. The bacterial cells play a role as a reductant agent as well as carrier matrix. The produced Pd nanoparticles were distributed in a carbon matrix. Immediately after adding the aqueous Na_2PdCl_4 solution into *Shewanella* cells, the color of mixture changed from yellow to black indicating Pd NPs formation. Ultimately, the mixture was carbonized at 420 °C, followed by adding KOH [117].

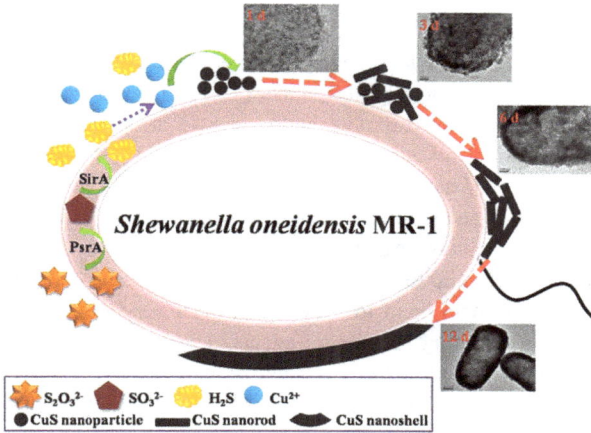

Fig. 6. Schematic and TEM of Self-assembly of complex hollow CuS nano/micro shell through extracellularly bio synthesized by Shewanellaoneidensis MR-1 at different time period.(Reproduced with permission from Ref.[116]).

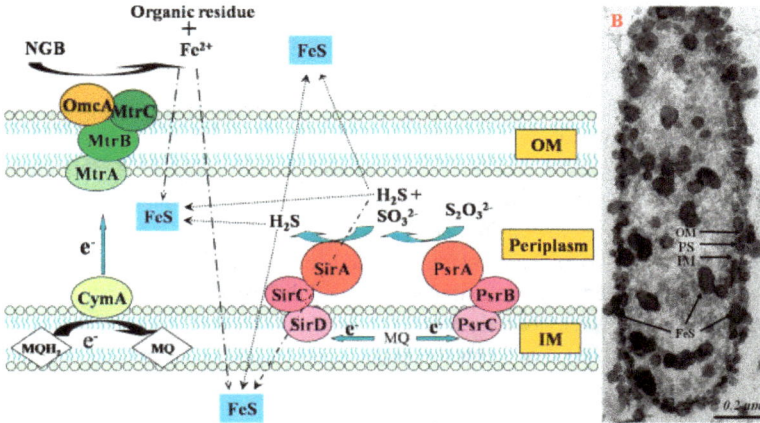

Fig. 7. A) Proposed mechanisms FeS nanoparticles biosynthesis coupled with degradation of NGB and thiosulfate by S. oneidensis MR-1 under anaerobic conditions. OM: outer membrane; IM: inner membrane; MQ: menquinone, B) thin sectional TEM images of S. oneidensis MR-1 after 5-d incubation with 300 mg/L NGB and 10mM thiosulfate which shows subcellular localization of FeS nanoparticles in cells. OM: outer membrane; IM: inner membrane; PS: periplasmic space. (Reproduced with permission from Ref.[104]).

2.2.2 Green quantum dot synthesis by algae

Similar to bacteria, algae can be found everywhere on the earth in many different verities and known as photoautotrophic organisms. They are rich in biochemicals which have various functional groups, e.g. carboxyl, hydroxyl, and amine [17, 128]. They can be used as reductant and stabilizer agents. Consequently, phyco-nanotechnology is emerged as a new promising field of nanotechnology [129]. Pugazhendhi et al. [130] synthesized Ag NPs by marine red algae Gelidiumamansii. By optimizing the concentration of silver and other reaction parameters, Ag NPs were synthesized and used against various pathogenic bacteria. According to FTIR spectrum, some bioactive molecules composited with Ag NPs which can help them to attach with other kind of cells and improve the biocompatibility, besides acting as capping agents. The strong bonds between Ag NPs and capping agents in the algae extract can stabilize nanoparticles and prevent them from aggregation for very long time [130, 131]. In a different study investigated by the same group, seaweed *E.compressa* was served to synthesize spherical shaped Ag NPs ranged from 4 to 24 nm. They displayed that many biochemicals in the seaweed extract were adsorbed on the Ag NP surface [128]. Sharma et al. [66], utilized *Chlorella pyrenoidosaextract to synthesized*TiO$_2$- graphene oxide (GO) nanocomposite. Titanium (VI) isopropoxide (TTIP) used as the Ti ions source was mixed by algae extract for 4 hours, then after drying at 60 °C, the TiO$_2$ NPs was annealed at 600 °C for 2 hours. Ultimately, TiO$_2$-GO nanocomposite was produced via hydrothermal method [132]. In an interesting study, polyesteramide polyol (PEA-RIC) was synthesized from algae oil and ricinoleic acid from castor oil. Afterwards they prepared silver doped hydroxyapatite nanoparticles using chicken egg-shells and combined them with PEA-RIC crosslinked with polyurethane to produce a nanocomposite coating. They claimed that the coating properties improved by increasing percentage of TiO$_2$ [133]. Iron oxide NPs were produced by brown (Colpomeniasinuosa) and red (Pterocladiacapillacea) seaweed separately. The seaweed aqueous extracts contained different biochemicals which played as capping agents as well as reductant. The particle sizes of NPs synthesized by *C. sinuosa* and *P. capillacea* aqueous extracts were 11.24–33.71 and 16.85–22.47 nm, respectively [134].

The water soluble polysaccharides extracted from *Pterocladiacapillacae, Janiarubins, Ulva faciata* and *Colpmeniasinusa* known as marine macro algae, were served as reductant and capping agents for producing Ag NPs. Afterwards, they were applied to cotton fabrics for the antimicrobial properties. Besides the stabilizer agents in the alga extracts, using citric acid and a binder enhanced the stability of Ag NPs on the fabric, therefore, the final product can be applied as a promising candidate for an antiseptic dressing or bandage [135]. Table 3 [128, 130, 135-143] represents a summary of some researches for green synthesized nanoparticles by using various algae.

Table 3 Green synthesized Quantum dots by using various algae[128, 130, 135-143].

Biogenic source	Metal/Metal Oxide	Application	Particle size (nm)	Shape	Ref.
Gelidiumamansii	Ag	Antimicrobial property	27-54	S	[130]
Enteromorphacompressa	Ag	Biomedical properties	4-24	S	[128]
Sargassumwightii	MgO	Anticancer, antimicrobial and photocatalytic activities	43	flower shaped	[136]
Ulva lactuca	Ag/ Au	Study for colon cancer cell	8-31	S	[137]
Gracilariaverrucosa	Au	Biocompatibility study	20-80	S and Tri	[138]
Dunaliellasalina	Au	Anticancer activity	5-45	S, Tri, and Hex	[139]
Egregia sp.	Au	Coated by biocompatible molecules	8	S	[140]
Enteromorphaflexuosa	Ag	Antimicrobial activities	2-32	S	[141]
Gracilariabirdiae	Ag	Antibacterial activity	20-94	S	[142]
Pterocladiacapillacae, Janiarubins, Ulva faciata, and *Colpmeniasinusa*	Ag	Antibacterial activity	7,7,12 and 20	S	[135]
Ulva lactuca seaweed	ZnO	Photocatalytic, antibiofilm and insecticidal activity	10-50	sponge-like asymmetrical	[143]

S: Spherical, R: Round, Hex: Hexagonal,Tri: Triangular.

2.2.3 Green quantum dots synthesis by fungus

Myconanotechnologyis a new field where fungus is served for synthesizing nanomaterials and nanostructures. However, myconanotechnology has recently emerged, and there are lot of potential applications in this field e.g. green synthesis quantum dots, green chemistry approaches for advancing food safety, etc. In addition, fungi are resistant to severe environments and harsh conditions. On the other hand, they mainly secrete reductive protein extracellularly which in turn results to easier downstream processes [144-146]. Therefore, several research groups have studied on fungus including filamentous fungus and yeast as biofactory for synthesizing nanostructure with desired morphology and size. Filamentous fungi are easy to handle, therefore, they are more preferred than other unicellular organism to synthesis metal/metal oxide composites [147, 148]. Jha et al. [75] compared using yeast with bacteria to synthesis of TiO_2 NPs. The particle size of TiO_2 NPs produced by yeast were smaller and mainly towards lower side which might be due to the fact that yeast is a eukaryote and they have a better level of organization at the cellular level [149]. Fungus also can secrete extracellular enzymes therefore, producing QDs via fungus is easy to scale up. Hulikere et al. [76] demonstrated an easy way to synthesis of Au NPs by endophytic fungus Cladosporiumcladosporioides which is isolated from seaweed [150]. They showed NADPH-dependent reductases possibly play a role in the metal NPs synthesis process in association with small phenolic molecules [150,151]. In another research, Singh et al. [78] synthesized silver nanoparticles by using endophytic fungi Alternaria sp. isolated from Raphanussativusleaves. They produced spherical Ag NPs with 10-30 nm size range [152]. Devi et al. [79] isolated three different endophytic fungi from an ethno-medicinal plant Potentillafulgens L. however, they had 98.7% similarity, the results were not the same. The TEM results revealed that all silver NPs synthesized by three funguses were same in shape, spherical, but not in particle size [153]. Yeast also can be used as a carrier for hosting NPs. They can absorb metal ions either in yeast cell envelope or inside yeast cells and reduce them in situ. The formation of NPs on the cell membrane or inside of the cells depends on the preparation methods. In Tollens's reaction, Glucan molecules at the cell membrane act as reductant agents and therefore, NPs were formed on the yeast cell membrane as a coverage. While NPs were synthesized inside of the cells by applying ultraviolet (UV) illumination on yeast [154]. Yamal et al. [81] applied yeast extract mannitol (YEM) medium to produce Au NPs. Mannitol, hexose sugar alcohol/polyol, help the yeast extract to reduce gold ions. So that Au NPs were synthesized at low temperature [155].

Overall, fungus and yeast are able to produce NPs extracellularly, which adds an essential advantage in the downstream processes as well as handling of biomass. In comparison with bacterial fermentation process, using fungus is more straightforward and easier. In addition, fungus produce more protein unlike bacteria, which in turn leads to gain higher yield by using fungus to synthesis NPs [146, 156, 157].

3. Green synthesis of quantum dots via wet chemical techniques

Wet chemical methods are one of the bottom-up approaches for synthesizing nanomaterials. It is called "wet chemical method, or aqueous method in case of using water as a solvent/media" since most of chemical reactions take place in a liquid phase/water. Due to multiple benefits, various methods have been established by many research groups in recent decades, to discover and develop an eco-friendly and economic perspective. The techniques are able to achieve desired structures, morphologies, sizes, and compounds, resulting in a wide range of properties. Wet chemical methods require a variety of reaction conditions such as concentration of components, different additives or surfactants, adding rate, temperature, heating method, heating/cooling rates, pH, type of solvent and it is a more challenging method. The composition, morphology, and particle size of the final product can be modified by adjusting the conditions. Therefore, it is very likely to synthesis products with different properties using just one process by changing parameters. It is well-known that changing particle size and morphology can alter the quantum dot properties. In results, modifying the condition can help to obtain the selective properties. On the other hand, playing with the wet chemical method condition is reasonably attainable, and the products are reproducible, then the methods can easily be scaled up. Therefore, many different wet chemical methods have been used for many industrial demand of quantum dot applications, especially as a continuous-flow synthesis technique [158].

Wet chemical method is one of the main techniques of the bottom-up approach. In the bottom-up approach nanomaterials/ nanostructures are fabricated by integration of atom by atom or molecular species together. The liquid surroundings in wet chemical methods produces a homogenous environment for atoms and molecules to meet each other freely and integrate in any possible way. Therefore, nanocomposites with different composition, morphology, and size are highly achievable. By adjusting different parameters, it is feasible to control shape, size, and structure of nanoparticles which in turn leads to modify quantum dot properties such as electronic, optical, magnetic, and surface properties. On the other hand, there are a lot of studies about the reactions in aqueous and the knowledge about the kinetic and thermodynamic parameters of this type of reactions are more accessible and easy to understand, due to researches in the past. Synthesizing nanoparticles via the bottom-top approach is established and well-known, therefore understanding the mechanism of synthesizing nanoparticles by applying wet chemical methods is less challenging than the biogenic methods such as plant-mediated methods, etc. In addition, the scalability of this method is more feasible since many industries are already adapted with wet chemical platform. The most common wet chemical methods for synthesis nanocomposites contains Sol-gel, solvo/hydrothermal, reverse micelle,

Materials Research Forum LLC
https://doi.org/10.21741/9781644901250-1

precipitation, thermal decomposition, microwave-assisted and ultrasonic-assisted synthesis methods.

Explaining all of these methods with their details is outside the scope of this chapter. Moreover, there are lots of books and review papers which lectured the techniques in details. The aim of this section is to address over view of each method and comparing together.

3.1 Sol-gel

The sol-gel method is one of the applicable techniques to synthesis quantum dot. In this method, a colloidal solution (sol) containing monomers convert to an integrated network (gel). This method has been discovered over 150 years ago when Graham studied on silica sols in 1864 [159]. He was the first person who used the term of "sol-gel". Due to the high abilities of this method, several research works have been published since then.

One of the major advantages of this method is the variety of product that can be produced by sol-gel process. Based on the application, the final product can be alerted from 0 D, e.g. powders and particles, to 2D, e.g. thin film and coating, 3D nanomaterials and composite, and nano-porous matrix. Fabricating thin film via sol-gel has several benefits including feasibility, low-temperature processing, accurate chemical and microstructural control. By this technique, sol/gel solution can be applied on any complex surfaces and spread it on the whole surface by spinning or spraying, Fig. 8 [86]. After sintering step, a thin uniform surface with high specific surface area is produced that has a wide range of applications such as optic, electronic, anti-bacterial, drug-delivery, etc. On the other hand, synthesizing different products with desired morphology, size, and composite is manageable via adjusting reaction conditions such as pH, time, temperature, heating rate, and additives. Metal alkoxides are typical precursors for sol-gel process such as tetraethyl orthosilicate (TEOS) and Titanium tetraisopropoxide which are used for synthesizing silica and titania composites via sol-gel method, respectively. Using alkoxide molecules facilitate production of hybrid organic-inorganic materials through combinations of inorganic and organic networks [160-162].

While sol-gel process is an effective method to produce nanocomposite exclusively as thin films, the metal alkoxides used as precursors are expensive, toxic, and very delicate and need to apply a special atmosphere, e.g. inert gas, for some materials. On the other hand, sol-gel is a method to synthesis metal oxide and it is rare to use for metal nanoparticles. Therefore, a modified sol-gel method has been developed, called Pechini method [163, 164]. In this method, different chelating agents such as EDTA, citric acid, etc. are introduced into ethylene glycol to form a polymer citrate gel. In case of using a glycol solvent, the technique is called Polyol method [161, 165]. After aging, the gel will be pyrolyzed. Nowadays, the Pechini process is broadly used in different industries such as fluorescent, magnetic, superconductors, catalysts, and coating. Compared to sol-gel, this method has relatively simple chemistry, more affordable and less toxic precursors [160, 161, 163, 165].

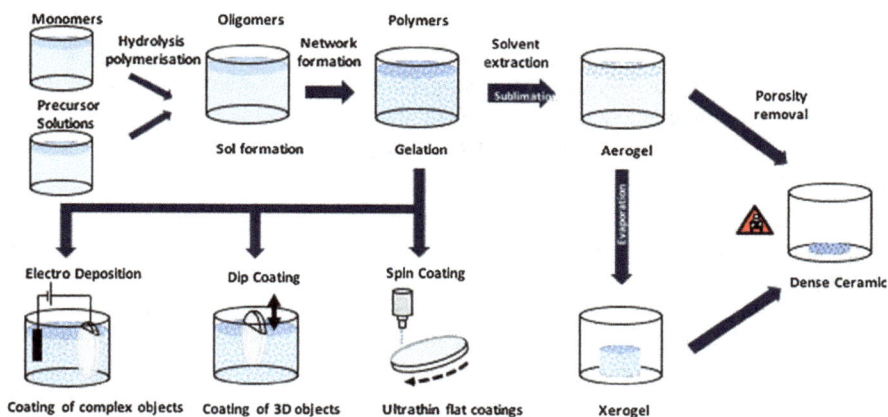

Fig. 8. Sol gel synthesis routes. Processes are defined as sol–gel by the transition of colloidal solution to an interconnected gel network (gelation). The further processing stages illustrated are non-redundant and may be combined depending on the specific needs of the application. (Reproduced with permission from Ref.[160]).

3.2 Solvo/hydrothermal

The solvo/hydrothermal takes place in an autoclave where the temperature is higher than RT/boiling point and the pressure is more than 1 atm. The autoclave consists of a steel pressure vessel which can tolerate supercritical condition of used solvent. In case of using water, hydrothermal, the supercritical point is around 647 K (or 374 °C) and 22.064 MPa (or 218 atm). One of the most important properties of supercritical fluid is the solubility. Therefore, even slightly soluble/insoluble materials can be conveniently dissolved in the supercritical fluid in the autoclave. Consequently, it is possible to synthesis any kind of nanomaterials with hydrothermal technique with high purity. On the other hand, due to using supercritical condition, it is able to apply water as a solvent instead of organic solvent. As a result, this method is very eco-friendly and green compare to other methods such as sol-gel, thermal decomposition, etc. Solvo/hydrothermal method is synthesized nanomaterials with high purity and quality, as well as with an arrow distribution particle size. Also, controlling morphology, size, and doping ions in the crystal lattice is easy to manage. Therefore, this technique is gained a good attention among researchers.

3.3   Microemulsions/ reverse micelle

Three ingredients required to synthesis quantum dots by microemulsions/ reverse micelle method include water, organic solvent (oil), and surfactant, in some cases also co-surfactant. Controlling the ratio of water: oil: surfactant is crucial to prepare a well-dispersed micelle solution with a desired shape. While water-in-oil microemulsions have been used to synthesis NPs for more than three decades, water/supercritical fluid microemulsions was recently applied for producing NPs with high ability to control in particle size of nanomaterials [166-168].

A supercritical fluid is created by keeping it above both its critical temperature (T_c) and critical pressure (P_c). In order to tailor size and morphology of NPs, density of solvent in the supercritical point can be changed through altering pressure and temperature. Nowadays, supercritical CO_2 is one of the particular interests and introduced as a novel green solvent. Using CO_2 as a supercritical environment can eliminate organic solvents. On the other hand, solvents can be removed by decreasing the pressure and releasing the CO_2 after the reaction is done. Therefore, besides eliminating wash steps, the surface of nanoparticles is completely free to conjugate with other molecules. In other words, the nature of this technique makes it a promising method to synthesis nanoparticles for potential commercial applications [166-168].

3.4   Thermal decomposition

Decomposition of inorganic/organic materials in the inert or particular atmosphere by increasing temperature is known as thermal decomposition. By increasing temperature, the molecules of the precursors decompose into metal, metal oxide, or other composite particles, in case of using inert, air, or particular atmosphere, respectively, and byproducts which evaporate or are washable after reaction [169, 170]. This method occurs in different phases, such as chemical gas phase deposition, pyrolysis in solid phase [98], and using organic solvent as a liquid environment. However, thermal decomposition in organic solvents is one of most renowned methods for synthesis of quantum dots. Appling an organic solvent can reduce the temperature of reaction. Also, liquid environment is liable to form quantum dots through self-assembly which in turn leads to synthesis nanoparticles with particular morphology and then specific properties. The other advantage of thermal decomposition in a liquid is having the possibility of using regular surfactant or capping agents to reduce growth rate of nanoparticles and prevent the surface from oxidation. Besides, organic coating assists binding of other molecules, such as biological molecules, on the particle surface for specific applications, for example drug delivery.

Among the impact factors on thermal decomposition method include temperature, precursors, concentration, and post processing, the type of surfactant and solvent is the most important parameter. By choosing an appropriate solvent and surfactant it is possible to control size and shape of nanoparticles. However, some solvents play as a surfactant/capping agent as well [171-174]. (For more details refer to [99]).

3.5 Precipitation

It is ensured that precipitation is one of the simplest methods among the NPs synthesis techniques. The precipitation method required neither particular equipment nor complex operation system. In addition, being budget friendly and having high yield without contamination risk results in the application of precipitation method as potential industrial application exclusively quantum dot industries. This method is a facile and green technique to prepare quantum dots. The mechanism behind this method is explained by particle nucleation and growth theory, where nuclei of NPs start to emerge at a certain concentration, supersaturation solution. Then additional atoms or molecules are deposited on the initial particles, known as growth step. The particle size of quantum dots can be manipulated by using surfactant, control temperature, time, or pH, which in turn leads to stop growth of particles, for more details refer to [80]. The products synthesized by precipitation are generally amorphous, especially if produced at low or room temperature. Therefore, in some cases, the precipitate from a reaction are post-annealed or calcined to improve crystallinity or purity.

In order to synthesize uniform quantum dots with desired particle size, the effect of operative factors on size and morphology should be studied and optimized [175, 176]. Besides a few common parameters such as precursor, concentration, pH, temperature, or reaction time, it is possible to assist precipitation by ultrasonication [177-182] or microwave [183-187] technique as a complementary method to promptly produce smaller NPs.

Microwave- and ultrasonication- assisted processing are considered as rapid methods superior to traditional heating. The temperature of the reaction can be rapidly increased even above the boiling point of the solvent, which in turn results in speeding the reaction speed up 10-1000 times, meaning reactions can be completed in a few minutes or seconds [188]. In addition, the particle size of synthesized QDs assisted by them are smaller with narrower particle size distribution. Also, the purity of the NPs is higher than other methods. As a results, the properties of NPs synthesis by these methods are enhanced and constant, comparing to other traditional methods [175, 189-192]. One of the other advantages of using microwave or ultrasonication is that they can effortlessly be adapted with other methods as a heat source such as hydrothermal, co-precipitation, and thermal decomposition [185, 189, 192-194]. That is why these methods have attracted a great deal of attention by researchers and industries.

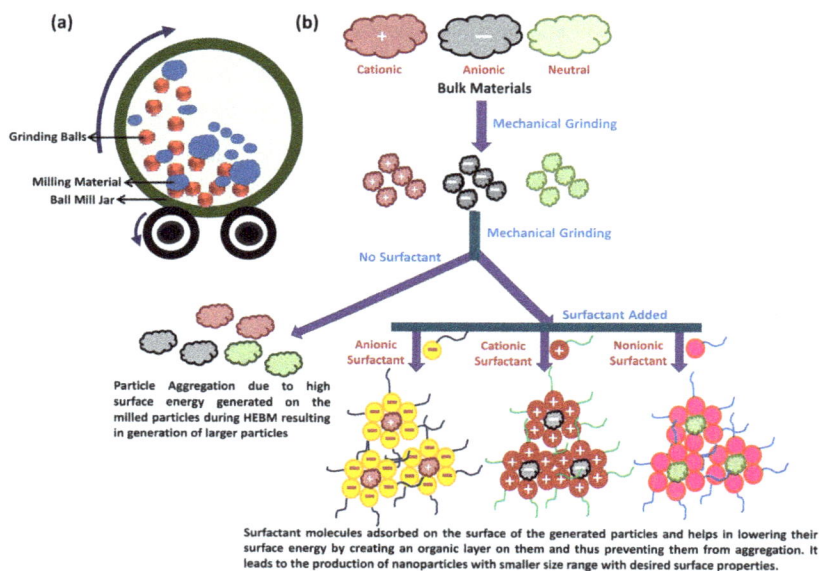

Fig. 9. (a) High energy ball milling system and (b) schematic representation of the NPs synthesis using HEBM method with and without surfactant. (Reproduced with permission from Ref.[199]).

4.   Solid-state techniques to synthesis quantum dots

Unlike many of the methods mentioned above, solid-state technique is operated on top-down approach. In this method developed by Benjamin for the production of super alloys [195], preliminary materials in micro size are fed to a mechanical mill. In a mill, balls transfer their kinetic energy to the raw materials between them which in turn leads to rupture materials into smaller pieces. Breaking particles results in creating a fresh surface to make a chemical reaction between precursors and consequently synthesize nanoparticles/composite. There are different types of mechanical mills in order to produce nanomaterials, such as, high energy ball mills, planetary ball mills, vibratory ball mills, low energy tumbling mills [196, 197]. The type of mill, speed of milling, temperature, weight ratio, time, particle size of precursors, and atmosphere [196, 198]. However, in most of them, the reaction would take place at room temperature, sometimes temperature or pressure were locally escalated by ball collisions which is considered as a mechano-chemical method, refer to Fig. 9a [125]. It is worth mentioning that researchers have been using surfactant- assisted ball mill to obtain an increase of the process efficiency, smaller particle size, and particular surface properties Fig. 9b [199].

One known method that avoids using solvent is solid-state techniques which is an easy scalable method to produce nanoparticles in the absence of solvents and capping agents. Therefore, this process is known as a green, eco-friendly technique to synthesis nanomaterials especially quantum dots [176, 197-200].

Conclusions

While nanocomposite synthesis via biogenic methods has become a popular topic recently, adapting these methods to commercial-scale has remained as a vital issue. In order to apply biological approaches for nanocomposite industries, it is necessary to evaluate the technologies from financial perspectives and reliability views. Owing to the fact that synthesizing NPs has emerged in the past few decades, there are plenty open challenges in this field. More noticeably, the yield of phytochemical reactions is very low, therefore, the cost of producing QDs with the methods increased dramatically. So that it would lose the interest of trade and industry. In addition, the nature of each method should be carefully distinguished. There are various unknown concerns in this field which should be addressed before scale them up. For example, what kind of active chemicals are responsible for the reaction? What is their individual quantity in the mixture? How does the condition of biological systems influence on the type and amount of components, e.g. bioactive components in plant extracts change by weather and time of harvest? Answering these questions not only helps to understand the mechanism behind each method, but also it can reconcile the existing green technology with industrial companies via producing QDs with higher efficacy. Overall, investigations into using microorganisms and plants to synthesis quantum dots are still at early stage and it needs considerable research.

On the other hand, synthesizing quantum dots via wet chemical/ solid-state techniques are well-known and researchers have used them to fabricating nanoparticles, especially for large scale application. Table 4 discusses the advantages and disadvantages of each method. Therefore, according to the type of QD and the facilities and equipment, researchers can decide which methods can be fitted in their lab. For example, sol-gel is a great method to synthesis metal oxide NPs or producing thin film/coating. However, it should be noted that this method needs high temperatures e. g. < 500 °C, hence the materials of NPs and the surface should be stable at high temperature. By using solvo/hydrothermal method, it is possible to synthesize nanocrystals with high purity but the reaction is happened in an autoclave at high pressure and temperature. Therefore, the safety of this method is low compare to others. Despite the fact that supercritical fluid technology is an expensive method to fabricate QDs, this method is a facile and rapid technology with controlling shape and size of NPs. this method can provide NPs with clean and bare surface which give them the ability of conjugating them with other molecules such as antibodies for therapeutic/imaging application. Although, Microemulsions/Reverse micelle method can be managed at RT and 1 atm, finding a proper surfactant and being costly can be considered as disadvantages of this method.

Table 4 Advantage and disadvantage of wet chemical techniques for synthesizing QDs.

Technique	Advantage	Disadvantage
Sol-Gel	Good method to synthesis metal oxide NPs. Scalable method High homogenous product Easy to dope ions/ atoms into crystal lattice Product diversity; NP, Thin film, coating, etc.	High temperature Synthesizing metal NPs is difficult. Expensive precursor, such as alkoxide precursors. Long procedure
Solvo/hydrothermal	High homogenous product high quality crystallized monodispersed nanocrystals. Varieties of product; meta, metal oxide, composite, etc. Eco-friendly, especially for hydrothermal. Adopted to using different precursors, even non-soluble materials.	High temperature and pressure Needs to optimize conditions for specific products
Microemulsions/Reverse micelle	Uniform NPs Very narrow size distribution Control shape with using different surfactant/condition	Finding a proper surfactant Expensive Needs specific concentration
Supercritical fluid technology	Eco-friendly Rapid Easy to remove solvent Facile nature Facilitate to collect NPs Control shape and particle size free of chemical contaminants	Expensive equipment Some case using high pressure and temperature Need high technology to keep CO_2 at critical point or above that
Precipitation	Simple mechanism Facile equipment Operative friendly	Low crystallinity Need post-processing

Materials Research Forum LLC
https://doi.org/10.21741/9781644901250-1

	Low cost	
	High yield and high purity	
	Compatible with many industries	
Thermal decomposition	Simple mechanism	High temperature
	Synthesis different product; M, MO, MS, etc.	Using organic solvent
	Control on shape and size	
Solid-state method, Ball milling	Solvent free,	Amorphous products
	RT, low pressure,	No control on shape
	Scalability	Particle size limitation
		Agglomeration
		Long process 12-24h

Precipitation is the simplest method to synthesis NPs. it requires inexpensive equipment and preliminary materials and can be operated at RT and 1 atm. Also, it is able to be compatible with pipeline equipment. Nonetheless, the purity would be an issue in some cases and the products need post-processing in term of purity and crystallinity. The other scalable method for QD synthesizing is Solid-state method e.g. Ball milling. In this method, having solvent is not requirement and the process happen at RT in 1 atm. In addition, adjusting temperature and pressure is feasible. Similar to precipitation method, the products from solid-state method needs post-processing due to amorphousness. Also, there is no control on shape and particle size.

Overall, finding a perfect method to synthesis QDs depends on many factors, such as feasibility, expenses, production and etc. however, researchers around the world try to find eco-friendly methods with high yield and purity. The ability of adjusting morphology and particle size is an enormous assistance since QD properties can be managed by their shape and size.

References

[1] T. Jamieson, R. Bakhshi, D. Petrova, R. Pocock, M. Imani, A.M. Seifalian, Biological applications of quantum dots, Biomaterials 28 (2007) 4717-4732. https://doi.org/10.1016/j.biomaterials.2007.07.014

[2] W.C.W. Chan, D.J. Maxwell, X. Gao, R.E. Bailey, M. Han, S. Nie, Luminescent quantum dots for multiplexed biological detection and imaging, Curr. Opin. Biotechnol. 13 (2002) 40-46. https://doi.org/10.1016/s0958-1669(02)00282-3.

[3] J. Xue, X. Wang, J.H. Jeong, X. Yan, Fabrication, photoluminescence and applications of quantum dots embedded glass ceramics, Chem. Eng. J. 383 (2020) 123082. https://doi.org/10.1016/j.cej.2019.123082

Materials Research Forum LLC
https://doi.org/10.21741/9781644901250-1

[4] A. Manikandan, Y.-Z. Chen, C.-C. Shen, C.-W. Sher, H.-C. Kuo, Y.-L. Chueh, A critical review on two-dimensional quantum dots (2d qds): From synthesis toward applications in energy and optoelectronics, Progress in Quantum Electronics 68 (2019) 100226. https://doi.org/10.1016/j.pquantelec.2019.100226

[5] V.G. Reshma, P.V. Mohanan, Quantum dots: Applications and safety consequences, J. Lumin. 205 (2019) 287-298. https://doi.org/10.1016/j.jlumin.2018.09.015

[6] J. Yao, P. Li, L. Li, M. Yang, Biochemistry and biomedicine of quantum dots: From biodetection to bioimaging, drug discovery, diagnostics, and therapy, Acta Biomater. 74 (2018) 36-55. https://doi.org/10.1016/j.actbio.2018.05.004

[7] R.S. Pawar, P.G. Upadhaya, V.B. Patravale, Quantum dots, (2018) 621-637. https://doi.org/10.1016/b978-0-12-813351-4.00035-3

[8] E.M. Egorova, A.A. Revina, Synthesis of metallic nanoparticles in reverse micelles in the presence of quercetin, Colloids Surf. Physicochem. Eng. Aspects 168 (2000) 87-96. https://doi.org/10.1016/s0927-7757(99)00513-0

[9] H. Uyama, M. Kuwabara, T. Tsujimoto, M. Nakano, A. Usuki, S. Kobayashi, Green nanocomposites from renewable resources: Plant oil–clay hybrid materials, Chem. Mater. 15 (2003) 2492-2494. https://doi.org/10.1021/cm0340227

[10] J.G. Parsons, J.R. Peralta-Videa, J.L. Gardea-Torresdey, Use of plants in biotechnology: Synthesis of metal nanoparticles by inactivated plant tissues, plant extracts, and living plants, 5 (2007) 463-485. https://doi.org/10.1016/s1474-8177(07)05021-8

[11] P. Kuppusamy, M.M. Yusoff, G.P. Maniam, N. Govindan, Biosynthesis of metallic nanoparticles using plant derivatives and their new avenues in pharmacological applications - an updated report, Saudi Pharm. J. 24 (2016) 473-484. https://doi.org/10.1016/j.jsps.2014.11.013

[12] A.K. Jha, K. Prasad, K. Prasad, A.R. Kulkarni, Plant system: Nature's nanofactory, Colloids Surf. B. Biointerfaces 73 (2009) 219-223. https://doi.org/10.1016/j.colsurfb.2009.05.018

[13] R. Zeiser, Advances in understanding the pathogenesis of graft-versus-host disease, British journal of haematology 187 (2019) 563-572. 10.1111/bjh.16190

[14] A. Altemimi, N. Lakhssassi, A. Baharlouei, D.G. Watson, D.A. Lightfoot, Phytochemicals: Extraction, isolation, and identification of bioactive compounds from plant extracts, Plants (Basel) 6 (2017). 10.3390/plants6040042

[15] M. Rai, A. Yadav, A. Gade, Current [corrected] trends in phytosynthesis of metal nanoparticles, Crit. Rev. Biotechnol. 28 (2008) 277-284. https://doi.org/10.1080/07388550802368903

[16] S. Rajeshkumar, L.V. Bharath, Mechanism of plant-mediated synthesis of silver nanoparticles - a review on biomolecules involved, characterisation and antibacterial activity, Chem. Biol. Interact. 273 (2017) 219-227. https://doi.org/10.1016/j.cbi.2017.06.019

[17] P.K. Gautam, A. Singh, K. Misra, A.K. Sahoo, S.K. Samanta, Synthesis and applications of biogenic nanomaterials in drinking and wastewater treatment, J. Environ. Manage. 231 (2019) 734-748. https://doi.org/10.1016/j.jenvman.2018.10.104

[18] M. Klekotko, K. Brach, J. Olesiak-Banska, M. Samoc, K. Matczyszyn, Popcorn-shaped gold nanoparticles: Plant extract-mediated synthesis, characterization and multiphoton-excited luminescence properties, Mater. Chem. Phys. 229 (2019) 56-60. https://doi.org/10.1016/j.matchemphys.2019.02.066

[19] G.A. Islan, S. Das, M.L. Cacicedo, A. Halder, A. Mukherjee, M.L. Cuestas, P. Roy, G.R. Castro, A. Mukherjee, Silybin-conjugated gold nanoparticles for antimicrobial chemotherapy against gram-negative bacteria, J. Drug Deliv. Sci. Technol. (2019) 101181. https://doi.org/10.1016/j.jddst.2019.101181

[20] I. Kumar, M. Mondal, V. Meyappan, N. Sakthivel, Green one-pot synthesis of gold nanoparticles using sansevieria roxburghiana leaf extract for the catalytic degradation of toxic organic pollutants, Mater. Res. Bull. 117 (2019) 18-27. https://doi.org/10.1016/j.materresbull.2019.04.029

[21] N. Thangamani, N. Bhuvaneshwari, Green synthesis of gold nanoparticles using simarouba glauca leaf extract and their biological activity of micro-organism, Chem. Phys. Lett. (2019). https://doi.org/10.1016/j.cplett.2019.07.015

[22] A. Boldeiu, M. Simion, I. Mihalache, A. Radoi, M. Banu, P. Varasteanu, P. Nadejde, E. Vasile, A. Acasandrei, R.C. Popescu, D. Savu, M. Kusko, Comparative analysis of honey and citrate stabilized gold nanoparticles: In vitro interaction with proteins and toxicity studies, J Photochem Photobiol B 197 (2019) 111519. https://doi.org/10.1016/j.jphotobiol.2019.111519

[23] M.K. Satheeshkumar, E.R. Kumar, C. Srinivas, N. Suriyanarayanan, M. Deepty, C.L. Prajapat, T.V.C. Rao, D.L. Sastry, Study of structural, morphological and magnetic properties of ag substituted cobalt ferrite nanoparticles prepared by honey assisted combustion method and evaluation of their antibacterial activity, J. Magn. Magn. Mater. 469 (2019) 691-697. https://doi.org/10.1016/j.jmmm.2018.09.039

[24] M.F. Zayed, W.H. Eisa, S.M. El-Kousy, W.K. Mleha, N. Kamal, Ficus retusa-stabilized gold and silver nanoparticles: Controlled synthesis, spectroscopic characterization, and sensing properties, Spectrochim. Acta A Mol. Biomol. Spectros. 214 (2019) 496-512. https://doi.org/10.1016/j.saa.2019.02.042

[25] M. Vinosha, S. Palanisamy, R. Muthukrishnan, S. Selvam, E. Kannapiran, S. You, N.M. Prabhu, Biogenic synthesis of gold nanoparticles from halymenia dilatata for pharmaceutical applications: Antioxidant, anti-cancer and antibacterial activities, Process Biochem. (2019). https://doi.org/10.1016/j.procbio.2019.07.013

[26] P.M. Anjana, M.R. Bindhu, R.B. Rakhi, Green synthesized gold nanoparticle dispersed porous carbon composites for electrochemical energy storage, Materials Science for Energy Technologies 2 (2019) 389-395. https://doi.org/10.1016/j.mset.2019.03.006

[27] I. Mohammad, Gold nanoparticle: An efficient carrier for mcp i of carica papaya seeds extract as an innovative male contraceptive in albino rats, J. Drug Deliv. Sci. Technol. 52 (2019) 942-956. https://doi.org/10.1016/j.jddst.2019.06.010

[28] S. Onitsuka, T. Hamada, H. Okamura, Preparation of antimicrobial gold and silver nanoparticles from tea leaf extracts, Colloids Surf. B. Biointerfaces 173 (2019) 242-248. https://doi.org/10.1016/j.colsurfb.2018.09.055

[29] Q. Zhou, M. Zhou, Q. Li, R. Wang, Y. Fu, T. Jiao, Facile biosynthesis and grown mechanism of gold nanoparticles in pueraria lobata extract, Colloids Surf. Physicochem. Eng. Aspects 567 (2019) 69-75. https://doi.org/10.1016/j.colsurfa.2019.01.039

[30] G.M. Asnag, A.H. Oraby, A.M. Abdelghany, Green synthesis of gold nanoparticles and its effect on the optical, thermal and electrical properties of carboxymethyl cellulose, Composites Part B: Engineering 172 (2019) 436-446. https://doi.org/10.1016/j.compositesb.2019.05.044

[31] A.I. Usman, A.A. Aziz, O.A. Noqta, Green sonochemical synthesis of gold nanoparticles using palm oil leaves extracts, Materials Today: Proceedings 7 (2019) 803-807. https://doi.org/10.1016/j.matpr.2018.12.078

[32] M.P. Patil, E. Bayaraa, P. Subedi, L.L.A. Piad, N.H. Tarte, G.-D. Kim, Biogenic synthesis, characterization of gold nanoparticles using lonicera japonica and their anticancer activity on hela cells, J. Drug Deliv. Sci. Technol. 51 (2019) 83-90. https://doi.org/10.1016/j.jddst.2019.02.021

[33] P. Boomi, R.M. Ganesan, G. Poorani, H. Gurumallesh Prabu, S. Ravikumar, J. Jeyakanthan, Biological synergy of greener gold nanoparticles by using coleus

aromaticus leaf extract, Mater. Sci. Eng. C Mater. Biol. Appl. 99 (2019) 202-210.
https://doi.org/10.1016/j.msec.2019.01.105

[34] S.K. Vemuri, R.R. Banala, S. Mukherjee, P. Uppula, S. Gpv, V.G. A, M. T, Novel
biosynthesized gold nanoparticles as anti-cancer agents against breast cancer:
Synthesis, biological evaluation, molecular modelling studies, Mater. Sci. Eng. C
Mater. Biol. Appl. 99 (2019) 417-429. https://doi.org/10.1016/j.msec.2019.01.123

[35] V. Sunderam, D. Thiyagarajan, A.V. Lawrence, S.S.S. Mohammed, A. Selvaraj, In-
vitro antimicrobial and anticancer properties of green synthesized gold nanoparticles
using anacardium occidentale leaves extract, Saudi J. Biol. Sci. 26 (2019) 455-459.
https://doi.org/10.1016/j.sjbs.2018.12.001

[36] T.S.K. Sharma, K. Selvakumar, K.Y. Hwa, P. Sami, M. Kumaresan, Biogenic
fabrication of gold nanoparticles using camellia japonica l. Leaf extract and its
biological evaluation, Journal of Materials Research and Technology 8 (2019) 1412-
1418. https://doi.org/10.1016/j.jmrt.2018.10.006

[37] S.E. Celik, B. Bekdeser, R. Apak, A novel colorimetric sensor for measuring
hydroperoxide content and peroxyl radical scavenging activity using starch-stabilized
gold nanoparticles, Talanta 196 (2019) 32-38.
https://doi.org/10.1016/j.talanta.2018.12.022

[38] M. Khoshnamvand, S. Ashtiani, C. Huo, S.P. Saeb, J. Liu, Use of alcea rosea leaf
extract for biomimetic synthesis of gold nanoparticles with innate free radical
scavenging and catalytic activities, J. Mol. Struct. 1179 (2019) 749-755.
https://doi.org/10.1016/j.molstruc.2018.11.079

[39] R.A. Zayadi, F. Abu Bakar, M.K. Ahmad, Elucidation of synergistic effect of
eucalyptus globulus honey and zingiber officinale in the synthesis of colloidal
biogenic gold nanoparticles with antioxidant and catalytic properties, Sustainable
Chemistry and Pharmacy 13 (2019) 100156. https://doi.org/10.1016/j.scp.2019.100156

[40] S. Rashid, M. Azeem, S.A. Khan, M.M. Shah, R. Ahmad, Characterization and
synergistic antibacterial potential of green synthesized silver nanoparticles using
aqueous root extracts of important medicinal plants of pakistan, Colloids Surf. B.
Biointerfaces 179 (2019) 317-325. https://doi.org/10.1016/j.colsurfb.2019.04.016

[41] S.S. Dakshayani, M.B. Marulasiddeshwara, M.N.S. Kumar, R. Golla, R.P. Kumar,
S. Devaraja, R. Hosamani, Antimicrobial, anticoagulant and antiplatelet activities of
green synthesized silver nanoparticles using selaginella (sanjeevini) plant extract, Int.
J. Biol. Macromol. 131 (2019) 787-797.
https://doi.org/10.1016/j.ijbiomac.2019.01.222

Materials Research Forum LLC
https://doi.org/10.21741/9781644901250-1

[42] S.H. Mohd Taib, K. Shameli, P. Moozarm Nia, M. Etesami, M. Miyake, R. Rasit Ali, E. Abouzari-Lotf, Z. Izadiyan, Electrooxidation of nitrite based on green synthesis of gold nanoparticles using hibiscus sabdariffa leaves, Journal of the Taiwan Institute of Chemical Engineers 95 (2019) 616-626. https://doi.org/10.1016/j.jtice.2018.09.021

[43] D. Tripathi, A. Modi, G. Narayan, S.P. Rai, Green and cost effective synthesis of silver nanoparticles from endangered medicinal plant withania coagulans and their potential biomedical properties, Mater. Sci. Eng. C Mater. Biol. Appl. 100 (2019) 152-164. https://doi.org/10.1016/j.msec.2019.02.113

[44] S. Francis, K.M. Nair, N. Paul, E.P. Koshy, B. Mathew, Green synthesized metal nanoparticles as a selective inhibitor of human osteosarcoma and pathogenic microorganisms, Materials Today Chemistry 13 (2019) 128-138. https://doi.org/10.1016/j.mtchem.2019.04.013

[45] P.M. Anjana, M.R. Bindhu, M. Umadevi, R.B. Rakhi, Antibacterial and electrochemical activities of silver, gold, and palladium nanoparticles dispersed amorphous carbon composites, Appl. Surf. Sci. 479 (2019) 96-104. https://doi.org/10.1016/j.apsusc.2019.02.057

[46] Z. Vaseghi, A. Nematollahzadeh, O. Tavakoli, Plant-mediated cu/cr/ni nanoparticle formation strategy for simultaneously separation of the mixed ions from aqueous solution, Journal of the Taiwan Institute of Chemical Engineers 96 (2019) 148-159. https://doi.org/10.1016/j.jtice.2018.10.020

[47] E.Y. Ahn, H. Jin, Y. Park, Assessing the antioxidant, cytotoxic, apoptotic and wound healing properties of silver nanoparticles green-synthesized by plant extracts, Mater. Sci. Eng. C Mater. Biol. Appl. 101 (2019) 204-216. https://doi.org/10.1016/j.msec.2019.03.095

[48] N. Ahmad, S. Sharma, M.K. Alam, V.N. Singh, S.F. Shamsi, B.R. Mehta, A. Fatma, Rapid synthesis of silver nanoparticles using dried medicinal plant of basil, Colloids Surf. B. Biointerfaces 81 (2010) 81-86. https://doi.org/10.1016/j.colsurfb.2010.06.029

[49] D. Sharma, M.I. Sabela, S. Kanchi, P.S. Mdluli, G. Singh, T.A. Stenstrom, K. Bisetty, Biosynthesis of zno nanoparticles using jacaranda mimosifolia flowers extract: Synergistic antibacterial activity and molecular simulated facet specific adsorption studies, J. Photochem. Photobiol. B 162 (2016) 199-207. https://doi.org/10.1016/j.jphotobiol.2016.06.043

[50] R. Dobrucka, J. Dlugaszewska, Biosynthesis and antibacterial activity of zno nanoparticles using trifolium pratense flower extract, Saudi J. Biol. Sci. 23 (2016) 517-523. https://doi.org/10.1016/j.sjbs.2015.05.016

[51] R. Yuvakkumar, J. Suresh, A.J. Nathanael, M. Sundrarajan, S.I. Hong, Novel green synthetic strategy to prepare zno nanocrystals using rambutan (nephelium lappaceum l.) peel extract and its antibacterial applications, Mater. Sci. Eng. C Mater. Biol. Appl. 41 (2014) 17-27. https://doi.org/10.1016/j.msec.2014.04.025

[52] D. Wang, H. Liu, Y. Ma, J. Qu, J. Guan, N. Lu, Y. Lu, X. Yuan, Recycling of hyper-accumulator: Synthesis of zno nanoparticles and photocatalytic degradation for dichlorophenol, J. Alloys Compd. 680 (2016) 500-505. https://doi.org/10.1016/j.jallcom.2016.04.100

[53] J. Fowsiya, G. Madhumitha, N.A. Al-Dhabi, M.V. Arasu, Photocatalytic degradation of congo red using carissa edulis extract capped zinc oxide nanoparticles, J. Photochem. Photobiol. B 162 (2016) 395-401. https://doi.org/10.1016/j.jphotobiol.2016.07.011

[54] N. Matinise, X.G. Fuku, K. Kaviyarasu, N. Mayedwa, M. Maaza, Zno nanoparticles via moringa oleifera green synthesis: Physical properties & mechanism of formation, Appl. Surf. Sci. 406 (2017) 339-347. https://doi.org/10.1016/j.apsusc.2017.01.219

[55] N. Mayedwa, N. Mongwaketsi, S. Khamlich, K. Kaviyarasu, N. Matinise, M. Maaza, Green synthesis of nickel oxide, palladium and palladium oxide synthesized via aspalathus linearis natural extracts: Physical properties & mechanism of formation, Appl. Surf. Sci. 446 (2018) 266-272. https://doi.org/10.1016/j.apsusc.2017.12.116

[56] T. Rasheed, F. Nabeel, M. Bilal, H.M.N. Iqbal, Biogenic synthesis and characterization of cobalt oxide nanoparticles for catalytic reduction of direct yellow-142 and methyl orange dyes, Biocatalysis and Agricultural Biotechnology 19 (2019) 101154. https://doi.org/10.1016/j.bcab.2019.101154

[57] S. Vasantharaj, S. Sathiyavimal, M. Saravanan, P. Senthilkumar, K. Gnanasekaran, M. Shanmugavel, E. Manikandan, A. Pugazhendhi, Synthesis of ecofriendly copper oxide nanoparticles for fabrication over textile fabrics: Characterization of antibacterial activity and dye degradation potential, J. Photochem. Photobiol. B 191 (2019) 143-149. https://doi.org/10.1016/j.jphotobiol.2018.12.026

[58] A.P. Angeline Mary, A. Thaminum Ansari, R. Subramanian, Sugarcane juice mediated synthesis of copper oxide nanoparticles, characterization and their antibacterial activity, Journal of King Saud University - Science (2019). https://doi.org/10.1016/j.jksus.2019.03.003

[59] S. Rajeshkumar, S. Menon, S. Venkat Kumar, M.M. Tambuwala, H.A. Bakshi, M. Mehta, S. Satija, G. Gupta, D.K. Chellappan, L. Thangavelu, K. Dua, Antibacterial and antioxidant potential of biosynthesized copper nanoparticles mediated through

cissus arnotiana plant extract, J. Photochem. Photobiol. B 197 (2019) 111531. https://doi.org/10.1016/j.jphotobiol.2019.111531

[60] N. Sripriya, S. Vasantharaj, U. Mani, M. Shanmugavel, R. Jayasree, A. Gnanamani, Encapsulated enhanced silver nanoparticles biosynthesis by modified new route for nano-biocatalytic activity, Biocatalysis and Agricultural Biotechnology 18 (2019) 101045. https://doi.org/10.1016/j.bcab.2019.101045

[61] K. Ranoszek-Soliwoda, E. Tomaszewska, K. Malek, G. Celichowski, P. Orlowski, M. Krzyzowska, J. Grobelny, The synthesis of monodisperse silver nanoparticles with plant extracts, Colloids Surf. B. Biointerfaces 177 (2019) 19-24. https://doi.org/10.1016/j.colsurfb.2019.01.037

[62] M. Chokkalingam, P. Singh, Y. Huo, V. Soshnikova, S. Ahn, J. Kang, R. Mathiyalagan, Y.J. Kim, D.C. Yang, Facile synthesis of au and ag nanoparticles using fruit extract of lycium chinense and their anticancer activity, J. Drug Deliv. Sci. Technol. 49 (2019) 308-315. https://doi.org/10.1016/j.jddst.2018.11.025

[63] S. Vasantharaj, S. Sathiyavimal, P. Senthilkumar, F. LewisOscar, A. Pugazhendhi, Biosynthesis of iron oxide nanoparticles using leaf extract of ruellia tuberosa: Antimicrobial properties and their applications in photocatalytic degradation, J. Photochem. Photobiol. B 192 (2019) 74-82. https://doi.org/10.1016/j.jphotobiol.2018.12.025

[64] D. Aksu Demirezen, Y.S. Yildiz, S. Yilmaz, D. Demirezen Yilmaz, Green synthesis and characterization of iron oxide nanoparticles using ficus carica (common fig) dried fruit extract, J. Biosci. Bioeng. 127 (2019) 241-245. https://doi.org/10.1016/j.jbiosc.2018.07.024

[65] L. Katata-Seru, T. Moremedi, O.S. Aremu, I. Bahadur, Green synthesis of iron nanoparticles using moringa oleifera extracts and their applications: Removal of nitrate from water and antibacterial activity against escherichia coli, J. Mol. Liq. 256 (2018) 296-304. https://doi.org/10.1016/j.molliq.2017.11.093

[66] S.N.A. Mohamad Sukri, K. Shameli, M. Mei-Theng Wong, S.-Y. Teow, J. Chew, N.A. Ismail, Cytotoxicity and antibacterial activities of plant-mediated synthesized zinc oxide (zno) nanoparticles using punica granatum (pomegranate) fruit peels extract, J. Mol. Struct. 1189 (2019) 57-65. https://doi.org/10.1016/j.molstruc.2019.04.026

[67] N. Shobha, N. Nanda, A.S. Giresha, P. Manjappa, S. P, K.K. Dharmappa, B.M. Nagabhushana, Synthesis and characterization of zinc oxide nanoparticles utilizing seed source of ricinus communis and study of its antioxidant, antifungal and anticancer

activity, Mater. Sci. Eng. C Mater. Biol. Appl. 97 (2019) 842-850.
https://doi.org/10.1016/j.msec.2018.12.023

[68] A. Thirumurugan, P. Aswitha, C. Kiruthika, S. Nagarajan, A.N. Christy, Green synthesis of platinum nanoparticles using azadirachta indica – an eco-friendly approach, Mater. Lett. 170 (2016) 175-178.
https://doi.org/10.1016/j.matlet.2016.02.026

[69] G. Sharma, R. Soni, N.D. Jasuja, Phytoassisted synthesis of magnesium oxide nanoparticles with swertia chirayaita, Journal of Taibah University for Science 11 (2018) 471-477. https://doi.org/10.1016/j.jtusci.2016.09.004

[70] J. Suresh, G. Pradheesh, V. Alexramani, M. Sundrarajan, S.I. Hong, Green synthesis and characterization of hexagonal shaped mgo nanoparticles using insulin plant (costus pictus d. Don) leave extract and its antimicrobial as well as anticancer activity, Adv. Powder Technol. 29 (2018) 1685-1694. https://doi.org/10.1016/j.apt.2018.04.003

[71] M.S. Al-Ruqeishi, T. Mohiuddin, L.K. Al-Saadi, Green synthesis of iron oxide nanorods from deciduous omani mango tree leaves for heavy oil viscosity treatment, Arabian Journal of Chemistry (2016). https://doi.org/10.1016/j.arabjc.2016.04.003

[72] S.S.U. Rahman, M.T. Qureshi, K. Sultana, W. Rehman, M.Y. Khan, M.H. Asif, M. Farooq, N. Sultana, Single step growth of iron oxide nanoparticles and their use as glucose biosensor, Results in Physics 7 (2017) 4451-4456.
https://doi.org/10.1016/j.rinp.2017.11.001

[73] P. Rajiv, B. Bavadharani, M.N. Kumar, P. Vanathi, Synthesis and characterization of biogenic iron oxide nanoparticles using green chemistry approach and evaluating their biological activities, Biocatalysis and Agricultural Biotechnology 12 (2017) 45-49.
https://doi.org/10.1016/j.bcab.2017.08.015

[74] Z. Izadiyan, K. Shameli, M. Miyake, H. Hara, S.E.B. Mohamad, K. Kalantari, S.H.M. Taib, E. Rasouli, Cytotoxicity assay of plant-mediated synthesized iron oxide nanoparticles using juglans regia green husk extract, Arabian Journal of Chemistry (2018). https://doi.org/10.1016/j.arabjc.2018.02.019

[75] Z. Izadiyan, K. Shameli, M. Miyake, S.Y. Teow, S.C. Peh, S.E. Mohamad, S.H.M. Taib, Green fabrication of biologically active magnetic core-shell fe3o4/au nanoparticles and their potential anticancer effect, Mater. Sci. Eng. C Mater. Biol. Appl. 96 (2019) 51-57. https://doi.org/10.1016/j.msec.2018.11.008

[76] G.W. Duan, J. Zhang, Y.M. Xu, Y.C. Zhang, Y.Z. Fu, Synthesis and characterization of Pt3Co@Pt nanocomposites using banana peel extract as novel surfactants, Synthesis and Reactivity in Inorganic, Metal-Organic, and Nano-Metal Chemistry 45 (2014) 203-209. https://doi.org/10.1080/15533174.2013.831452

[77] G.J. Zhou, S.H. Li, Y.C. Zhang, Y.Z. Fu, Biosynthesis of cds nanoparticles in banana peel extract, J Nanosci Nanotechnol 14 (2014) 4437-4442. 10.1166/jnn.2014.8259

[78] O.J. Nava, C.A. Soto-Robles, C.M. Gómez-Gutiérrez, A.R. Vilchis-Nestor, A. Castro-Beltrán, A. Olivas, P.A. Luque, Fruit peel extract mediated green synthesis of zinc oxide nanoparticles, J. Mol. Struct. 1147 (2017) 1-6. https://doi.org/10.1016/j.molstruc.2017.06.078

[79] S. Sangar, S. Sharma, V.K. Vats, S.K. Mehta, K. Singh, Biosynthesis of silver nanocrystals, their kinetic profile from nucleation to growth and optical sensing of mercuric ions, Journal of Cleaner Production 228 (2019) 294-302. https://doi.org/10.1016/j.jclepro.2019.04.238

[80] N.T. Thanh, N. Maclean, S. Mahiddine, Mechanisms of nucleation and growth of nanoparticles in solution, Chem. Rev. 114 (2014) 7610-7630. https://doi.org/10.1021/cr400544s

[81] P. Kuppusamy, S. Ilavenil, S. Srigopalram, G.P. Maniam, M.M. Yusoff, N. Govindan, K.C. Choi, Treating of palm oil mill effluent using commelina nudiflora mediated copper nanoparticles as a novel bio-control agent, Journal of Cleaner Production 141 (2017) 1023-1029. https://doi.org/10.1016/j.jclepro.2016.09.176

[82] M. Sathishkumar, K. Sneha, I.S. Kwak, J. Mao, S.J. Tripathy, Y.S. Yun, Phyto-crystallization of palladium through reduction process using cinnamom zeylanicum bark extract, J. Hazard. Mater. 171 (2009) 400-404. https://doi.org/10.1016/j.jhazmat.2009.06.014

[83] H.Y. El-Kassas, M.M. El-Sheekh, Cytotoxic activity of biosynthesized gold nanoparticles with an extract of the red seaweed corallina officinalis on the mcf-7 human breast cancer cell line, Asian Pac J Cancer Prev 15 (2014) 4311-4317. 10.7314/apjcp.2014.15.10.4311

[84] A. Cassidy, C. Kay, Phytochemicals: Classification and occurrence, (2013) 39-46. https://doi.org/10.1016/b978-0-12-375083-9.00226-9

[85] K. Heneman, S. Zidenberg-Cherr, Nutrition and health info sheet: Phytochemicals, 8313 (2008).

[86] N. Sahu, D. Soni, B. Chandrashekhar, D.B. Satpute, S. Saravanadevi, B.K. Sarangi, R.A. Pandey, Synthesis of silver nanoparticles using flavonoids: Hesperidin, naringin and diosmin, and their antibacterial effects and cytotoxicity, International Nano Letters 6 (2016) 173-181. https://doi.org/10.1007/s40089-016-0184-9

[87] S. Fahimirad, F. Ajalloueian, M. Ghorbanpour, Synthesis and therapeutic potential of silver nanomaterials derived from plant extracts, Ecotoxicol. Environ. Saf. 168 (2019) 260-278. https://doi.org/10.1016/j.ecoenv.2018.10.017

[88] S. C.G. Kiruba Daniel, K. Nehru, M. Sivakumar, Rapid biosynthesis of silver nanoparticles using eichornia crassipes and its antibacterial activity, Current Nanoscience 8 (2012) 125-129. 10.2174/157341371120801012 5

[89] S. Mukherjee, C.R. Patra, Biologically synthesized metal nanoparticles: Recent advancement and future perspectives in cancer theranostics, Future Sci OA 3 (2017) FSO203. 10.4155/fsoa-2017-0035

[90] M.S. Akhtar, J. Panwar, Y.-S. Yun, Biogenic synthesis of metallic nanoparticles by plant extracts, ACS Sustainable Chemistry & Engineering 1 (2013) 591-602. https://doi.org/10.1021/sc300118u

[91] M.I. Din, A.G. Nabi, A. Rani, A. Aihetasham, M. Mukhtar, Single step green synthesis of stable nickel and nickel oxide nanoparticles from calotropis gigantea : Catalytic and antimicrobial potentials, Environmental Nanotechnology, Monitoring & Management 9 (2018) 29-36. https://doi.org/10.1016/j.enmm.2017.11.005

[92] J. Saikia, N.g.B. Allou, S. Sarmah, P. Bordoloi, C. Gogoi, R.L. Goswamee, Reductant free synthesis of silver-carbon nanocomposite using low temperature carbonized ipomoea carnea stem carbon and study of its antibacterial property, Journal of Environmental Chemical Engineering 6 (2018) 4226-4235. https://doi.org/10.1016/j.jece.2018.06.025

[93] L. Castro, M.L. Blázquez, J.A. Muñoz, F. González, C. García-Balboa, A. Ballester, Biosynthesis of gold nanowires using sugar beet pulp, Process Biochem. 46 (2011) 1076-1082. https://doi.org/10.1016/j.procbio.2011.01.025

[94] M.P. Desai, G.M. Sangaokar, K.D. Pawar, Kokum fruit mediated biogenic gold nanoparticles with photoluminescent, photocatalytic and antioxidant activities, Process Biochem. 70 (2018) 188-197. https://doi.org/10.1016/j.procbio.2018.03.027

[95] A.K. Singh, O.N. Srivastava, One-step green synthesis of gold nanoparticles using black cardamom and effect of ph on its synthesis, Nanoscale Res Lett 10 (2015) 1055. 10.1186/s11671-015-1055-4

[96] V. Kumar, S.C. Yadav, S.K. Yadav, Syzygium cumini leaf and seed extract mediated biosynthesis of silver nanoparticles and their characterization, Journal of Chemical Technology & Biotechnology 85 (2010) 1301-1309. https://doi.org/10.1002/jctb.2427

[97] N.A. Bouqellah, M.M. Mohamed, Y. Ibrahim, Synthesis of eco-friendly silver nanoparticles using allium sp . And their antimicrobial potential on selected vaginal bacteria, Saudi Journal of Biological Sciences (2018). https://doi.org/10.1016/j.sjbs.2018.04.001

[98] M. Salavati-Niasari, F. Davar, Z. Fereshteh, Synthesis of nickel and nickel oxide nanoparticles via heat-treatment of simple octanoate precursor, J. Alloys Compd. 494 (2010) 410-414. https://doi.org/10.1016/j.jallcom.2010.01.063

[99] Z. Fereshteh, M. Salavati-Niasari, Effect of ligand on particle size and morphology of nanostructures synthesized by thermal decomposition of coordination compounds, Adv. Colloid Interface Sci. 243 (2017) 86-104. https://doi.org/10.1016/j.cis.2017.03.001

[100] E. Ismail, M. Khenfouch, M. Dhlamini, S. Dube, M. Maaza, Green palladium and palladium oxide nanoparticles synthesized via aspalathus linearis natural extract, J. Alloys Compd. 695 (2017) 3632-3638. https://doi.org/10.1016/j.jallcom.2016.11.390

[101] V. Helan, J.J. Prince, N.A. Al-Dhabi, M.V. Arasu, A. Ayeshamariam, G. Madhumitha, S.M. Roopan, M. Jayachandran, Neem leaves mediated preparation of nio nanoparticles and its magnetization, coercivity and antibacterial analysis, Results in Physics 6 (2016) 712-718. https://doi.org/10.1016/j.rinp.2016.10.005

[102] P. Anbu, S.C.B. Gopinath, H.S. Yun, C.-G. Lee, Temperature-dependent green biosynthesis and characterization of silver nanoparticles using balloon flower plants and their antibacterial potential, J. Mol. Struct. 1177 (2019) 302-309. https://doi.org/10.1016/j.molstruc.2018.09.075

[103] R. Prasad, R. Pandey, I. Barman, Engineering tailored nanoparticles with microbes: Quo vadis?, Wiley Interdiscip Rev Nanomed Nanobiotechnol 8 (2016) 316-330. https://doi.org/10.1002/wnan.1363

[104] X. Xiao, W.-W. Zhu, H. Yuan, H.-W. Li, Q. Li, H.-Q. Yu, Biosynthesis of fes nanoparticles from contaminant degradation in one single system, Biochem. Eng. J. 105 (2016) 214-219. https://doi.org/10.1016/j.bej.2015.09.022

[105] D.H. Kim, R.A. Kanaly, H.G. Hur, Biological accumulation of tellurium nanorod structures via reduction of tellurite by shewanella oneidensis mr-1, Bioresour. Technol. 125 (2012) 127-131. https://doi.org/10.1016/j.biortech.2012.08.129

[106] S. He, Z. Guo, Y. Zhang, S. Zhang, J. Wang, N. Gu, Biosynthesis of gold nanoparticles using the bacteria rhodopseudomonas capsulata, Mater. Lett. 61 (2007) 3984-3987. https://doi.org/10.1016/j.matlet.2007.01.018

[107] M. Lengke, G. Southam, Bioaccumulation of gold by sulfate-reducing bacteria cultured in the presence of gold(i)-thiosulfate complex, Geochim. Cosmochim. Acta 70 (2006) 3646-3661. https://doi.org/10.1016/j.gca.2006.04.018

[108] Y. Konishi, T. Tsukiyama, T. Tachimi, N. Saitoh, T. Nomura, S. Nagamine, Microbial deposition of gold nanoparticles by the metal-reducing bacterium shewanella algae, Electrochim. Acta 53 (2007) 186-192. https://doi.org/10.1016/j.electacta.2007.02.073

[109] P. Manivasagan, K.H. Kang, D.G. Kim, S.K. Kim, Production of polysaccharide-based bioflocculant for the synthesis of silver nanoparticles by streptomyces sp, Int. J. Biol. Macromol. 77 (2015) 159-167. https://doi.org/10.1016/j.ijbiomac.2015.03.022

[110] N. Faghri Zonooz, M. Salouti, Extracellular biosynthesis of silver nanoparticles using cell filtrate of streptomyces sp. Eri-3, Sci. Iranica 18 (2011) 1631-1635. https://doi.org/10.1016/j.scient.2011.11.029

[111] C. Saravanan, R. Rajesh, T. Kaviarasan, K. Muthukumar, D. Kavitake, P.H. Shetty, Synthesis of silver nanoparticles using bacterial exopolysaccharide and its application for degradation of azo-dyes, Biotechnol Rep (Amst) 15 (2017) 33-40. https://doi.org/10.1016/j.btre.2017.02.006

[112] D.A. Bazylinski, R.B. Frankel, H.W. Jannasch, Anaerobic magnetite production by a marine, magnetotactic bacterium, Nature 334 (1988) 518-519. https://doi.org/10.1038/334518a0

[113] R.Y. Sweeney, C. Mao, X. Gao, J.L. Burt, A.M. Belcher, G. Georgiou, B.L. Iverson, Bacterial biosynthesis of cadmium sulfide nanocrystals, Chem. Biol. 11 (2004) 1553-1559. https://doi.org/10.1016/j.chembiol.2004.08.022

[114] L. Wang, S. Chen, Y. Ding, Q. Zhu, N. Zhang, S. Yu, Biofabrication of morphology improved cadmium sulfide nanoparticles using shewanella oneidensis bacterial cells and ionic liquid: For toxicity against brain cancer cell lines, J. Photochem. Photobiol. B 178 (2018) 424-427. https://doi.org/10.1016/j.jphotobiol.2017.11.007

[115] X. Xiao, X.B. Ma, H. Yuan, P.C. Liu, Y.B. Lei, H. Xu, D.L. Du, J.F. Sun, Y.J. Feng, Photocatalytic properties of zinc sulfide nanocrystals biofabricated by metal-reducing bacterium shewanella oneidensis mr-1, J. Hazard. Mater. 288 (2015) 134-139. https://doi.org/10.1016/j.jhazmat.2015.02.009

[116] X. Xiao, Q.-Y. Liu, X.-R. Lu, T.-T. Li, X.-L. Feng, Q. Li, Z.-Y. Liu, Y.-J. Feng, Self-assembly of complex hollow cus nano/micro shell by an electrochemically active bacterium shewanella oneidensis mr-1, Int. Biodeterior. Biodegrad. 116 (2017) 10-16. https://doi.org/10.1016/j.ibiod.2016.09.021

[117] L. Xiong, J.-J. Chen, Y.-X. Huang, W.-W. Li, J.-F. Xie, H.-Q. Yu, An oxygen reduction catalyst derived from a robust pd-reducing bacterium, Nano Energy 12 (2015) 33-42. https://doi.org/10.1016/j.nanoen.2014.11.065

[118] Y.F. Guan, B.C. Huang, C. Qian, L.F. Wang, H.Q. Yu, Improved pvdf membrane performance by doping extracellular polymeric substances of activated sludge, Water Res. 113 (2017) 89-96. https://doi.org/10.1016/j.watres.2017.01.057

[119] N.Q. Zhou, L.J. Tian, Y.C. Wang, D.B. Li, P.P. Li, X. Zhang, H.Q. Yu, Extracellular biosynthesis of copper sulfide nanoparticles by shewanella oneidensis mr-1 as a photothermal agent, Enzyme Microb. Technol. 95 (2016) 230-235. https://doi.org/10.1016/j.enzmictec.2016.04.002

[120] Z.Y. Yan, Q.Q. Du, J. Qian, D.Y. Wan, S.M. Wu, Eco-friendly intracellular biosynthesis of cds quantum dots without changing escherichia coli's antibiotic resistance, Enzyme Microb. Technol. 96 (2017) 96-102. https://doi.org/10.1016/j.enzmictec.2016.09.017

[121] A.K. Suresh, M.J. Doktycz, W. Wang, J.W. Moon, B. Gu, H.M. Meyer, 3rd, D.K. Hensley, D.P. Allison, T.J. Phelps, D.A. Pelletier, Monodispersed biocompatible silver sulfide nanoparticles: Facile extracellular biosynthesis using the gamma-proteobacterium, shewanella oneidensis, Acta Biomater. 7 (2011) 4253-4258. https://doi.org/10.1016/j.actbio.2011.07.007

[122] Y.C. Huo, W.W. Li, C.B. Chen, C.X. Li, R. Zeng, T.C. Lau, T.Y. Huang, Biogenic fes accelerates reductive dechlorination of carbon tetrachloride by shewanella putrefaciens cn32, Enzyme Microb. Technol. 95 (2016) 236-241. https://doi.org/10.1016/j.enzmictec.2016.09.013

[123] Y. Roh, R.J. Lauf, A.D. McMillan, C. Zhang, C.J. Rawn, J. Bai, T.J. Phelps, Microbial synthesis and the characterization of metal-substituted magnetites, Solid State Commun. 118 (2001) 529-534. https://doi.org/10.1016/s0038-1098(01)00146-6

[124] G. Liu, H. Yu, N. Wang, R. Jin, J. Wang, J. Zhou, Microbial reduction of ferrihydrite in the presence of reduced graphene oxide materials: Alteration of fe(iii) reduction rate, biomineralization product and settling behavior, Chem. Geol. 476 (2018) 272-279. https://doi.org/10.1016/j.chemgeo.2017.11.023

[125] N. Naimi-Shamel, P. Pourali, S. Dolatabadi, Green synthesis of gold nanoparticles using fusarium oxysporum and antibacterial activity of its tetracycline conjugant, J. Mycol. Med. 29 (2019) 7-13. https://doi.org/10.1016/j.mycmed.2019.01.005

[126] J.H. Lee, D.W. Kennedy, A. Dohnalkova, D.A. Moore, P. Nachimuthu, S.B. Reed, J.K. Fredrickson, Manganese sulfide formation via concomitant microbial manganese

Materials Research Forum LLC
https://doi.org/10.21741/9781644901250-1

oxide and thiosulfate reduction, Environ. Microbiol. 13 (2011) 3275-3288. 10.1111/j.1462-2920.2011.02587.x

[127] M.J. Marshall, A.S. Beliaev, A.C. Dohnalkova, D.W. Kennedy, L. Shi, Z. Wang, M.I. Boyanov, B. Lai, K.M. Kemner, J.S. McLean, S.B. Reed, D.E. Culley, V.L. Bailey, C.J. Simonson, D.A. Saffarini, M.F. Romine, J.M. Zachara, J.K. Fredrickson, C-type cytochrome-dependent formation of u(iv) nanoparticles by shewanella oneidensis, PLoS Biol. 4 (2006) e268. 10.1371/journal.pbio.0040268

[128] V.S. Ramkumar, A. Pugazhendhi, K. Gopalakrishnan, P. Sivagurunathan, G.D. Saratale, T.N.B. Dung, E. Kannapiran, Biofabrication and characterization of silver nanoparticles using aqueous extract of seaweed enteromorpha compressa and its biomedical properties, Biotechnol Rep. (Amst) 14 (2017) 1-7. https://doi.org/10.1016/j.btre.2017.02.001

[129] T.N.V.K.V. Prasad, V.S.R. Kambala, R. Naidu, Phyconanotechnology: Synthesis of silver nanoparticles using brown marine algae cystophora moniliformis and their characterisation, J. Appl. Phycol. 25 (2012) 177-182. https://doi.org/10.1007/s10811-012-9851-z

[130] A. Pugazhendhi, D. Prabakar, J.M. Jacob, I. Karuppusamy, R.G. Saratale, Synthesis and characterization of silver nanoparticles using gelidium amansii and its antimicrobial property against various pathogenic bacteria, Microb. Pathog. 114 (2018) 41-45. https://doi.org/10.1016/j.micpath.2017.11.013

[131] A.M. Awwad, N.M. Salem, A.O. Abdeen, Green synthesis of silver nanoparticles using carob leaf extract and its antibacterial activity, Int. J. Ind. Chem. 4 (2013) 29. 10.1186/2228-5547-4-29

[132] M. Sharma, K. Behl, S. Nigam, M. Joshi, Tio2-go nanocomposite for photocatalysis and environmental applications: A green synthesis approach, Vacuum 156 (2018) 434-439. https://doi.org/10.1016/j.vacuum.2018.08.009

[133] C.K. Patil, H.D. Jirimali, J.S. Paradeshi, B.L. Chaudhari, V.V. Gite, Functional antimicrobial and anticorrosive polyurethane composite coatings from algae oil and silver doped egg shell hydroxyapatite for sustainable development, Prog. Org. Coat. 128 (2019) 127-136. https://doi.org/10.1016/j.porgcoat.2018.11.002

[134] D.M.S.A. Salem, M.M. Ismail, M.A. Aly-Eldeen, Biogenic synthesis and antimicrobial potency of iron oxide (fe3o4) nanoparticles using algae harvested from the mediterranean sea, egypt, The Egyptian Journal of Aquatic Research (2019). https://doi.org/10.1016/j.ejar.2019.07.002

[135] H.M. El-Rafie, M.H. El-Rafie, M.K. Zahran, Green synthesis of silver nanoparticles using polysaccharides extracted from marine macro algae, Carbohydr. Polym. 96 (2013) 403-410. https://doi.org/10.1016/j.carbpol.2013.03.071

[136] A. Pugazhendhi, R. Prabhu, K. Muruganantham, R. Shanmuganathan, S. Natarajan, Anticancer, antimicrobial and photocatalytic activities of green synthesized magnesium oxide nanoparticles (mgonps) using aqueous extract of sargassum wightii, J. Photochem. Photobiol. B 190 (2019) 86-97. https://doi.org/10.1016/j.jphotobiol.2018.11.014

[137] N. Gonzalez-Ballesteros, M.C. Rodriguez-Arguelles, S. Prado-Lopez, M. Lastra, M. Grimaldi, A. Cavazza, L. Nasi, G. Salviati, F. Bigi, Macroalgae to nanoparticles: Study of ulva lactuca l. Role in biosynthesis of gold and silver nanoparticles and of their cytotoxicity on colon cancer cell lines, Mater. Sci. Eng. C Mater. Biol. Appl. 97 (2019) 498-509. https://doi.org/10.1016/j.msec.2018.12.066

[138] C. Chellapandian, B. Ramkumar, P. Puja, R. Shanmuganathan, A. Pugazhendhi, P. Kumar, Gold nanoparticles using red seaweed gracilaria verrucosa: Green synthesis, characterization and biocompatibility studies, Process Biochem. 80 (2019) 58-63. https://doi.org/10.1016/j.procbio.2019.02.009

[139] A.K. Singh, R. Tiwari, V.K. Singh, P. Singh, S.R. Khadim, U. Singh, Laxmi, V. Srivastava, S.H. Hasan, R.K. Asthana, Green synthesis of gold nanoparticles from dunaliella salina, its characterization and in vitro anticancer activity on breast cancer cell line, J. Drug Deliv. Sci. Technol. 51 (2019) 164-176. https://doi.org/10.1016/j.jddst.2019.02.023

[140] J.A. Colin, I.E. Pech-Pech, M. Oviedo, S.A. Águila, J.M. Romo-Herrera, O.E. Contreras, Gold nanoparticles synthesis assisted by marine algae extract: Biomolecules shells from a green chemistry approach, Chem. Phys. Lett. 708 (2018) 210-215. https://doi.org/10.1016/j.cplett.2018.08.022

[141] M. Yousefzadi, Z. Rahimi, V. Ghafori, The green synthesis, characterization and antimicrobial activities of silver nanoparticles synthesized from green alga enteromorpha flexuosa (wulfen) j. Agardh, Mater. Lett. 137 (2014) 1-4. https://doi.org/10.1016/j.matlet.2014.08.110

[142] A.P. de Aragão, T.M. de Oliveira, P.V. Quelemes, M.L.G. Perfeito, M.C. Araújo, J.d.A.S. Santiago, V.S. Cardoso, P. Quaresma, J.R. de Souza de Almeida Leite, D.A. da Silva, Green synthesis of silver nanoparticles using the seaweed gracilaria birdiae and their antibacterial activity, Arabian Journal of Chemistry (2016). https://doi.org/10.1016/j.arabjc.2016.04.014

[143] R. Ishwarya, B. Vaseeharan, S. Kalyani, B. Banumathi, M. Govindarajan, N.S. Alharbi, S. Kadaikunnan, M.N. Al-Anbr, J.M. Khaled, G. Benelli, Facile green synthesis of zinc oxide nanoparticles using ulva lactuca seaweed extract and evaluation of their photocatalytic, antibiofilm and insecticidal activity, J. Photochem. Photobiol. B 178 (2018) 249-258. https://doi.org/10.1016/j.jphotobiol.2017.11.006

[144] P.L. Kashyap, S. Kumar, A.K. Srivastava, A.K. Sharma, Myconanotechnology in agriculture: A perspective, World J. Microbiol. Biotechnol. 29 (2013) 191-207. https://doi.org/10.1007/s11274-012-1171-6

[145] A. Gade, A. Ingle, C. Whiteley, M. Rai, Mycogenic metal nanoparticles: Progress and applications, Biotechnol. Lett. 32 (2010) 593-600. https://doi.org/10.1007/s10529-009-0197-9

[146] M.H. Hanafy, Myconanotechnology in veterinary sector: Status quo and future perspectives, Int J Vet Sci Med 6 (2018) 270-273. https://doi.org/10.1016/j.ijvsm.2018.11.003

[147] K. Kalishwaralal, V. Deepak, S. Ramkumarpandian, H. Nellaiah, G. Sangiliyandi, Extracellular biosynthesis of silver nanoparticles by the culture supernatant of bacillus licheniformis, Mater. Lett. 62 (2008) 4411-4413. https://doi.org/10.1016/j.matlet.2008.06.051

[148] K. AbdelRahim, S.Y. Mahmoud, A.M. Ali, K.S. Almaary, A.E. Mustafa, S.M. Husseiny, Extracellular biosynthesis of silver nanoparticles using rhizopus stolonifer, Saudi J. Biol. Sci. 24 (2017) 208-216. https://doi.org/10.1016/j.sjbs.2016.02.025

[149] A.K. Jha, K. Prasad, A.R. Kulkarni, Synthesis of tio2 nanoparticles using microorganisms, Colloids Surf. B. Biointerfaces 71 (2009) 226-229. https://doi.org/10.1016/j.colsurfb.2009.02.007

[150] M. Hulikere, Manjunath, C.G. Joshi, A. Danagoudar, J. Poyya, A.K. Kudva, D. Bl, Biogenic synthesis of gold nanoparticles by marine endophytic fungus-cladosporium cladosporioides isolated from seaweed and evaluation of their antioxidant and antimicrobial properties, Process Biochem. 63 (2017) 137-144. https://doi.org/10.1016/j.procbio.2017.09.008

[151] M. Manjunath Hulikere, C.G. Joshi, Characterization, antioxidant and antimicrobial activity of silver nanoparticles synthesized using marine endophytic fungus-cladosporium cladosporioides, Process Biochem. 82 (2019) 199-204. https://doi.org/10.1016/j.procbio.2019.04.011

[152] T. Singh, K. Jyoti, A. Patnaik, A. Singh, R. Chauhan, S.S. Chandel, Biosynthesis, characterization and antibacterial activity of silver nanoparticles using an endophytic

fungal supernatant of raphanus sativus, J. Genet. Eng. Biotechnol. 15 (2017) 31-39. https://doi.org/10.1016/j.jgeb.2017.04.005

[153] L.S. Devi, S.R. Joshi, Ultrastructures of silver nanoparticles biosynthesized using endophytic fungi, J. Microsc. Ultrastruct. 3 (2015) 29-37. https://doi.org/10.1016/j.jmau.2014.10.004

[154] J. Li, G. Ma, H. Liu, H. Liu, Yeast cells carrying metal nanoparticles, Mater. Chem. Phys. 207 (2018) 373-379. https://doi.org/10.1016/j.matchemphys.2018.01.001

[155] G. Yamal, P. Sharmila, K.S. Rao, P. Pardha-Saradhi, Yeast extract mannitol medium and its constituents promote synthesis of au nanoparticles, Process Biochem. 48 (2013) 532-538. https://doi.org/10.1016/j.procbio.2013.02.011

[156] S. Gupta, K. Sharma, R. Sharma, Myconanotechnology and application of nanoparticles in biology, Recent Research in Science and Technology 4 (2012) 36-38

[157] S. Menon, R. S, V.K. S, A review on biogenic synthesis of gold nanoparticles, characterization, and its applications, Resource-Efficient Technologies 3 (2017) 516-527. https://doi.org/10.1016/j.reffit.2017.08.002

[158] A.V. Nikam, B.L.V. Prasad, A.A. Kulkarni, Wet chemical synthesis of metal oxide nanoparticles: A review, CrystEngComm 20 (2018) 5091-5107. https://doi.org/10.1039/c8ce00487k

[159] T. Graham, XXXV.—on the properties of silicic acid and other analogous colloidal substances, J. Chem. Soc. 17 (1864) 318-327. https://doi.org/10.1039/js8641700318

[160] G.J. Owens, R.K. Singh, F. Foroutan, M. Alqaysi, C.-M. Han, C. Mahapatra, H.-W. Kim, J.C. Knowles, Sol–gel based materials for biomedical applications, Prog. Mater Sci. 77 (2016) 1-79. https://doi.org/10.1016/j.pmatsci.2015.12.001

[161] F. Bensebaa, Wet production methods, 19 (2013) 85-146. https://doi.org/10.1016/b978-0-12-369550-5.00002-1

[162] M. Farhadi-Khouzani, Z. Fereshteh, M.R. Loghman-Estarki, R.S. Razavi, Different morphologies of zno nanostructures via polymeric complex sol–gel method: Synthesis and characterization, J. Sol-Gel Sci. Technol. 64 (2012) 193-199. https://doi.org/10.1007/s10971-012-2847-y

[163] L. Dimesso, Pechini processes: An alternate approach of the sol–gel method, preparation, properties, and applications, (2016) 1-22. https://doi.org/10.1007/978-3-319-19454-7_123-1

[164] M.P. Pechini, Method of preparing lead and alkaline earth titanates and niobates and coating method using the same to form a capacitor.

[165] Z. Fereshteh, R. Rojaee, A. Sharifnabi, Effect of different polymers on morphology and particle size of silver nanoparticles synthesized by modified polyol method, Superlattices Microstruct. 98 (2016) 267-275. https://doi.org/10.1016/j.spmi.2016.08.034

[166] J. Eastoe, M.J. Hollamby, L. Hudson, Recent advances in nanoparticle synthesis with reversed micelles, Adv. Colloid Interface Sci. 128-130 (2006) 5-15. https://doi.org/10.1016/j.cis.2006.11.009

[167] A.S. Deshmukh, P.N. Chauhan, M.N. Noolvi, K. Chaturvedi, K. Ganguly, S.S. Shukla, M.N. Nadagouda, T.M. Aminabhavi, Polymeric micelles: Basic research to clinical practice, Int. J. Pharm. 532 (2017) 249-268. https://doi.org/10.1016/j.ijpharm.2017.09.005

[168] S. Asgari, A.H. Saberi, D.J. McClements, M. Lin, Microemulsions as nanoreactors for synthesis of biopolymer nanoparticles, Trends Food Sci. Technol. 86 (2019) 118-130. https://doi.org/10.1016/j.tifs.2019.02.008

[169] F. Davar, Z. Fereshteh, M. Salavati-Niasari, Nanoparticles ni and nio: Synthesis, characterization and magnetic properties, J. Alloys Compd. 476 (2009) 797-801. https://doi.org/10.1016/j.jallcom.2008.09.121

[170] Z. Fereshteh, M. Salavati-Niasari, K. Saberyan, S.M. Hosseinpour-Mashkani, F. Tavakoli, Synthesis of nickel oxide nanoparticles from thermal decomposition of a new precursor, J. Cluster Sci. 23 (2012) 577-583. https://doi.org/10.1007/s10876-012-0477-8

[171] M. Salavati-Niasari, Z. Fereshteh, F. Davar, Synthesis of cobalt nanoparticles from [bis(2-hydroxyacetophenato)cobalt(ii)] by thermal decomposition, Polyhedron 28 (2009) 1065-1068. https://doi.org/10.1016/j.poly.2009.01.012

[172] M. Salavati-Niasari, F. Davar, Z. Fereshteh, Synthesis and characterization of zno nanocrystals from thermolysis of new precursor, Chem. Eng. J. 146 (2009) 498-502. https://doi.org/10.1016/j.cej.2008.09.042

[173] M. Salavati-Niasari, Z. Fereshteh, F. Davar, Synthesis of oleylamine capped copper nanocrystals via thermal reduction of a new precursor, Polyhedron 28 (2009) 126-130. https://doi.org/10.1016/j.poly.2008.09.027

[174] F. Davar, M. Salavati-Niasari, Z. Fereshteh, Synthesis and characterization of sno2 nanoparticles by thermal decomposition of new inorganic precursor, J. Alloys Compd. 496 (2010) 638-643. https://doi.org/10.1016/j.jallcom.2010.02.152

[175] C. Qiu, Y. Hu, Z. Jin, D.J. McClements, Y. Qin, X. Xu, J. Wang, A review of green techniques for the synthesis of size-controlled starch-based nanoparticles and their

applications as nanodelivery systems, Trends Food Sci. Technol. 92 (2019) 138-151. https://doi.org/10.1016/j.tifs.2019.08.007

[176] P.G. Jamkhande, N.W. Ghule, A.H. Bamer, M.G. Kalaskar, Metal nanoparticles synthesis: An overview on methods of preparation, advantages and disadvantages, and applications, J. Drug Deliv. Sci. Technol. 53 (2019) 101174. https://doi.org/10.1016/j.jddst.2019.101174

[177] S.A. Arote, A.S. Pathan, Y.V. Hase, P.P. Bardapurkar, D.L. Gapale, B.M. Palve, Investigations on synthesis, characterization and humidity sensing properties of zno and zno-zro2 composite nanoparticles prepared by ultrasonic assisted wet chemical method, Ultrason. Sonochem. 55 (2019) 313-321. https://doi.org/10.1016/j.ultsonch.2019.01.012

[178] D.E. Fouad, C. Zhang, T.D. Mekuria, C. Bi, A.A. Zaidi, A.H. Shah, Effects of sono-assisted modified precipitation on the crystallinity, size, morphology, and catalytic applications of hematite (α-fe2o3) nanoparticles: A comparative study, Ultrason. Sonochem. 59 (2019) 104713. https://doi.org/10.1016/j.ultsonch.2019.104713

[179] H. Jian-feng, Z. Xie-rong, C. Li-yun, X. Xin-bo, Preparation of y2bacuo5 nanoparticles by a co-precipitation process with the aid of ultrasonic irradiation, J. Mater. Process. Technol. 209 (2009) 2963-2966. https://doi.org/10.1016/j.jmatprotec.2008.07.001

[180] V. Mohanraj, R. Jayaprakash, R. Robert, J. Balavijayalakshmi, S. Gopi, Effect of particle size on optical and electrical properties in mixed cds and nis nanoparticles synthesis by ultrasonic wave irradiation method, Mater. Sci. Semicond. Process. 56 (2016) 394-402. https://doi.org/10.1016/j.mssp.2016.08.014

[181] B. Pohl, R. Jamshidi, G. Brenner, U.A. Peuker, Experimental study of continuous ultrasonic reactors for mixing and precipitation of nanoparticles, Chem. Eng. Sci. 69 (2012) 365-372. https://doi.org/10.1016/j.ces.2011.10.058

[182] D. Gopi, J. Indira, L. Kavitha, M. Sekar, U.K. Mudali, Synthesis of hydroxyapatite nanoparticles by a novel ultrasonic assisted with mixed hollow sphere template method, Spectrochim. Acta A Mol. Biomol. Spectrosc. 93 (2012) 131-134. https://doi.org/10.1016/j.saa.2012.02.033

[183] S. Guru, A.K. Bajpai, S.S. Amritphale, Influence of nature of surfactant and precursor salt anion on the microwave assisted synthesis of barium carbonate nanoparticles, Mater. Chem. Phys. 241 (2020) 122377. https://doi.org/10.1016/j.matchemphys.2019.122377

[184] D. MubarakAli, Microwave irradiation mediated synthesis of needle-shaped hydroxyapatite nanoparticles as a flocculant for chlorella vulgaris, Biocatalysis and Agricultural Biotechnology 17 (2019) 203-206. https://doi.org/10.1016/j.bcab.2018.11.025

[185] H.N. Deepak, K.S. Choudhari, S.A. Shivashankar, C. Santhosh, S.D. Kulkarni, Facile microwave-assisted synthesis of cr2o3 nanoparticles with high near-infrared reflection for roof-top cooling applications, J. Alloys Compd. 785 (2019) 747-753. https://doi.org/10.1016/j.jallcom.2019.01.254

[186] D.S. Chauhan, C.S.A. Gopal, D. Kumar, N. Mahato, M.A. Quraishi, M.H. Cho, Microwave induced facile synthesis and characterization of zno nanoparticles as efficient antibacterial agents, Materials Discovery 11 (2018) 19-25. https://doi.org/10.1016/j.md.2018.05.001

[187] O. Reyes, M. Pal, J. Escorcia-García, R. Sánchez-Albores, P.J. Sebastian, Microwave-assisted chemical synthesis of Zn_2SnO_4 nanoparticles, Mater. Sci. Semicond. Process. 108 (2020) 104878. https://doi.org/10.1016/j.mssp.2019.104878

[188] B.A. Roberts, C.R. Strauss, Toward rapid, "green", predictable microwave-assisted synthesis, Acc. Chem. Res. 38 (2005) 653-661. https://doi.org/10.1021/ar040278m

[189] X.H. Zhu, Q.M. Hang, Microscopical and physical characterization of microwave and microwave-hydrothermal synthesis products, Micron 44 (2013) 21-44. https://doi.org/10.1016/j.micron.2012.06.005

[190] A. Mirzaei, G. Neri, Microwave-assisted synthesis of metal oxide nanostructures for gas sensing application: A review, Sensor. Actuator. B Chem. 237 (2016) 749-775. https://doi.org/10.1016/j.snb.2016.06.114

[191] M. Hujjatul Islam, M.T.Y. Paul, O.S. Burheim, B.G. Pollet, Recent developments in the sonoelectrochemical synthesis of nanomaterials, Ultrason. Sonochem. 59 (2019) 104711. https://doi.org/10.1016/j.ultsonch.2019.104711

[192] S. Rahemi Ardekani, A. Sabour Rouh Aghdam, M. Nazari, A. Bayat, E. Yazdani, E. Saievar-Iranizad, A comprehensive review on ultrasonic spray pyrolysis technique: Mechanism, main parameters and applications in condensed matter, J. Anal. Appl. Pyrolysis 141 (2019) 104631. https://doi.org/10.1016/j.jaap.2019.104631

[193] N.N. Huy, V.T. Thanh Thuy, N.H. Thang, N.T. Thuy, L.T. Quynh, T.T. Khoi, D. Van Thanh, Facile one-step synthesis of zinc oxide nanoparticles by ultrasonic-assisted precipitation method and its application for H_2S adsorption in air, J. Phys. Chem. Solids 132 (2019) 99-103. https://doi.org/10.1016/j.jpcs.2019.04.018

Materials Research Forum LLC
https://doi.org/10.21741/9781644901250-1

[194] L.S.K. Achary, P.S. Nayak, B. Barik, A. Kumar, P. Dash, Ultrasonic-assisted green synthesis of β-amino carbonyl compounds by copper oxide nanoparticles decorated phosphate functionalized graphene oxide via mannich reaction, Catal. Today (2019). https://doi.org/10.1016/j.cattod.2019.07.050

[195] J.S. Benjamin, Dispersion strengthened superalloys by mechanical alloying, Metallurgical Transactions 1 (1970) 2943–2951. https://doi.org/10.1007/bf03037835

[196] M.S. El-Eskandarany, Introduction, (2001) 1-21. https://doi.org/10.1016/b978-081551462-6.50003-2

[197] A.R. Jones, Mechanical alloying, (2001) 1-5. https://doi.org/10.1016/b0-08-043152-6/00912-8

[198] G.A. Marcelo, C. Lodeiro, J.L. Capelo, J. Lorenzo, E. Oliveira, Magnetic, fluorescent and hybrid nanoparticles: From synthesis to application in biosystems, Mater. Sci. Eng. C Mater. Biol. Appl. 106 (2020) 110104. https://doi.org/10.1016/j.msec.2019.110104

[199] C. Dhand, N. Dwivedi, X.J. Loh, A.N. Jie Ying, N.K. Verma, R.W. Beuerman, R. Lakshminarayanan, S. Ramakrishna, Methods and strategies for the synthesis of diverse nanoparticles and their applications: A comprehensive overview, RSC Adv. 5 (2015) 105003-105037. https://doi.org/10.1039/c5ra19388e

[200] Z. Fereshteh, M. Fathi, R. Mozaffarinia, Synthesis and characterization of fluorapatite nanoparticles via a mechanochemical method, J. Cluster Sci. 26 (2014) 1041-1053. https://doi.org/10.1007/s10876-014-0793-2

Quantum Dots – Properties and Applications
Materials Research Foundations **96** (2021) 53-80

Materials Research Forum LLC
https://doi.org/10.21741/9781644901250-2

Chapter 2

Fabrication Techniques for Quantum Dots

Jyoti Patel, Bhawana Jain, Ajaya Kumar Singh*

Department of Chemistry, Govt. V.Y.T. PG. Autonomous, College, Durg, Chhattisgarh, 491001, India

ajayaksingh_au@yahoo.co.in

Abstract

The nanotechnological expansion involves the innovation and designing of materials at the nanoscale regime with controlled properties. Production of nanomaterials with good crystallinity, shape control, and narrow distribution of size plays a significant role in QD-based devices and applications. Various strategies ranging from simple wet chemical methods to advanced atomic layer deposition strategies have been employed for the production of QDs. In this chapter, a prominent and detailed discussion of conventional techniques in addition to the up-to-date development in the synthesis of recent QDs is given. Synthesis routes based on the microwave or ultrasonically assisted and cluster-seed process are of great significance.

Keywords

Quantum Dots, Lithography, Etching, Microemulsion, Epitaxy, Ultrasonic, Microwave, Hydrothermal, Solvothermal

Contents

Materials Research Forum LLC
https://doi.org/10.21741/9781644901250-2

1. Introduction

In the last two decades, the nanostructured materials or quantum dots (QDs) have acquired enormous consideration owing to their characteristic properties transitional between the bulk and molecular levels [1]. QDs also known as quasi zero-dimensional structures or artificial atoms are semiconductor nanocrystals that show discrete levels of energy moreover their bandgaps are adjustable by variation in the sizes [2]. QDs are generally made of elements of groups II-VI or III-V. These are defined as nanocrystals having their substantial dimensions comparable to or lesser than the Bohr exciton radius [3]. When the dimension of the crystals are reduced to nanoscales, they become "molecule-like" giving rise to unique properties differing greatly from their bulk

Materials Research Forum LLC
https://doi.org/10.21741/9781644901250-2

counterparts [4,5] possessing a wide range of engineering applications. They are photostable and mostly showing a broad absorption nevertheless, their specific wavelengths of emission are useful for chemical sensing [6,7], biological imaging [8,9], screening of cancer cells and drug delivery [10], organic dye removal [11], solar paints [12], anti-fake labeling [13,14], LEDs and photovoltaic solar cells [15,16].

QDs are the most often reported nanomaterials during the last few years both fundamentally as well as promising resources for varied functions [17,18]. Besides, they are among the earliest nanomaterials incorporated with biology [19,20], extensively projected for their application in number of marketable and scientific commodities [21]. For instance, CdSe/ZnS QDs at present are the chief accessible nanomaterials as conjugated secondary antibodies [22]. All these demands for simplistic, straightforward and economical synthetic routes for fabrication of nanoscale materials for an easy manufacturing process on a commercial scale.

2. Synthesis or fabrication techniques of nanostructures

In nanotechnology, the synthesis of nanoscale materials with monitored size and shape plays a vital role. Since last few years, attention is given to devising newer methods for the synthesis of nanomaterials advantageous over the conventional ones. For the regular synthesis of QDs, meticulous mechanistic exploration has been going on [23,24]. Typically, QDs preparation consists of nucleation and the growth process. The saturation of monomers follows the nucleation process where consumption of the extra monomers in the system takes place. QDs with varying shapes and size distributions can be prepared based on the monomer concentration. The nanocrystal growth is kinetically and thermodynamically controlled. For instance, minute trait ratio is observed during the deliberate growth and development under thermodynamically managed conditions whereas highly anisotropic shapes necessitate a kinetically controlled environment. Alivisatos et al. [25] proposed that almost round particles are obtained with large size regimes at low monomer concentration. Due to uninterrupted development of large constituents and depletion of smaller elements, Ostwald ripening is observed. Moreover, when the concentration of monomer enhances, the small trait ratios are still obtained with narrow size division. Thus, the size determination process takes place as a result of the faster growth rate of small constituents than larger ones consequently resulting in monodispersed nanocrystals.

During synthesis, the size and the conductive properties of the crystals can be circumspectly controlled. Some of the excellent review articles [26-29], interesting papers [30-34], and revised books [35-38] have came out dealing with meticulous physicochemical features, fabrication techniques as well as applications. This book chapter presents a comprehensive view of the conventional literature techniques and focuses on the recent system for research of stable as well as non-toxic QDs citing significant examples.

3. Fabrication techniques of QDs

The fabrication of QDs can be attained chiefly by the following techniques, the first is known as top-down synthesis method and second as bottom-up method.

3.1 Top-down methods

In this approach, the bulk materials are thinned to prepare QDs. Different techniques involved in this method are:

3.1.1 Electron beam lithography

3.1.2 Focused ion beam methods

3.1.3 Etching techniques

3.1.1 Electron beam lithography

Electron beam lithography (also known as EBL or e-beam lithography) originated from scanning electron micrograph, was first given by Buck and Shoulders (1958), moreover, Mollenstedt and Speidel first demonstrated EBL (1960) [39]. Since many years, a variety of particle beams were used in lithography. These techniques were immensely studied to level down lithography to nanoscale regimes (Brodie, Muray [40], Timp [41]) Due to the exceptionally elevated diffraction limited resolution, the electron source could be applied for transferring patterns of nanometer size regime.

It is the practice of transmitting patterns or an outline over the face of an object by scanning a sheet of a film known as resist by a specifically and closely focused electron beam known as exposure, followed by vigilantly eliminating the unexposed or exposed areas using a developing solvent. The method permits synthesis of extremely minute structures, from submicrometer to a few nanometer regimes, by casing the chosen regions with the help of resist or revealing the resist-coated parts. The exposed regions are then again developed for thin-film deposition or etching whereas the covered parts are secluded throughout these processes [42].

Usually, there are two methods in EBL, projection printing and direct writing [43]. Recently, EBL has turned out to be the conventional selection in synthesis of nanoscale structures, equally by direct writing as well as projection printing techniques. Moreover, it has consistently been applied to create master masks as well as reticles from computer-assisted drawing archives in the semiconductor industry [44]. The masks are generally applied during projection printing for replication of outlines over the silicon films. It has also been applied in direct writing, in which a focused electron beam directly encroaches over the resist to perform various activities. Versatile pattern formations with extremely high resolutions are the chief advantages of EBL over photolithography.

Zheng et al. [45] recently demonstrated a kirigami-inspired (paper cutting art) EBL procedure for metallic composition with considerably improved precision and efficiency. The multiscale structures are made by initially generating nanotrench outlines over a metal layer by EBL and then selective peeling of the film outside the contours as diagramed in Fig.1 [45]. In conventional methods, exposure of the whole pattern is required, and in nanotrench contours, very less exposure area is desired, hence less duration for exposure and increased statistical accuracy is obtained. The process permits multiscale compositions with nanoregime characteristics, providing potential uses in anti-counterfeiting and gap-plasmon-enhanced spectroscopy.

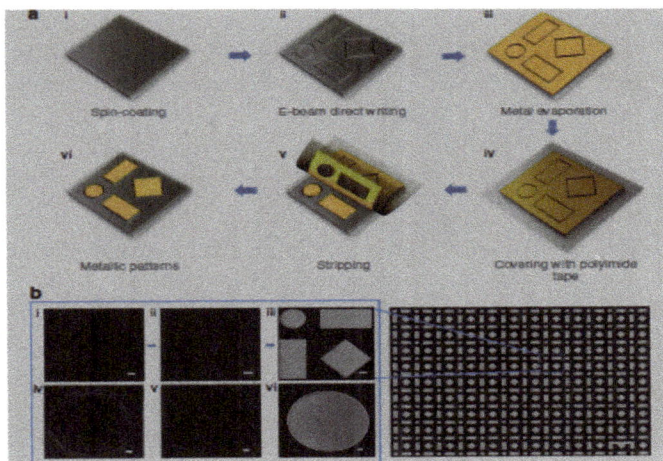

Fig. 1(a) Flowchartshowing kirigami-inspired EB-Patterning for metallic structures. (i)silica substrate with spin-coated resist (ii) outline by EBL (iii)deposition of metal by EB-evaporator; (iv) Covering with Polyimide tape; (v) the peeling process, detachment (vi) patterns achieved by removing the tape. (b) Scanning electron micrographs of the array. The unit cell entails a diamond, two rectangles, and a circle. Expanded image of unit cell previous (i) and subsequent to (ii) metal evaporation, and (iii) subsequent to stripping. Particulars for individual circle from (iv) to (vi). Reprinted with permission from ref. [45]. Copyright 2018 Springer Nature.

a)   Projection printing

Projection EBL (P-EBL) development was first of all started by Bell Laboratories in 1989 by the contraption of scattering with angular limitation in P-EBL (SCALPEL) method [46]. In 1980, IBM put the basis for projection reduction exposure with variable axis immersion lenses (PREVAIL) system [47,48]. Both the hypothesis constructs a tiny field image of a mask on a film to produce nano-regime sub-patterns.

A slender mask can be employed or cutouts enclosed template mask with the help of which electron beams could surpass is desirable. A slit situated at the reverse focal plane obstructs the robustly scattered electrons, whereas the ones surpassing the membrane undergo slight variation in the path and progress all the way through slit. Consequently, the undisturbed electrons pass the slit forming a higher contrast illustration over the substrate plane or wafer.

To evade the extreme distortion owing to field irregularities, the size of the beam is set relatively undersized nearly 1mm. The miniature size of field and printing region is an important concern in SCALPEL method. For imaging a 300 mm film, the complete film needs to get uncovered serially by covering these tiny fields collectively with good precision. Moreover, projection EBL never became a practical tool for nanofabrication. Problems like mask difficulties together with patching up number of sub-patterns in a particular pattern, the irregularity constraints, and extreme thermal expansion and absorption, along with other problems prevented the P-EBL from being a realistic gizmo [43].

The small parts in the film, as well as reticle are sutured by the grouping of electronic-beam scanning (high-speed) and mechanical scanning (moderate-speed) in PREVAIL method. A scheme with adjustable-axis lenses is the foundation behind this approach which allows movement of the optical axis to a fixed curvature, turning the electron axis to specifically track the curved changeable beam with the intention that the beam efficiently stay on the axis, removing the entire off-axis irregularities. The work is executed by the superposition of various magnetically deflected areas [48]. A PREVAIL method has lately accomplished an improved printing region of 10×10 mm [49]. The outcome made the PREVAIL system hopeful. However, some of the improvement is desirable as large scan areas are essential for the projection of the reticle, additionally the improved outline area of 10-mm is of lesser scope as well as magnitude than the present 300-mm film size [43].

b)   Direct patterning

QDs patterning is significant for their incorporation into the solid-state equipments. EBL is a prominent lithographic method for direct patterning (DP), allocating production of nanoscale structures with accurateness [50]. DP through EBL presents a simple and accurate approach used to pattern Au nanostructures [51]. DP, unlike usual lithography, provides a significant generalization in fabrication by means of the operational constituent working as resist. DP prevents tricky circumstances of resist deposit, carving, and elimination, as a result, evades resist/film adhesion problems, extended handing times and irregularity during generation at the etching stage.

V. Nandwana et al. [52] proposed a simple technique for the DP of QDs nanostructures by way of EBL. Initial, CdSe/ZnS QDs functionalized with trioctylphosphine oxide film were deposit over silicon substrate covered with gold. Subsequently, the outline was directly written over the film by EBL. Lastly, unrevealed QDs are detached by rinsing with toluene. The bare regions remain affixed with substrate attributable to electron-axis influenced cross relations between organic ligands. The optical features, lifetimes as well as fluorescence properties were untouched previously as well as subsequent to exposure.

c)   Multi-column (MC) × multi-beam (MB) technique of EBL

Parker et al. [53] proposed a novel multiple-beam theory utilizing many columns with many beams producing a huge number of analogous beams for writing. This conquers the restrictions of low voltage operation as well as space-charge interactions. Also, validation of the optimal amount of columns and beams for each column was given. With this method, fabrication of ≥60 wafers per hour can be obtained.

Conventional sole column, probe-making or contoured-beam methods are not creditable to produce owing to small throughput, restricted by overall current, for a certain resolution, caused by space-charge property. Moreover, structures writing with sole-pixel beams possess other intricacy of elevated data rate. Space-charge property could be lowered by introducing current in detached beams, for instance the solo-beam micro-columns [54] or the dispersed beams [55]. The troubles related to the rate of data are resolved by revealing numerous analogous pixels for example contoured beams [56], cell projection [57]. Various number of beams between 10 to 1000 were positioned in single column by various groups [58,59] for equivalent exposure but experienced the space charge trouble. Thus, Parker et al. [53] combined these two approaches in an (MC) × (MB) EBL system thus providing a scheme of physical space necessities, preserving the miniature dot size range on the film.

3.1.2   Focused ion beam methods

Focused ion beams (FIBs) generate minute beams of several positive ions (from sub-nanometer to a few micrometers) energy (between a few keV to ~ 150 keV) to produce the exclusion of sample substance by sputtering at currents between sub-pA to μA [60].

The technique is used in the manufacturing of semiconductors. The method is similar to that of scanning electron microscope (SEM) though FIBs utilize a thinly focused ions beam controlled at high or low beam currents while an electron beam is employed in SEM. The size of the ion beam decides the various dimensional properties of the QDs. A diameter beam between 8–20 nm suitable for laboratory and marketable requirements is reported, assigning the QDs etching to sizes < 100 nm [61].

The majority of FIBs device use metal ion in liquid state for instance gallium. Occasionally gold in addition to iridium are employed as well. In case of gallium, the metal is positioned close to a tungsten pointer which when warmed up moists the pointer and runs to the tip. The tip radius is extremely small of the order ~2 nm. Higher electric field at the tip results in ionization of the gallium as well as field emission. Ions source then picks up speed up to 40 keV, centered over sample using variable lens. A recent FIB could convey several nano amperes of current. But expensive instruments during the process results in low yield and cause residual surface damage [62].

Friedensen et al. [63] successfully created nanowires by thinning and cutting from larger structures using Ga^+-based FIB milling, for controlling the thickness of Bi_2Se_3. The Bi_2Se_3 nanowires prepared by FIBs showed enhanced room temperature photoconductivity, preserving the originality of their topological surface states at smaller temperatures. Fig.2 [63] shows a Ga^+-based FIB milling.

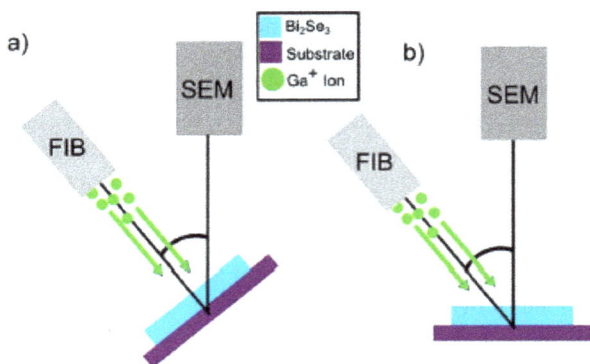

Fig. 2 Figure showing arrangement of FIB milling with flake of Bi_2Se_3. (a) straight-on milling. (b) angled milling. Reprinted with permission from ref. [63]. Copyright 2018 Springer Nature.

3.1.3 Etching techniques

The etching is a renowned process and plays a vital role in the nano-syntheses. In dried up etching, the compartment is filled with a gas and plasma is produced using controlled radiofrequency voltage, which further assists in breaking the gaseous molecules down to reactive residues. These molecules with elevated kinetic energy then hit the plane producing a volatile substituent to etch a particular outlined model. When the active constituents are ions, the method is identified to be reactive ion etching. With a covering design, discriminatory etching of constituents is obtained. With this technique, using boron trichloride and argon, GaAs/AlGaAs QDs ~ 40 nm is reported. ZnTe QDs close-packed array having interdot space approx. 190 - 350 nm was synthesized using methane and hydrogen by etching technique [64].

Chen et al. [65] proposed the HF/H_2O_2 based anisotropic etching of Si with the aid of Cu nanoparticles due to the Si crystal with anisotropic electrochemical nature. The rate of etching, mechanism, and ratios were studied. The Cu assisted etching produced inverted pyramid arrays for controlling the reflections in Si solar cells whereas the upright pyramids are obtained from the anisotropic and localized etching process. Fig. 3 [65] illustrates the exterior morphology of (100) Si etched pyramid structures.

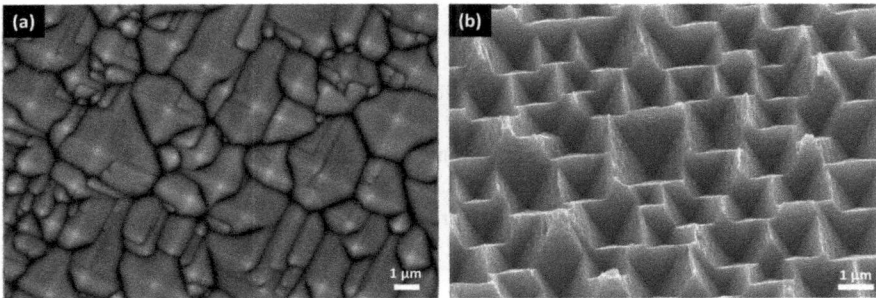

Fig. 3 Exterior morphology of Si (a) upright pyramidal structures etched in alkaline solution (b) Cu based acid-etched inverted pyramidal structures. Reprinted with permission from ref. [65]. Copyright 2018 Springer Nature.

3.2 Bottom-up methods

In this approach, different physical and chemical methods are strapped up to form clusters and nanostructures. These are of two types-

a) Wet-chemical techniques

b) Vapor-phase techniques

Vapor-phase techniques occur by molecular beam epitaxy, physical or chemical vapor deposition, and sputtering, etc. while in wet-chemical process solution phase methods like sol-gel method, microemulsion, hot-solution techniques are significant.

3.2.1 Wet-chemical techniques

Wet-chemical techniques are traditional methods of precipitation in which a single solution or mixtures are the precursors of the chemical reactions giving rise to zero-dimensional QDs with cautious control over various parameters. The QDs of specific dimensions can be obtained by regulating the temperature, pH, and ratios of cationic to anionic precursors. Capping agents, micellar dispersions or stabilizers, surfactants with different amount of precursors also have prominent results in these routes. The method consist nucleation and controlled development of nanocrystals. Nucleation takes place homogeneously when constituents of the solute unite and attain a critical dimension without the aid of an offered crossing point [66]. Different wet chemical techniques are discussed below.

3.2.1.1 Sol-gel process

It is a powerful method which is an amalgamation of three major steps, comprising hydrolysis, sol development or condensation, along with subsequent gel development [67,68]. In a usual method, a sol, dissolved in solvent, is controlled via metals (normally metal chloride and alkoxides) as precursor. The precursor undergoes hydrolysis meanwhile condenses giving rise to sol, subsequently polymerization occurs producing gel. Thus, the sol advances enroute for the development of a gel-type diphasic organization that consists of solid as well as liquid-phase with morphologies varying between isolated molecules to continuous polymers. Afterward residual solvents were eliminated with a considerable shriveling and densification [62].

The rate of solvent removal is eventually determined by the allocation of gel porosity. Thermal treatment is often followed for polycondensation as well as to improve stability and mechanical property by final sintering, densification, and grain development. For instance, ZnO QDs has been prepared by adding the precursors, zinc-acetate in ethanol, sodium hydroxide, and subsequent open air aging [69]. By sol-gel technique, Park et al. [70] synthesized ZnO thin films, and used it as n-type-heterojunction QDs-solar cells. The process requires low temperature, but it bears an unprecise size distribution and high defect concentrations.

Materials Research Forum LLC
https://doi.org/10.21741/9781644901250-2

3.2.1.2 Microemulsion technique

Microemulsions are colloidal surfactant film stabilized 'nano-dispersions' of water in oil (or oil in water). These are considered as true nanoreactors that are thermodynamically stable and can accomplish chemical reactions for the synthesis of nanomaterials. It is an adaptable method that permits the synthesis of a huge variety of nanomaterials alone or in grouping with other methods. The scheme of working is that by suitably controlling the synthesis parameters, nanoreactors could produce tailor-made yields to the nanoscale regime with innovative and exceptional properties. The specific control of various parameters resulting in the final particle sizes and shapes is still a confront, but the affluence of the potential which can be achieved from nano to macroscale regime is a big drive to work with these miniature dispersed particles [71].

The dimensions of QDs could be altered by adjusting the surfactant to water molar proportions with fine size distributions. Major drawbacks of the process include integration of impurities, low yield, and defects. The reverse micelle is an accepted method for preparing QDs, where nanosized globules of water in solution of n-alkane are obtained using surfactants, like triton-X, cetylpyridinium chloride or cetyltrimethyl-ammonium chloride. The hydrophobic as well as the hydrophilic groups are present equally at the reverse ends, large number of small drops known as micelles are produced. The vigorous stirring of micellar solutions occur with exchange of reactant molecules, resulting in dynamic collisions. The development of the resulting QDs is constrained by micelle size which in turn is adjusted by molar proportion of water and surfactant. The technique is inured to produce QDs of II-VI groups, like CdS [72,73] core/shell CdS: Mn/ZnS [74,75] and carbon QDs using hydride reducing agents [76]. A one-step simple synthesis technique for single QDs with spherical silica nanoparticles using cyclohexane and synperonic NP-5 as oil phase and surfactant respectively was given by Darbandi et al. [77].

3.2.1.3 Hot-solution decomposition techniques

Pyrolysis of the organometallic complexes at very high temperatures is known to be as Hot-solution decomposition process. It was first given by Bawendi and co-workers in 1993 [78]. The method entails an initial degassing and drying of the solvent to ~300°C in a vacuum. A moisture-free mixture of precursor and solvent is organized and introduced energetically with stirring in a container, causing a uniform seeding process producing QDs, with consequent development throughout the Ostwald ripening. The high free energy of small sized crystals allows them to drop weights to greater size causing gradual enhancement of QDs size at ~250°C. The QDs dispersion is stabilized by the coordinating solvent thus improving passivation of the surface, providing an adsorption barricade to lower the QDs growth. At regular intervals, aliquots may be taken to attain a preferred particle size. The scheme has been widely used for preparing II-VI [79], IV-VI [80] and III-V [81] QDs. The purity of coordinating solvent also possesses a chief task in managing the dimensionality. Disadvantage of process lies in its higher expenses owing to the higher temperature, the organometallic solvents toxicity and reduced solubility in water [62].

3.2.2 Vapour-phase techniques

In this technique, layers or banded stratum of QDs are grown through the atom-next to-atom manner by hetero-epitaxial development of tensed objects [82,83]. Usually, the layered materials grow as homogeneous, epitaxial layers [84] tracked by nucleation and development of islands [85] or as zero-dimensional nanostructures straightly on the substrate (*Stranski-Krastonow* mode) [86]. Based on lattice strain or interfacial energies the growth modes are monitored. For instance, an overlayered substrate with huge lattice mismatch and aptly minute interface energies, growth occur layer-by-layer. Conversely, for a thick film with monolayers, and huge strain power, the arrangement decreases the entire free energy through intruding the layer to secluded islands or QDs formation takes place. *Stranski-Krastonow* development is seen with a material having fine lattice equivalent with components, however, surface energy is smaller compared to the interfacial energy sum between overlayer and substrate [87]. Moreover, the self-assembling of QDs by these methods is an efficient one in producing QDs arrays without a template.

3.2.2.1 Molecular beam epitaxy (MBE)

This technique is an atomic layer by layer growing method for crystals, depending on the reaction of heated crystalline substrate with atomic or molecular beams, at ultra-high vacuum condition. Thus, a significant concern in epitaxial growth of nanocrystal is the facade conditions of preliminary substrate along with succeeding stratums. MBE is distinctive in two respects: it is based on the reaction of *electron axis* with crystalline face, depending on various *kinetic systems*, secondly performed in ultra-high vacuum conditions [88]. It is used for growing III-V [89,90] and II-VI semiconductors nanocrystals [91,92]. MBE has lower throughput than other epitaxial forms. During the process, reflection with elevated energy diffraction of electrons is frequently employed for observing development of layers.

Lei et al. [93] prepared to conduct $LaAlO_3/SrTiO_3$ interfaces with atom-next to-atom layer laser MBE technique at elevated oxygen levels showing no oxygen vacancies. Fig.4 [93] shows an MBE arrangement of atom-next to-atom layer. The targets SrO and TiO_2

are interchangeably ablated by an ultra-violet beam, the reflection with elevated energy diffraction of electrons arrangement is obtained and real-time analysis is done. The $SrTiO_3$ film is created in an atom-next to-atom layer style by MBE.

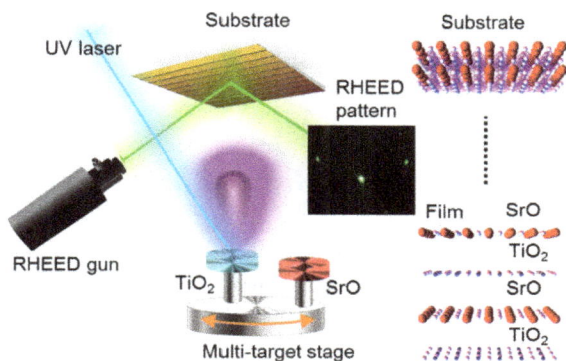

Fig. 4 Diagramatic representation of ALL-Laser MBE setup and synthesis of SrTiO3 film assembled in an atom-next to-atom layer style [93]. Reprinted with permission from ref. [93]. Copyright 2017 Springer Nature.

3.2.2.2 Physical vapor deposition (PVD) technique

The PVD technique is characterized by a method in which the material gets converted from a solid or liquid phase to vapor and again to a nanoscale or thin film solid phase. In other words, the material is vaporized as atom or molecules, carried in the vapor form through a vacuum or low-pressure gas. In industries, the frequently used PVD technique is sputtering, in which the exterior of the sputtering object is attacked with gaseous ion at high voltage acceleration. The momentum thus gets conveyed from the incoming particle to the atom/molecules. Arc vapor deposition, vacuum deposition, pulsed laser ablation, and ion plating are some of the commonly used PVD process [94,95]. PVD technique for the synthesis of Nb_2O_5 and CdSe/CdTe QDs are previously reported [96-98].

3.2.2.3 Chemical vapor deposition (CVD) technique

In typical CVD technique, the precursors in gaseous form get converted by a chemical reaction into solid film or powder and get deposited over the substrate at higher temperatures. Volatile reactions occur during the process, unused precursors and end products are attained, eliminated by the gas surge inside reaction compartment. CVD are of different types, for instance (i) Vapour phase epitaxy when the technique is employed for deposition of only one crystal, (ii) Metal-organic deposition technique where metal-organic species are the precursors, (iii) Plasma enhanced CVD where the plasma enhances the reaction and (iv) Low-pressure CVD where the decomposition takes place at low pressure. Likewise photochemical, atmospheric pressure and laser CVD are eminent as well [94,95]. Kim et al. [99] prepared ZnSe/ZnS QDs by metal-organic CVD with the *Stranski Krastanov* growth mode. Dhawan et.al [100] synthesized InAs QDs over germanium by the metal-organic deposition method.

3.3   Other synthetic techniques

3.3.1   Using ultrasonic or microwave irradiation

In current years, the exceptional features of the microwaves or ultrasonic techniques has appeared as probable substitutes for preparing very small crystals like QDs. Preparation of QDs between 1-5 nm by an aqueous mixture of precursors have been reported [101-103]. Chemical synthesis by ultrasonic irradiation is an advantageous and approving technique considering the temperature, time for synthesis and the homogeneity of the particles size [104]. In the sonochemical process, the acoustic cavitations occur in contact with ultrasonic waves (30 kHz–10 MHz); where the temperature gets raised up to 5000 Kelvin within nanoseconds, higher pace of cooling (1012 Ks^{-1}) along with pressures ~1700 kPa. The developed bubbles got collapsed, during cavitations, resulting in shock-waves; altering the ultrasound-irradiated matter structurally [105]. Also, by ultrasonic-assisted methods, the surface defects, and morphology of QDs can be restricted [106]. Rajabi et al. [11] prepared ZnS QDs with average particle size 5 nm where zinc nitrate was dispersed in a solution and L-cysteine and sodium sulphide were added and sonicated at 40°C.

3.3.2   Hydrothermal synthesis technique

Hydrothermal synthesis method has been developed to synthesize high-quality, water-soluble QDs [107,108]. The process involves inorganic salts crystallization, directed by temperature and pressure, in aqueous solution. The solubility characteristically decreases as the pressure and/or temperature is decreased, resulting in the formation of nanocrystals as precipitate. Variation in the temperature, reaction, pressure, aging time and precursors, results in different sizes of the QDs. For QDs synthesis, other wet-chemical techniques were also reported. To prepare sulfide-based QDs, H$_2$S gas has been passed to precursors [109,110].

3.3.3 Solvothermal technique

The technique is analogous to hydrothermal synthetic method, in which the reactions are accomplished in an autoclave. The variations in the methods are the precursors which are generally non-aqueous in the solvothermal method. The word "solvothermal" is generally depicts organics employed as solvents. The method is employed to produce diverse morphologies of TiO_2 nanocrystals [90] and nanosheets of carbon employing sodium and ethanol, which when react produce an intermediate. It is followed by pyrolysis, and graphene sheets are obtained [91].

3.3.4 From microorganism bio-template

Bio-fabrication means a set of techniques attempted to engineer intricate constructs (living and non-living) with predetermined biological characteristics employing biomaterials, extracellular matrices, fragments or intact living cells, etc. [113]. Microbes such as bacteria, algae, virus and fungi possessing a good adaptation capacity, variety in shapes and sizes, existence of various functional groups at the surfaces, biological characters, basic cell wall porosity, and simplicity of handling facilitate them to be employed as templates for biosynthesis of diverse bionanomaterials based on a bottom-up approach [114]. By immobilization and other technique, living microbes could be preset on a chip to produce micro and nanostructured arrays. By manipulating microbial cell movement, commanding various factors such as galvanotaxis, chemotaxis, phototaxis, and magnetotaxis offers insight for the improvement of novel fabrication methods utilizing the microbes as a smart cell factory for producing functional metabolites.

Li et al. [115] developed a uniform sized CdSe QDs in intracellular redox environment of yeasts, demonstrating a prospective for cellular biosynthesis of nanoparticles. Bai et al. [116] used Rhodobacter sphaeroides to illustrate an enhancement in the dimension of CdS nanocrystals with increasing duration for incubation of cells.

3.3.5 Electrochemical assembly

By means of electrochemical techniques, a highly ordered array of QDs could be self-assembled. In this method, template is produced through an ionic reaction over a metal-electrolyte crossing point resulting in an assemblage of nanocrystals or QDs, over the metal surface. After that, it will be employed as mask used in mesa-etching of the nanoscale structures over a selected substrate. The method is comparatively cheap and also there is absence of any radiation harm unlike beam lithography, which comprises a good yield appropriate for large production. Bandyopadhyay et al. [117,118] illustrated an electrochemically self-assembled route for well structured partial periodic assortment of QDs and nonlinear optical properties of self-assembled CdS QDs. The same group also

Materials Research Forum LLC
https://doi.org/10.21741/9781644901250-2

reported Raman study of self-assembled CdS QDs quasi periodic arrays over an anodized alumina film [119].

3.3.6 Cluster-seed method

For producing huge quantities of QDs in a single batch, Nigel Pickett [120] developed the new method known to be as Cluster-Seed method. It was already known that cluster of $Cd_{10}Se_4(SPh)_{16}$ germs CdSe QDs formation moreover quantity of QDs obtained is equivalent to the clusters added in a batch process. Jawaid et al. [121] synthesized CdSe QDs by subsequently enhancing the clusters where it was observed that the nanocrystals obtained are proportional to clusters. Moreover, using clusters resulted in the development of smaller QDs due to the struggle between huge numbers of dots for fewer precursors; also no QDs were obtained in the absence of clusters. Fig. 5 [121] shows synthesis of QDs by cluster seed method. Besides the group proposed a scheme producing lone CdSe QD employing $[Na(H_2O)_3]_2[Cu_4(SPh)_6]$ as germ seed. This technique was derived from the germ-seed scheme, where organometallic groups act as nucleating core for QDs.

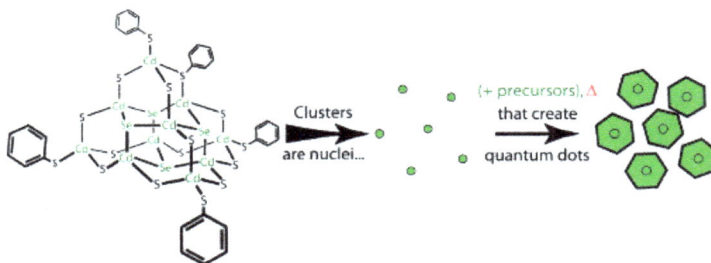

Fig. 5 QDs synthesis by Cluster-seed method. The organometallic clusters work as nucleation points producing equal amount of QDs. Reprinted with permission from ref. [121]. Copyright 2017 American Chemical Society.

3.4 Bulk-manufacturing technique

For the fabrication of an extensive assortment of QDs sizes along with compositions, the traditional "high-temperature dual injection" small scale production method is frequently used for II-VI group materials. But for III-V group QDs, the nature of bonding is more covalent so it becomes quite intricated to part the nucleation and development of nanoparticle by this method. Besides at a higher scale, quick injections of the huge

Materials Research Forum LLC
https://doi.org/10.21741/9781644901250-2

amounts of solution into one another causes temperature differential resulting in extensive size divisions.

Thus, a reiterating technique, synthesizing huge extent of good-quality, consistent QDs entail precursors in the company of a molecular cluster where the integrity of the cluster are retained acting as a preassembled core template. Each molecule of the cluster compounds work as a nucleation centre over which nanoparticles develop. Thus, the nucleation methods at elevated-temperatures could be avoided for the nanocrystals development as appropriate nucleating centres are offered by the molecular cluster [122]. The growth of the particle is sustained by the intermittent incorporation of precursors at reasonable temperatures until the preferred particle dimensions are attained [121].

4.   Perspective

Since the last decade, the recent trend in exploration of science is intended towards nanotechnological growth and its application in different fields. The recent lighting, LEDs and photovoltaic solar cells, electronic gadgets and digital display utilize energy efficiently from bright QDs. Besides QDs have applications in biological imaging, screening of cancer cells, drug delivery, chemical sensing, labeling, solar paints, organic dye removal. All these stipulate a simple and economical route for the production of nanomaterials through a trouble-free manufacturing method at marketable scales. This chapter presents the conventional top-down as well as bottom-up techniques for QDs synthesis. Also the article brings some up to date advances of the recent times with the intention that the current developments in the field could be trailed with no trouble.

Acknowledgement

Jyoti Patel is thankful to DST, New Delhi, India for Research fellowship under Women Scientist Scheme (SR/WOS-A/CS-82/2018).

References

[1]   S. Raj, J.H. Yun, G.R. Adilbish, K.Ch, I.H. Lee, M.S. Lee, Y.T.Yu, Formation of core@multi-shell CdSe@CdZnS–ZnS quantum dot heterostructure films by pulse electrophoresis deposition, Superlattices Microstruct. 83 (2015) 618-626. https://doi.org/10.1016/j.spmi.2015.03.043

[2]   P. Matagne, J.P. Leburton, Quantum Dots: Artificial Atoms and Molecules, in: H.S. Nalwa, S. Bandyopadhyay (Eds.) Quantum Dots and Nanowires, American Scientific Publishers Stevenson Ranch, California, 2003, pp. 2-66.

[3]   W.C. Chan, D.J. Maxwell, X.Gao, R.E. Bailey, M. Han, Luminescent quantum dots for multiplexed biological detection and imaging, Curr. Opin. Biotechnol. 13 (2002) 40–46. https://doi.org/10.1016/S0958-1669(02)00282-3

[4]   A.P. Alivisatos, Perspectives on the physical chemistry of semiconductor nanocrystals, J. Phys. Chem., 100 (1996) 13226-13239. https://doi.org/10.1021/jp9535506

[5]   A.M. Smith, S. Nie, Semiconductor nanocrystals: structure, properties, and band gap engineering, Acc. Chem. Res. 43 (2010) 190-200. https://doi.org/10.1021/ar9001069

[6]   L.Pang, Y.Zhou, W.Gao, J. Zhang, H. Song, X. Wang, Y. Wang, X. Peng, Curcuminbased fluorescent and colorimetric probe for detecting cysteine in living cells and zebrafish, Ind. Eng. Chem. Res. 56 (2017) 7650−7655. https://doi.org/10.1021/acs.iecr.7b02133

[7]   F. Cai, D. Wang, M. Zhu, S. He, Pencil-like imaging spectrometer for biosamples sensing, Biomed. Opt. Express 8 (2017) 5427−5436. https://doi.org/10.1364/BOE.8.005427

[8]   D. Zhang, N. Xu, H. Li, Q. Yao, F. Xu, J. Fan, J. Du, X .Peng, Probing thiophenol pollutant in solutions and cells with bodipy based fluorescent probe, Ind. Eng. Chem. Res. 56 (2017) 9303−9309. https://doi.org/10.1021/acs.iecr.7b02557

[9]   Y. Chen, T. Wei, Z. Zhang, T. Chen, J. Li, J. Qiang, J. Lv, F. Wang, X. Chen, A Benzothiazole-based fluorescent probe for ratiometric detection of Al3+ in aqueous medium and living cells, Ind. Eng. Chem. Res. 56 (2017) 12267−12275. https://doi.org/10.1021/acs.iecr.7b02979

[10]   J. Pardo, Z. Peng, R.M. Leblanc, Cancer targeting and drug delivery using carbon-based quantum dots and nanotubes, Molecules 23 (2018) 378-398. https://doi.org/10.3390/molecules23020378

[11]   H.R. Rajabi, F. Karimia,; H. Kazemdehdashti, L. Kavoshi, Fast sonochemically-assisted synthesis of pure and doped zinc sulfide quantum dots and their applicability in organic dye removal from aqueous media, J. Photochem. Photobiol. 181 (2018) 98–105. https://doi.org/10.1016/j.jphotobiol.2018.02.016

[12]   G. Shen, Z. Du, Z. Pan, J. Du, X. Zhong, Solar paint from TiO2 particles supported quantum dots for photoanodes in quantum dot−sensitized solar cells, ACS Omega 3 (2018) 1102–1109. https://doi.org/10.1021/acsomega.7b01761

[13] K. Jiang, L. Zhang, J. Lu, C. Xu, C. Cai, H. Lin, Triple-mode emission of carbon dots: applications for advanced anti-counterfeiting, Angew. Chem. Int. Ed. 55 (2016) 7231–7235. https://doi.org/10.1002/anie.201602445

[14] M. You, M. Lin, S. Wang, X. Wang, G. Zhang, Y. Hong, Y. Dong, G. Jin, F. Xu, Three-dimensional quick response code based on inkjet printing of upconversion fluorescent nanoparticles for drug anti-counterfeiting, Nanoscale 8 (2016) https://doi.org/10096–10104. 10.1039/c6nr01353h

[15] Y.W. Zhang, G. Wu, H. Dang, K. Ma, S. Chen, Multicolored mixed-organic-cation perovskite quantum dots (FAxMA1-xPbX3,X=Br and I) for white light-emitting diode, Ind. Eng. Chem. Res. 56 (2017) 10053–10059. https://doi.org/10.1021/acs.iecr.7b02309

[16] D.V. Talapin, J.S. Lee, M.V. Kovalenko, E.V. Shevchenko, Prospects of colloidal nanocrystals for electronic and optoelectronic applications, Chem. Rev. 110 (2010) 389–458. https://doi.org/10.1021/cr900137k

[17] U. Resch-genger, M. Grabolle, S. Cavaliere-Jaricot, R. Nitschke, T. Nann, Quantum dots versus organic dyes as fluorescent labels, Nat. Methods 5 (2008) 763–775. https://doi.org/10.1038/nmeth.1248

[18] X. Michalet, F.F. Pinaud, L.A. Bentolila, J.M. Tsay, S. Doose, J.J. Li, G. Sundaresan, A.M. Wu, S.S. Gambhir, S. Weiss, Quantum dots for live cells, in vivo imaging, and diagnostics, Science 307 (2005) 538–544. https://doi.org/10.1126/science.1104274

[19] M. Bruchez Jr, M. Moronne, P. Gin, S. Weiss, A.P. Alivisatos, Semiconductor nanocrystals as fluorescent biological labels, Science 281 (1998) 2013–2016. https://doi.org/10.1126/science.281.5385.2013

[20] W.C. Chan, S. Nie, Quantum dot bioconjugates for ultrasensitive nonisotopic detection, Science 281 (1998) 2016–2018. https://doi.org/10.1126/science.281.5385.2016

[21] H.M. Azzazy, M.M. Mansour, S.C. Kazmierczak, From diagnostics to therapy: prospects of quantum dots, Clin. Biochem. 40 (2007) 917–927. https://doi.org/10.1016/j.clinbiochem.2007.05.018

[22] T.J. Deerinck, The application of fluorescent quantum dots to confocal, multiphoton, and electron microscopic imaging. Toxicol. Pathol. 36 (2008) 112–116. https://doi.org/10.1177/0192623307310950

[23]   Y. Yin, A.P. Alivisatos, Colloidal nanocrystal synthesis and the organic– inorganic interface, Nature 437 (2005) 664–670. https://doi.org/10.1038/nature04165

[24]   Z.A. Peng, X.G. Peng, Nearly monodisperse and shape-controlled CdSe nanocrystals via alternative routes: nucleation and growth, J. Am. Chem. Soc. 124 (2002) 3343–3353. https://doi.org/10.1021/ja0173167

[25]   A.P. Alivisatos, Semiconductor clusters, nanocrystals, and quantum dots, Science, 271 (1996) 933–937. https://doi.org/10.1126/science.271.5251.933

[26]   R. Dhar, Deepika, Synthesis and current applications of quantum dots: A Review, Nanosci. Nanotech. An Inter. J. 4 (2014) 32-38.

[27]   S. Mishra, B. Panda, S.S. Rout, An elaboration of quantum dots and its applications, Inter. J. of Adv. in Engin. Technol. 5 (2012) 141-145.

[28]   E. Petryayeva, W.R. Algar, I.L. Medintz, Quantum dots in bioanalysis: a review of applications across various platforms for fluorescence spectroscopy and imaging. Appl.Spectrosc. 67 (2013) 215-252. https://doi.org/10.1366/12-06948

[29]   X. Cheng, S.B. Lowe, P.J. Reece, J.J. Gooding, Colloidal silicon quantum dots: from preparation to the modification of self-assembled monolayers (SAMs) for bio-applications. Chem. Soc. Rev. 43 (2014) 2680-2700. https://doi.org/10.1039/c3cs60353a

[30]   P. Mulpur, T.M. Rattan, V. Kamisetti, One-step synthesis of colloidal quantum dots of iron selenide exhibiting narrow range fluorescence in the green region. J.Nanosci. 2013 (2013) 1-5. https://doi.org/10.1155/2013/804290

[31]   H. Zhang, Y. Li, X. Liu, P. Liu, Y. Wang, T. An, H. Yang, D. Jing, H. Zhao, Determination of iodide via direct fluorescence quenching at nitrogen-doped carbon quantum dot fluorophores, Environ. Sci. Technol. Lett. 1 (2014) 87-91. https://doi.org/10.1021/ez400137j

[32]   K. Hoshino, A. Gopal, M.S. Glaz, D.A. Vanden Bout, X. Zhang, Nanoscale fluorescence imaging with quantum dot near-field electroluminescence, Appl. Phys. Lett. 101 (2012) 043118. https://doi.org/10.1063/1.4739235.

[33]   J. Zhang, W. Sun, L. Yin, X. Miao, D. Zhang, One-pot synthesis of hydrophilic CuInS2 and CuInS2-ZnS colloidal quantum dots, J. Mater. Chem. C 2 (2014) 4812-4817. https://doi.org/10.1039/C3TC32564D

[34]   M. Bottrill, M. Green, Some aspects of quantum dot toxicity, Chem. Commun. 47 (2011) 7039-7050. https://doi.org/10.1039/c1cc10692a

[35]   A. Kitai, Luminescent Materials and Applications, John Wiley & Sons Ltd. West Sessex, England, 2008.

[36]   D. Bera, L. Qian, P.H. Holloway, Semiconducting Quantum Dots for Bioimaging, in:Y.Pathak, D.Thassu (Eds.), Drug Delivery Nanoparticles Formulation and Characterization, first ed., CRC Press, UK, 2009, pp. 349-366.

[37]   S. Bandyopadhyay, H.S. Nalwa, Quantum Dots and Nanowires. American ScientificPublishers, USA, 2003.

[38]   H.S. Nalwa, Nanostructured Materials and Nanotechnology, Academic Press: San Diego, California, USA, 2002.

[39]   T.R. Groves, Electron beam lithography, in: M. Feldman (Ed.), Nanolithography: The Art of Fabricating Nanoelectronic and Nanophotonic devices and systems, Woodhead Publishing, Philadelphia, USA, 2014, pp. 80-115.

[40]   I. Brodie, J.J. Muray, The Physics of Micro/Nano-Fabrication, Plenum Press, New York, USA, 1992.

[41]   G. Timp, Nanotechnology, Springer-Verlag, American Institute of Physics, New York,USA, 1999.

[42]   N. Pala, M. Karabiyik, Electron Beam Lithograph, in: B. Bhushan (Eds.), Encyclopedia of Nanotechnology. Springer, Dordrecht, New York, 2016, pp. 19-77.

[43]   A.A. Tseng, K. Chen, C.D. Chen, K.J. Ma, Electron beam lithography in nanoscale fabrication: recent development, IEEE Trans. Electron. Packag. Manuf. 26 (2003) 141-149. https://doi.org/10.1109/TEPM.2003.817714

[44]   S.K. Ghandhi, VLSI Fabrication Principles: Silicon and Gallium Arsenide, second ed. Wiley, New York, 1994.

[45]   M. Zheng, Y. Chen, Z. Liu, Y. Liu, Y. Wang, P. Liu, Q. Liu, K. Bi, Z.Shu,Y. Zhang, H.Duan, Kirigami-inspired multiscale patterning of metallic structures via predefined nanotrench templates,Microsyst. Nanoeng. 5 (2019) 1-11. https://doi.org/10.1038/s41378-019-0100-3

[46]   S.D. Berger,J.M.Gibson, New approach to projection electron lithography with demonstrated 0.1 micron linewidth, Appl. Phys. Lett., 57 (1990) 153–155.

[47]   H.C. Pfeiffer, Advanced e-beam systems for manufacturing, in: M. Peckerar, (Ed.), Proceedings Electron-Beam, X-Ray, Ion Beam Submicrometer Lithographies for Manufacturing II, Microlithography CA, USA, 1671 (1992) 100–110.

[48]   Prevail: IBM's e-beam technology for next-generation lithography, IBM MicroNews, 6 (2000) 41–44. https://doi.org/10.1117/12.390056

[49]   R.S. Dhaliwal, W.A. Enichen, S.D. Golladay, R.A. Kendall, J.E. Lieberman, H.C. Pfeiffer, D.J. Pinckney, C.F. Robinson, J.D. Rockrohr, W. Stickle, and E.V. Tressler, Prevail: Electron projection technology approach for next-generation lithography, IBM J. Res. Develop., 45(2000) 615–638. https://doi.org/10.1147/rd.455.0615

[50]   A. del Campo, E. Arzt, Fabrication approaches for generating complex micro- and nanopatterns on polymeric surfaces, Chem. Rev. 108 (2008) 911. https://doi.org/10.1021/cr050018y

[51]   M.H.V. Werts, M. Lambert, J.P. Bourgoin and M. Brust, Nanometer scale patterning of langmuir–blodgett films of gold nanoparticles by electron beam lithography, Nano Lett. 2 (2002) 43-47. https://doi.org/10.1021/nl015629u

[52]   V. Nandwana, C. Subramani, Y.C. Yeh, B. Yang, S. Dickert, M.D. Barnes, M.T. Tuominen, V.M. Rotello, Direct patterning of quantum dot nanostructures via electron beam lithography, J. Mater. Chem. 21 (2011) 16859-16862. https://doi.org/10.1039/C1jm11782c

[53]   N.W. Parker, A.D. Brodie, J.H. McCoy, A high throughput NGL electron beam direct-write lithography system, SPIE Emerg.Lithograph. Technol.3997 (2000)

[54]   T.H.P. Chang, D.P. Kern, L.P. Muray, Arrayed miniature electron beam columns for high throughput sub-100 nm lithography, J. Vac. Sci. Technol. B 10 (1992) 2743-2748. https://doi.org/10.1116/1.585994

[55]   T.R. Groves, R.A. Kendall, Distributed multiple variable shaped electron beam column for high throughput maskless lithography, J. Vac. Sci. Technol. B 16 (1998) 3168-3173. https://doi.org/10.1116/1.590458

[56]   H.C. Pfeiffer, Variable spot shaping for electron-beam lithography, J. Vac. Sci. Technol. 15 (1978) 887-890. https://doi.:10.1116/1.569621

[57]   K. Hattori, Electron-beam direct writing system EX-8D employing character projection exposure method, J. Vac. Sci. Technol. B 11 (1993)2346-2351. https://doi.org/10.1116/1.586984

[58]   G.I. Winograd, L. Han, M.A. McCord, R.F.W. Pease, V. Krishnamurthi, Multipexed blanker array for parallel electron beam lithography, J. Vac. Sci. Technol. B 16 (1998) 3174-3176. https://doi.org/10.1116/1.590345

[59]   J. Schneider, Patterned negative affinity photocathodes for maskless electron beam lithography, J. Vac. Sci. Technol. B 16 (1998)3192-3196. https://doi.org/10.1116/1.590349

[60]   M. Utlaut, Focused Ion Beams For Nano-Machining And Imaging, in: M. Feldman (Ed.), Nanolithography: The Art of Fabricating Nanoelectronic and Nanophotonic devices and systems, Woodhead Publishing, Philadelphia, USA, 2014, pp. 116-157.

[61]   E. Chason, S.T. Picraux, J.M. Poate, J.O. Borland, M.I. Current, T.D. delaRubia, D.J. Eaglesham, O.W. Holland, M.E. Law, C.W. Magee, J.W. Mayer, J. Melngailis, A.F. Tasch, Ion beams in silicon processing and characterization,J. Appl. Phys. 81 (1997) 6513-6561. https://doi.org/10.1063/1.365193

[62]   R. Karmakar, Quantum dots and it method of preparations-revisited, Prajnan O Sadhona. 2 (2015) 116-142.

[63]   S. Friedensen, J.T. Mlack, M. Drndic, Materials analysis and focused ion beam nanofabrication of topological insulator Bi2Se3, Sci. Rep. 7 (2017) 13466-13473. https://doi.org/10.1038/s41598-017-13863-6

[64]   K. Tsutsui, E.L. Hu, C.D.W. Wilkinson, Reactive ion etched II-VI quantum dots: dependence of etched profile on pattern geometry. Jpn. J. Appl. Phys. 32 (1993) 6233-6236. https://doi.org/10.1143/JJAP.32.6233

[65]   W. Chen, Y. Liu, L. Yang, J. Wu, Q. Chen, Y. Zhao, Y. Wang, X. Du, Difference in anisotropic etchingcharacteristics of alkaline andcopper based acid solutions for single-crystalline Si, 8 (2018)3408-3416. https://doi.org/10.1038/s41598-018-21877-x

[66]   C. Burda, X.B. Chen, R. Narayanan, M.A. El-Sayed, Chemistry and properties of nanocrystals of different shapes, Chem. Rev. 105 (2005) 1025–1102. https://doi.org/10.1021/cr030063a

[67]   B. Mashford, J. Baldauf, T.L. Nguyen, A.M. Funston, P. Mulvaney, Synthesis of quantum dot doped chalcogenide glasses via sol-gel processing. J. Appl. Phys. 109 (2011) 94305-94312. https://doi.org/10.1063/1.3579442

[68]   L. Korala, Z. Wang, Y. Liu, S. Maldonado, S.L. Brock, Uniform thin films of CdSe and CdSe(ZnS) core shell) quantum dots by sol–gel assembly: enabling photoelectrochemical characterization and electronic applications, ACS Nano 7 (2013), 1215-1223. https://doi.org/10.1021/nn304563j

[69]   J. Bang, H. Yang, P.H. Holloway, Enhanced and stable green emission of ZnO nanoparticles by surface segregation of Mg. Nanotechnology 17 (2006), 973-978. https://doi.org/10.1088/0957-4484/17/4/022

[70]   H.Y. Park, I. Ryu, J. Kim, S. Jeong, S. Yim, S.Y. Jang, PbS quantum dot solar cells integrated with sol–gel-derived ZnO as an n-type charge-selective layer,J. Phys. Chem. C. 118 (2014), 17374-17382. https://doi.org/10.1021/jp504156c

[71]   M.A. Lopez-Quintela, Synthesis of nanomaterials in microemulsions: formation mechanisms and growth control, Curr. Opin. Colloid Interface Sci. 8 (2003) 137-144. https://doi.org/10.1016/S1359-0294(03)00019-0

[72]   V.L. Colvin, A.N. Goldstein, A.P. Alivisatos, Semiconductor nanocrystals covalently bound to metal-surfaces with self-assembled monolayers, J. Am. Chem. Soc. 114 (1992) 5221-5230. https://doi.org/10.1021/ja00039a038

[73]   K. Lemke, J. Koetz, Polycation-capped CdS quantum dots synthesized in reverse microemulsions. J. Nanomater. 10 (2012)1-10. https://doi.org/10.1155/2012/478153

[74]   H.Yang, P.H. Holloway, Enhanced photoluminescence from CdS:Mn/ZnS core/shell quantum dots. Appl. Phys. Lett. 82 (2003) 1965-1967. https://doi.org/10.1063/1.1563305

[75]   H.Yang, P.H. Holloway, G. Cunningham, K.S. Schanze, CdS:Mn nanocrystals passivated by ZnS: Synthesis and luminescent properties, J. Chem. Phys. 121 (2004) 10233-10240. https://doi.org/10.1063/1.1808418

[76]   K. Linehan, H. Doyle, Size controlled synthesis of carbon quantum dots using hydride reducing agents, J. Mater. Chem. C 2 (2014) 6025-6031. https://doi.org/10.1039/C4TC00826J

[77]   M. Darbandi, R. Thomann, T. Nann, Single quantum dots in silica spheres by microemulsion synthesis. Chem. Mater. 17 (2005) 5720-5725. https://doi.org/10.1021/cm051467h

[78]   C.B. Murray, D.J. Norris, M.G. Bawendi, Synthesis and characterization of nearly monodisperseCdE (E = sulfur, selenium, tellurium) semiconductor nanocrystallites. J. Am. Chem. Soc. 115 (1993) 8706-8715. https://doi.org/10.1021/ja00072a025

[79]   H. Lee, P.H. Holloway, H. Yang, Synthesis and characterization of colloidal ternary ZnCdSe semiconductor nanorods. J. Chem. Phys. 125 (2006) 2363181–2363189. https://doi.org/10.1063/1.2363181

[80]   L. Bakueva, S. Musikhin, M.A. Hines, T.W.F. Chang, M. Tzolov, G.D. Scholes, E.H. Sargent, Size-tunable infrared (1000–1600 nm) electroluminescence from PbS quantum-dotnanocrystals in a semiconducting polymer. Appl. Phys. Lett. 82 (2003) 2895–2897. https://doi.org/10.1063/1.1570940

Materials Research Forum LLC
https://doi.org/10.21741/9781644901250-2

[81]   D. Battaglia, X.G. Peng, Formation of high quality InP and InAsnanocrystals in a noncoordinating solvent. Nano Lett. 2 (2002) 1027–1030. https://doi.org/10.1021/nl025687v

[82]   E. Kurtz, J. Shen, M. Schmidt, M. Grun, S.K. Hong, D. Litvinov, D. Gerthsen, T. Oka, T. Yao, C. Klingshirn, Formation and properties of self-organized II–VI quantum islands, Thin Solid Films 367 (2000) 68-74. https://doi.org/10.1016/S0040-6090(00)00665-9

[83]   M.T. Swihart, Vapor-phase synthesis of nanoparticles. Curr. Opin. Colloid Interface Sci. 8 (2003) 127-133. https://doi.org/10.1016/S1359-0294(03)00007-4

[84]   F.C. Frank, J.H. van der Merwe, One-Dimensional Dislocations. I. Static Theory. Proceedings of the Royal Society A, London, 1949.

[85]   M. Volmer, A.Z. Weber, Nucleus formation in supersaturated systems, J. Phys. Chem. 119 (1926) 277-301.

[86]   I.N. Stranski, V.L. Krastanow, Self-assembled quantum dots. Akad. Wiss. Lit. Mainz Math.-Natur. KI. IIb 146 (1939) 797.

[87]   D.J. Eaglesham, M. Cerullo, Dislocation-free Stranski-Krastanow growth of Ge on Si(100). Phys. Rev. Lett. 64 (1990) 1943-1946. https://doi.org/10.1103/PhysRevLett.64.1943

[88]   K. Alavi, Molecular Beam Epitaxy, Encyclopedia of Materials: Science and Technology, Elsevier, Second Ed. 2001

[89]   G. Trevisi, L. Seravalli, P. Frigeri, C. Bocchi, V. Grillo, L. Nasi, I. Suarez, D. Rivas, G. Munoz-Matutano, J. Martínez-Pastor, MBE growth and properties of low-density InAs/GaAs quantum dot structures, Cryst. Res. Technol. 46 (2011) 801-804. https://doi.org/10.1002/crat.201000622

[90]   Y. Ma, S. Huang, C. Zeng, T. Zhou, Z. Zhong, T. Zhou, Y. Fan, X. Yang, J. Xia, Z. Jiang, Towards controllable growth of self-assembled SiGe single and double quantum dot nanostructures. Nanoscale 6 (2014) 3941-3948. https://doi.org/10.1039/C3NR04114J

[91]   S.H. Xin, P.D. Wang, A. Yin, C. Kim, M. Dobrowolska, J.L. Merz, J.K. Furdyna, Formation of self-assembling CdSe quantum dots on ZnSe by molecular beam epitaxy. Appl. Phys. Lett. 69 (1996) 3884-3886. https://doi.org/10.1063/1.117558

[92]   S. Nakamura, K. Kitamura, H. Umeya, A. Jia, M. Kobayashi, A.Yoshikawa, M. Shimotomai, S. Nakamura, K. Takahashi, Bright electroluminescence from CdS

quantum dot LED structures. Electron. Lett. 34 (1998) 2435-2436.
https://doi.org/10.1049/el:19981668

[93]   Q. Lei, M.Golalikhani, B.A. Davidson, G. Liu, D.G. Schlom, Q. Qiao, Y. Zhu,
R.U. Chandrasena, W. Yang, A.X. Gray, E.Arenholz, A.K. Farrar, D.A. Tenne, M. Hu,
J.Guo, R.K. Singh, X. Xi, Constructing oxide interfaces and heterostructures by atomic
layer-by-layer laser molecular beam epitaxy,npj Quant. Mater. 10 (2017) 1-7.
https://doi.org/10.1049/el:19981668

[94]   E. Comini, One and Two Dimentional Metal Oxide Nano Structures for Chemical
Sensing,in: R. Jaaniso, O.K. Tan, (Eds.), Semiconductor Gas Sensors, Woodhead
Publising Limited, 2013, 299-310.

[95]   D.M. Mattox, Hand book of Physical Vapor Deposition (PVD) Processing,second
ed. Elsevier Inc., USA, 2010.

[96]   C. Burda, X.B. Chen, R. Narayanan, M.A. El-Sayed, Chemistry and properties of
nanocrystals of different shapes. Chem. Rev. 105 (2005) 1025-1102.
https://doi.org/10.1021/cr030063a

[97]   S. Dhawan, T. Dhawan, A.G. Vedeshwar, Growth of Nb2O5 quantum dots by
physical vapor deposition, Mater. lett. 126 (2014) 32-35.
https://doi.org/10.1016/j.matlet.2014.03.107

[98]   M.M.D. Kumar, S. Devadason, S. Rajesh, Formation of CdSe/CdTe quantum dots
in multilayer thin films using PVD method, AIP Conf. Proc. 1451 (2012) 176-178.
https://doi.org/10.1063/1.4732406

[99]   Y.G. Kim, Y.S. Joh, J.H. Song, K.S. Baek, S.K. Chang, E.D. Sim, Temperature-
dependent photoluminescence of ZnSe/ZnS quantum dots fabricated under the
Stranski–Krastanov mode. Appl. Phys. Lett. 83 (2003) 2656-2658.
https://doi.org/10.1063/1.1612898

[100] T. Dhawan, R.Tyagi, R.K. Bag, M. Singh, P. Mohan, T. Haldar, R. Murlidharan,
R.P. Tandon, Growth of InAs quantum dots on germanium substrate using metal
organic chemical vapor deposition technique, Nanoscale Res. Lett. 5 (2010) 31-37.
https://doi.org/10.1007/s11671-009-9439-y

[101] M.H. Entezari, N. Ghows, Micro-emulsion under ultrasound facilitates the fast
synthesis of quantum dots of CdS at low temperature, Ultrason. Sonochem. 18 (2011)
127-134. https://doi.org/10.1016/j.ultsonch.2010.04.001

[102] H.F. Qian, L. Li, J. Ren, One-step and rapid synthesis of high quality alloyed
quantum dots (CdSe–CdS) in aqueous phase by microwave irradiation with

controllable temperature. Mater. Res. Bull. 40 (2005) 1726-1736.
https://doi.org/10.1016/j.materresbull.2005.05.022

[103] H. Gao, C. Xue, G. Hu, K. Zhu, Production of graphene quantum dots by ultrasoundassisted exfoliation in supercritical CO2/H2O medium, Ultrason. Sonochem. 37 (2017) 120–127. https://doi.org/10.1016/j.ultsonch.2017.01.001

[104] W. Yang, B. Zhang, N. Ding, W. Ding, L. Wang, M. Yu, Q. Zhang, Fast synthesize ZnO quantum dots via ultrasonic method, Ultrason. Sonochem. 30 (2016) 103–112. https://doi.org/10.1016/j.ultsonch.2015.11.015

[105] J.C. Wang, Y. Zhou, Suppressing bubble shielding effect in shock wave lithotripsy by low intensity pulsed ultrasound, Ultrasonics 55 (2015) 65–74. https://doi.org/10.1016/j.ultras.2014.08.004

[106] E. Moghaddam, A.A. Youzbashi, A. Kazemzadeh, M.J. Eshraghi, Preparation ofsurface-modified ZnO quantum dots through an ultrasound assisted sol–gel process,Appl. Surf. Sci. 346 (2015) 111–114. https://doi.org/10.1016/j.apsusc.2015.03.207

[107] L.C. Wang, L.Y. Chen, T. Luo, Y.T. Qian, A hydrothermal method to prepare the spherical ZnS and flower-like CdSmicrocrystallites, Mater. Lett. 60 (2006) 3627-3630. https://doi.org/10.1016/j.matlet.2006.03.072

[108] H. Yang, W. Yin, H. Zhao, R. Yang, Y. Song, A complexant-assisted hydrothermal procedure for growing well-dispersed InPnanocrystals, J. Phys. Chem. Solids 69 (2008) 1017-1022. https://doi.org/10.1016/j.jpcs.2007.11.017

[109] V. Sankaran, C.C. Cummins, R.R. Schrock, R.E. Cohen, R.J. Silbey, Small lead sulfide (PbS) clusters prepared via ROMP block copolymer technology, J. Am. Chem. Soc. 112 (1990) 6858–6859. https://doi.org/10.1021/ja00175a019

[110] J.G. Winiarz, L.M. Zhang, J. Park, P.N. Prasad, Inorganic: organic hybrid nanocomposites for photorefractivity at communication wavelengths, J. Phys. Chem. B 106 (2002) 967–970. https://doi.org/10.1021/jp013805h

[111] R.C. Xie, J.K. Shang, Morphological control in solvothermal synthesis of titanium oxide, J. Mater. Sci. 42 (2007) 6583-6589. https://doi.org/10.1007/s10853-007-1506-0

[112] C. Mohammad, T. Pall, J.A. Stride, Gram-scale production of graphene based on solvothermal synthesis and sonication, Nat. Nano. 4 (2009) 30-33. https://doi.org/10.1038/nnano.2008.365

[113] Y. Liu, E. Kim, R. Ghodssi, G.W.Rubloff, J.N. Culver, W.E. Bentley, G.F. Payne,Biofabrication to build the biology-device interface, Biofabrication 2(2010) 1-21. https://doi.org/10.1088/1758-5082/2/2/022002

[114] M.W. Ullah, Z. Shi, X. Shi, D. Zeng, S. Li, G. Yang, Microbes as structural templates in biofabrication: study of surface chemistry and applications, ACS Sustain. Chem. Eng. 5 (2017) 11163-11175. https://doi.org/ 10.1021/acssuschemeng.7b02765

[115] 54   Y. Li, R. Cui, P. Zhang, B.B. Chen, Z.Q. Tian, L. Li, B. Hu, D.W. Pang, Z.X. Xie, Mechanism-oriented controllability of intracellular quantum dots formation: The role of glutathione metabolic pathway, ACS Nano, 7 (2013) 2240–2248. https://doi.org/10.1021/nn305346a

[116] H. Bai, Z. Zhang, Y. Guo, W. Jia, Biological synthesis of size-controlled cadmium sulfide nanoparticles using immobilizedrhodobactersphaeroides, Nanoscale Res. Lett. 4 (2009) 717–723. https://doi.org/10.1007/s11671-009-9303-0

[117] S. Bandyopadhyay, A.E. Miller, H.C. Chang, G. Banerjee, V. Yuzhakov, D.F. Yue, R.E. Ricker, S.Jones, J.A. Eastman, E. Baugher, M. Chandrasekhar, Electrochemically assembled quasi-periodic quantum dot arrays, Nanotechnology.7 (1996) 360. https://doi.org/10.1088/0957-4484/7/4/010

[118] S. Bandyopadhyay, L. Menon, N. Kouklin, H. Zeng, D.J. Sellmyer, Electrochemically self-assembled quantum dot arrays. J. Electron. Mater. 28 (1999) 515-519. https://doi.org/10.1007/s11664-999-0104-0

[119] A. Balandin, K.L. Wang, N. Kouklin, S. Bandyopadhyay, Raman spectroscopy of electrochemically self-assembled CdS quantum dots, Appl. Phys. Lett. 76 (2000) 137-139. https://doi.org/10.1063/1.125681

[120] N. Pickett, Controlled preparation of nanoparticle materials U.S. Patent 7,867,556 B2.

[121] A.M. Jawaid, S.Chattopadhyay, D.J. Wink, L.E. Page, P.T. Snee, Cluster-seeded synthesis of doped CdSe:Cu4 quantum dots, ACS Nano 7 (2013) 3190-3197. https://doi.org/10.1021/nn305697q

[122] N.L. Pickett, O. Masala, J. Harris, Commercial volumes of quantum dots: controlled nanoscale synthesis and micron-scale applications. Material Matters 3 (2008) 24-26.

Quantum Dots – Properties and Applications
Materials Research Foundations **96** (2021) 81-94

Materials Research Forum LLC
https://doi.org/10.21741/9781644901250-3

Chapter 3

Green and One-Pot Synthesis of Mint Derived Carbon Quantum Dots for Metal Ion Sensing

Hasan ESKALEN[1]*, Serhan URUŞ[2], Şükrü ÖZĞAN[3], Beyhan TAHTA[2], Ali Burak SÜNBÜL[2]

[1]Vocational School of Health Services, Department of Opticianry, Kahramanmaraş Sütçu İmam University, Kahramanmaraş, Turkey

[2] Department of Chemistry, Faculty of Art and Sciences, *Kahramanmaraş Sütçü İmam University, 46100, Kahramanmaraş*, 46100, Turkey

[3]Department of Physics, Faculty of Art and Sciences, *Kahramanmaraş Sütçü İmam University, 46100, Kahramanmaraş*, 46100, Turkey

* eskalen@ksu.edu.tr, heskalen@gmail.com

Abstract

A green and simple synthesis of carbon quantum dots (CQDs) was derived from dried mint leaves by hydrothermal method. Crystalline structure of the synthesized CQDs was characterized with X-ray diffraction (XRD) method. The morphological properties of the CQDs were investigated with transmission electron microscopy (TEM). The optical behaviors of the CQDs were examined with fourier transfom infrared spectrophotometer (FT-IR), ultraviolet visible (UV-Vis) and photoluminescence spectrophotometer techniques. Crystalline structure of the CQDs was found as amorphous in nature and the average diameter of the CDs was calculated as 8.13 nm from TEM study. According to the fluorescence emission spectra of the samples, synthesized CQDs was sensitive to mainly Ag(I), Cr(III) and Fe(III) ions. Especially, Ag(I) was the most sensible compared to other metal ions. Quenching effect of the CQDs was also evaluated by using ascorbic acid to metal ions added CQDs samples. Ascorbic acid showed the quenching effect for all the metal ion added samples except Sn(II) ion.

Keywords

Carbon Quantum Dots, Green Synthesis, Hydrothermal Method, Quenching, Fluorescent Sensing

Contents

1. Introduction

Carbon quantum dots (CQDs) or simple carbon dots (CDs) have gained considerable attention for their superior properties such as low toxicity, small sizes, water soluble, biocompatibility and good photoluminescence properties [1]. Moreover, CQDs can be simply synthesized with low cost methods that puts these a step ahead to some complex, toxic, time consuming and high cost synthesis methods [2]. Today CQDs have been examined from very different perspectives for their potential applications in chemistry, sensing and biosensors, in vivo and in vitro bioimaging, energy and super capacitors, photo catalysis, drug carrier and cancer therapy [3-13].

In recent years, carbon quantum dots from various sources have been used for chemical sensing application. Wang et al. [14], synthesized carbon quantum dots from liquid fuels and examined in vivo sensing for Cu^{2+} in the rat brain. Fe^{3+} ions detection by using one pot hydrothermally synthesized CDs was studied by Sachdev et al. [15]. Recently, *Catharanthusroseus* leaves without any chemical used to synthesized carbon quantum dots with approximately 5 nm size and 28.2 % quantum yield. The synthesized CDs were used to examine multi ion sensing and selectivity (Fe^{3+} and Al^{3+} ions) [16]. Moreover, other than ions sensing, carbon quantum dots have been used as sensors for picric acid [17], histamine [18], insulin [19] and clonazepam [20].

Silver nanoparticles have wide spread application areas including, cosmetics, textile and painting [21]. Moreover, silver metal has high electrical and thermal conductivity and its antibacterial properties have been widely investigated [22]. According to drinking water quality standards of WHO (World Health Organization) the Ag^+ ion concentration of less than 0.1 ppm will be considered as drinkable however, above this level, the Ag^+ ions might get stored in human skeletal and liver [23]. To sum up, Ag^+ ion sensing is important for medical, environmental and industrial aspects.

In the present work, green, one pot and simple hydrothermal method was used to synthesize carbon quantum dots using dried mint as the carbon source. Blue emissive water dispersible carbon quantum dots were obtained. The morphology, crystalline structure, spectroscopic and optical properties were analyzed in detail. The prepared and analyzed carbon quantum dots are utilized for silver ion sensing.

2. Experimental

2.1 Chemicals and materials

Iron(III) chloride hexahydrate ($FeCl_3.6H_2O$), manganase (II) chloride tetrahydrate ($MnCl_2.4H_2O$), barium chloride dihydrate ($BaCl_2.2H_2O$), silver(I) nitrate ($AgNO_3$), chromium (III) chloride ($CrCl_3$), zinc(II) chloride ($ZnCl_2$), aluminium chloride ($AlCl_3$), calcium chloride ($CaCl_2$), magnesium chloride hexahydrate ($MgCl_2·6H_2O$), nickel(II) chloride hexahydrate ($NiCl_2·6H_2O$), cobalt(II) chloride ($CoCl_2$), copper(II) chloride dihydrate $CuCl_2·2H_2O$, lead(II) chloride ($PbCl_2$), mercury(II) chloride ($HgCl_2$), sodium chloride ($NaCl$), potassium chloride (KCl), tin(II) chloride dihydrate ($SnCl_2.2H_2O$) were purchased from Isolab, Sigma Aldrich, Merck, Alfa Aesar, Abcr and AFG Bioscience. Ultra-pure water (Millipore) system was applied to execute the all experimental works.

2.2 Equipments

Transmission electron microscopy (TEM) images of the carbon dots (CDs) were recorded by using Hitachi HT7700 with EXALENS, 120 kV, electron microscope. Photoluminescence (PL) spectra were collected on a Varian Cary Eclipse spectrometer. UV-Vis absorption spectra were performed by using Shimadzu-1800 UV-Visible spectrometer. FT-IR spectra were obtained with Perkin-Elmer Spectrum 400 system attached ATR apparatus in therange 4000–400 cm^{-1}. X-ray diffraction (XRD) pattern of samples was recorded by using a Philips X`Pert PRO XRD withCu Kα radiation (λ = 0.154056 nm, set at 40 kV and 30 mA).

Materials Research Forum LLC
https://doi.org/10.21741/9781644901250-3

2.3   Synthesis of carbon quantum dots

The dried mint leaves bought from local market were ground with mortar, then sieved to reduce size of raw particles. 1 g of mint leaves was dissolved in 50 mL pure water and ultra-sonicated for 15 minutes. The prepared solution was heated at 200 °C for 12 hours in a teflon lined stainless steel autoclave. After heat treatment the solution was naturally cooled down to room temperature. The obtained solution was centrifuged at 14000 rpm for 20 minutes.

2.4   Fluorescent sensing for metal ions

The 0.01 M standard stock solution of each metal ion was freshly prepared and then diluted to 0.1 µM in order to inverstigate the sensing properties of carbon quantum dots. For each fluorescent analysis, 1 mL of 0.1 µM metal ion solution was added to 1 mL (0.1 mg/mL) CQDs solutions separately for 1 hour and then measured in photoluminescence instrument. Photoluminescence spectra of the metal ion added carbon quantum dot samples were recorded between 390-620 nm with the excitation wavelength of 275 nm with 5 nm and 10 nm excitation and emission slit widths respectively. The changings in the emissions were collected at 550 nm in the photo luminescence spectra. Ascorbic acid showed the quenching properties for the metal ion added carbon quantum dot samples except Sn(II) added carbon quantum dot sample. For investigating the quenching tests, 100 µL (5.68 mM) ascorbic acid was added to each metal ion added carbon quantum dot samples.

3.   Result and discussion

UV-Visible spectrum of carbon dot mixture was recorded between 200-600 nm in order to evaluate the optical behaviours of C-dot mixture in water. π-π^*transitionsof C=C in graphitic sp^2 domains were observed as a sharp peak at 275 nm. A peak about 335 nm was becauseof n-π^*transitions in C=O substituents from alcoholic and carboxylic structures in carbon quantum dots (Fig. 1) [24].

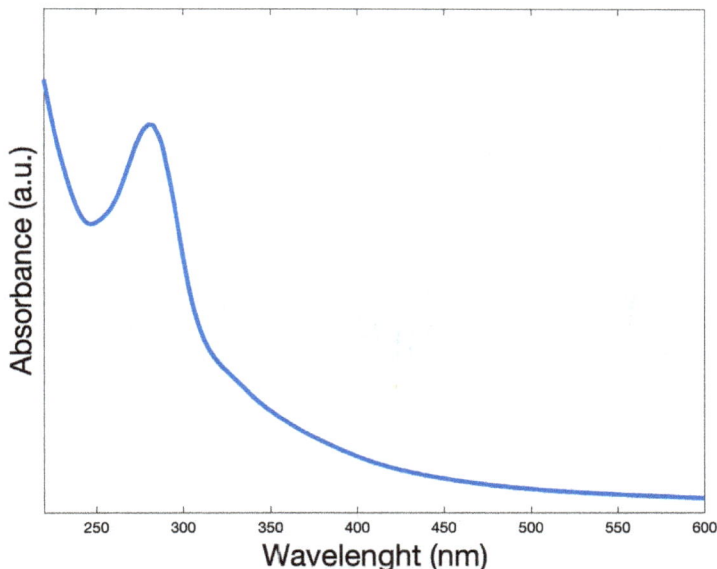

Fig. 1 UV-Vis absorption spectrum of CDs derived from mint leaves [24].

X-ray diffraction (XRD) was used to investigate crystallite structure of the synthesized CDs. The obtained XRD graph is given in Fig. 2 and a board diffraction peak centered at 25.04 degree (2θ) is observed. This implies that the synthesized carbon dots were amorphous in nature. The transmission electron microscopy (TEM) of the dried mint derived CDs was examined as shown in Fig. 3. As seen the particles are generally uniformly distributed. The diameter of synthesized CDs is further investigated by selecting 35 different CDs and it is illustrated in Fig. 4. The average CDs diameter are calculated as 8.13 nm with standard deviation of 2.26 nm.

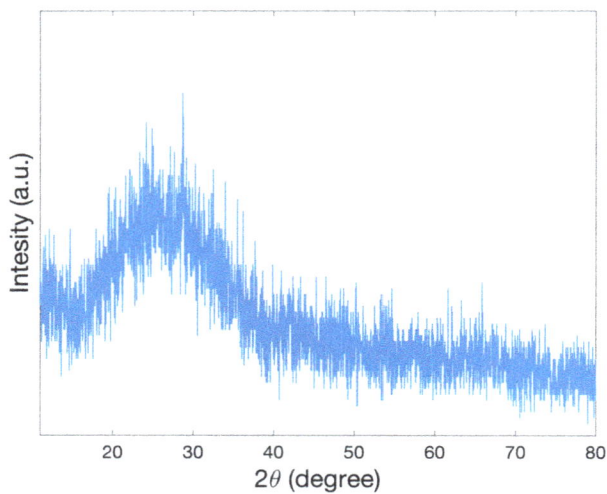

Fig. 2 X-ray diffraction pattern of synthesized CDs.

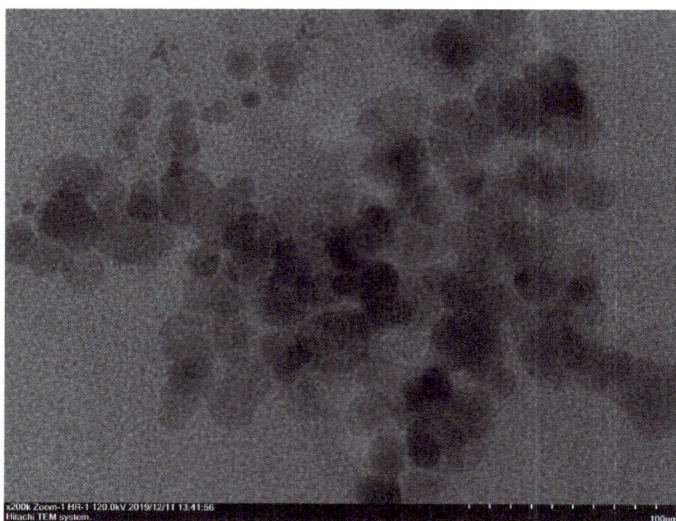

Fig. 3 TEM morphologies of synthesized CDs.

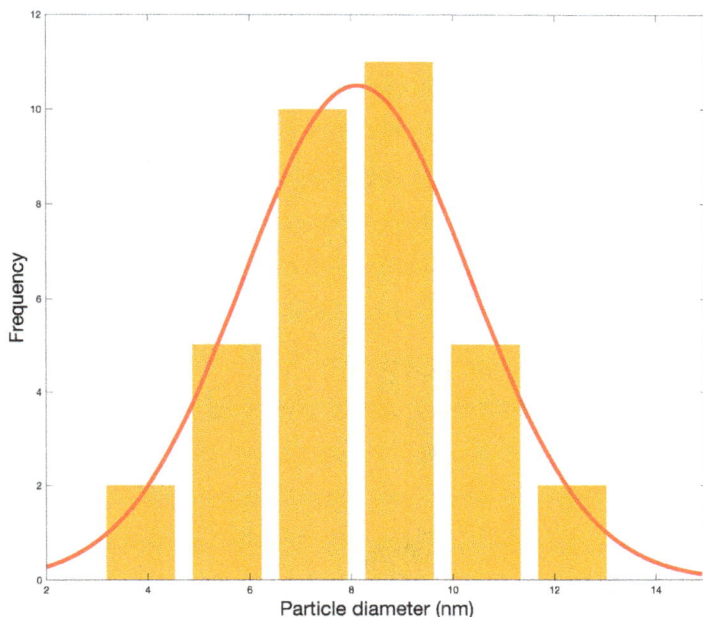

Fig. 4 Particle diameters distribution of synthesized CDs.

Photoluminescence spectra of aqueous mixture of carbon quantum dot were obtained between 390-620 nm (Fig. 5). When carbon quantum dots were excited at 428 nm (λex), the highest fluorescence intensity was recorded at 486 nm (λem) in the emission spectra. Carbon quantum dots showed the turquoise color in UV cabinet using 365 nm lamp (Fig. 6). Because of having the mono-disperse particle distribution, while observing slightly bathochromic shifts, intensities increased in the emission specta when the excitation wavelengths increased [25, 26].

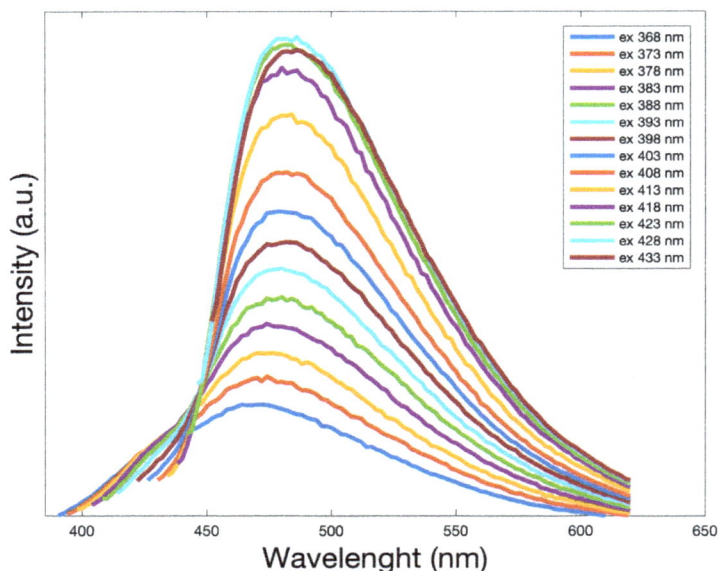

Fig. 5 Excitation dependent photoluminescence spectra of synthesized CDs.

Aqueous medium covered some of the stretches in FT-IR spectrum of carbon quantum dots. The band of OH stretches from alcoholic, phenolic and carboxylic structures observed around 3750-2800 cm^{-1} indicating the existence of large numbers of hydroxyl groups coated the aliphatic and aromatic C-H stretches in FT-IR spectrum of carbon quantum dot (Fig. 6). Additionally, C=O and C=C stretches were recorded at 1640 cm^{-1} as a single band [24-26].

Materials Research Forum LLC
https://doi.org/10.21741/9781644901250-3

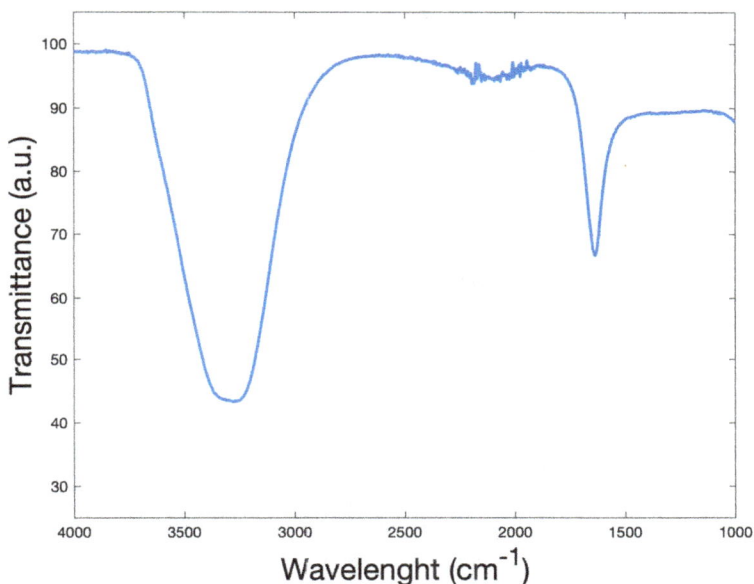

Fig. 6 FT-IR spectrum of synthesized CDs.

3.1   Fluorescent sensing

The excitation wavelength dependent fluorescence emission spectra of the metal ion added carbon quantum dot samples were recorded between 390-620 nm with the excitation wavelength of 275 nm. The changings in the emissions were collected at 550 nm in the photoluminescence spectra. According to the I/I_0 ratio calculated from the fluorescence emission spectra of the samples, the intensities were remarkably increased for Ag(I) and slightly increased for Cr(III) and Fe(III) ions as compared the other metal ions (Fig. 7). It can be said that the synthesized carbon quantum dots from dried mint leaves are very sensitive for Ag(I) ion in water. The sensitivity of carbon quantum dots to Ag(I) is possibly because of the coordination of alcoholic, phenolic and carboxylic structures to metal cores.

Quantum Dots – Properties and Applications   Materials Research Forum LLC
Materials Research Foundations **96** (2021) 81-94   https://doi.org/10.21741/9781644901250-3

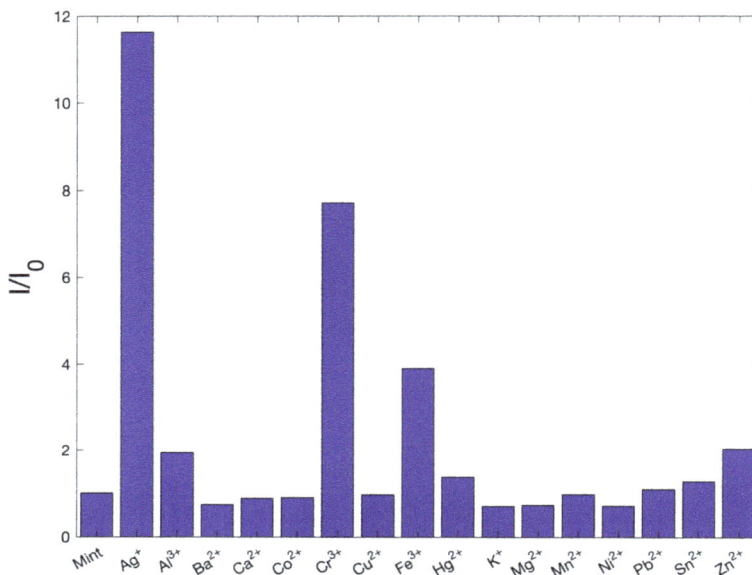

Fig. 7 Metal ions sensing properties of the CDs.

In order to examine the quenching tests, ascorbic acid solution was used for each metal ions added carbon quantum dot samples. The intensities of the metal ions added carbon quantum dot samples decreased when ascorbic acid was added to metal ion added carbon quantum dot samples except Sn(II) added sample. It can be said that ascorbic acid showed the quenching properties for all the metal ions added to carbon quantum dot samples except Sn(II) sample (Fig. 8).

Materials Research Forum LLC
https://doi.org/10.21741/9781644901250-3

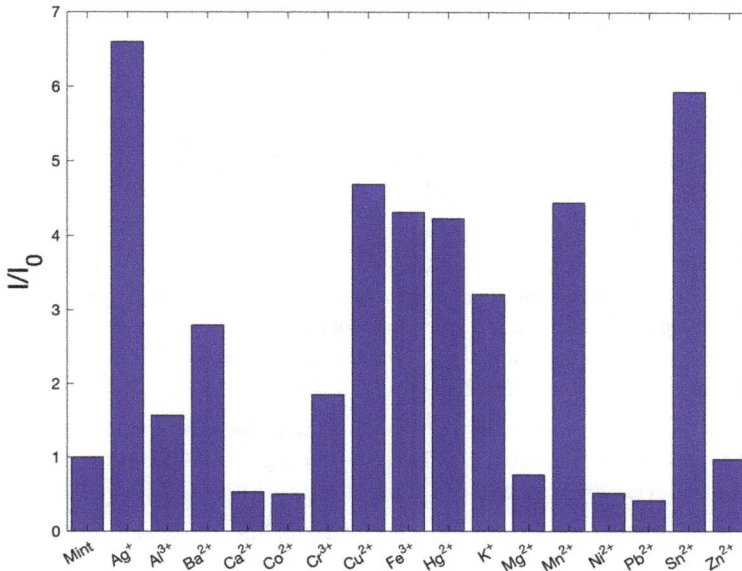

Fig. 8 Quenching effect of ascorbic acid on metal ions added CDs samples.

Conclusions

To sum up, carbon quantum dots derived from dried mint leaves were synthesized by hydrothermal method without using any chemicals. The synthesized CDs was characterized with XRD, TEM, FT-IR, UV-Vis and photoluminescence techniques. The XRD shows the amorphous nature of the CDs and the average diameter of the CDs were calculated as 8.13 nm from TEM observation. Moreover, uniform size distribution of the CDs also revealed from excitation dependent photoluminescence spectra with slightly bathochromic shift. Metal ions sensing properties of the CDs were also investigated. The synthesized CDs were sensitive to Ag(I) Cr(III) and Fe(III) ions, with sensitivity highest for Ag(I) ion. Ascorbic acid showed a quenching effect for all the metal ion added CDs samples except Sn(II) sample.

Acknowledgments

This work was financially supported by Kahramanmaras Sutcu Imam University, (KSU) Scientific Research Projects Coordination Department, under Project No. 2019/1-26M.

References

[1] N. Gao, L. Huang, T. Li, J. Song, H. Hu, Y. Liu, S. Ramakrishna, Application of carbon dots in dye-sensitized solar cells:, J. Appl. Polym. Sci 137 (2020) 48443. https://doi.org/10.1002/app.48443

[2] S. Liu, J. Tian, L. Wang, Y. Zhang, X. Qin, Y. Luo, A.M. Asiri, A.O. Al-Youbi, X. Sun, Hydrothermal treatment of grass: a low-cost, green route to nitrogen-doped, carbon-rich, photoluminescent polymer nanodots as an effective fluorescent sensing platform for label-free detection of Cu (II) ions, Adv. Mater. 24 (2012) 2037-2041. https://doi.org/10.1002/adma.201200164

[3] S. Sagbas, N. Sahiner, Carbon dots: preparation, properties, and application, Nanocarbon and its Composites, Duxford, United Kingdom Woodhead Publishing, an imprint of Elsevier 2019, pp. 651-676.https://doi.org/10.1016/C2017-0-01980-8

[4] S. Rai, B.K. Singh, P. Bhartiya, A. Singh, H. Kumar, P.K. Dutta, G.K. Mehrotra, Lignin derived reduced fluorescence carbon dots with theranostic approaches: Nano-drug-carrier and bioimaging, J. Lumin. 190 (2017) 492-503. https://doi.org/10.1016/j.jlumin.2017.06.008

[5] N. Sahiner, S.S. Suner, M. Sahiner, C. Silan, nitrogen and sulfur doped carbon dots from amino acids for potential biomedical applications, J. Fluoresc.29 (2019) 1191-1200. https://doi.org/10.1007/s10895-019-02431-y

[6] M.O. Alas, A. Güngör, R. Genc, E. Erdem, Feeling the power: Robust supercapacitor from nanostructured conductive polymer fostered with Mn^{+2} and carbon dots, Nanoscale 11 (2019) 12804-12816.https://doi.org/10.1039/C9NR03544C

[7] S. Dinç, M. Kara, M.D. Kars, F. Aykül, H. Çiçekci, M. Akkuş, Biocompatible yogurt carbon dots: evaluation of utilization for medical applications, Appl. Phys. A 123 (2017) 572. https://doi.org/10.1007/s00339-017-1184-y

[8] T. Guner, H. Yuce, D. Tascioglu, E. Simsek, U. Savaci, A. Genc, S. Turan, M.M. Demir, Optimization and performance of nitrogen-doped carbon dots as a color conversion layer for white-LED applications, Beilstein J. Nanotechnol. 10 (2019) 2004-2013. https://doi.org/10.3762/bjnano.10.197

[9] E. Yavuz, S. Dinc, M. Kara, Effects of endogenous molasses carbon dots on macrophages and their potential utilization as anti-inflammatory agents, Appl. Phys. A 126 (2020) 22. https://doi.org/10.1007/s00339-020-3318-x

[10] R. Kandra, S. Bajpai, Synthesis, mechanical properties of fluorescent carbon dots loaded nano composites chitosan film for wound healing and drug delivery, Arabian J. Chem.13 (2020) 4882-4894.https://doi.org/10.1016/j.arabjc.2019.12.010

[11] V. Ansi, K. Vijisha, K. Muraleedharan, N. Renuka, fluorescent carbon nanodots as an efficient nitro aromatic sensor-analysis based on computational perspectives, Sens. Actuators. A302 (2020) 111817.https://doi.org/10.1016/j.sna.2019.111817

[12] L. Mohammadi, M.M. Heravi, S. Sadjadi, M. Malmir, Hybrid of graphitic carbon nitride and palladated magnetic carbon dot: An efficient catalyst for coupling reaction, Chem. Select 4 (2019) 13404-13411. https://doi.org/10.1002/slct.201903078

[13] S. Raina, A. Thakur, A. Sharma, D. Pooja, A.P. Minhas, Bactericidal activity of Cannabis sativa phytochemicals from leaf extract and their derived carbon dots and Ag@ Carbon Dots, Mater. Lett. 262 (2019) 127122.https://doi.org/10.1016/j.matlet.2019.127122

[14] C. Wang, C. Pan, X. Wei, F. Yang, W. Wu, L. Mao, Emissive carbon dots derived from natural liquid fuels and its biological sensing for copper ions, Talanta 208 (2020) 120375.https://doi.org/10.1016/j.talanta.2019.120375

[15] A. Sachdev, P. Gopinath, Green synthesis of multifunctional carbon dots from coriander leaves and their potential application as antioxidants, sensors and bioimaging agents, Analyst 140 (2015) 4260-4269. https://doi.org/10.1039/C5AN00454C

[16] T. Arumugham, M. Alagumuthu, R.G. Amimodu, S. Munusamy, S.K. Iyer, A sustainable synthesis of green carbon quantum dot (CQD) from Catharanthus roseus (white flowering plant) leaves and investigation of its dual fluorescence responsive behavior in multi-ion detection and biological applications, Sustain. Mater. Technol. 23 (2020) e00138.https://doi.org/10.1016/j.susmat.2019.e00138

[17] Q. Niu, K. Gao, Z. Lin, W. Wu, Amine-capped carbon dots as a nanosensor for sensitive and selective detection of picric acid in aqueous solution via electrostatic interaction, Anal. Methods 5 (2013) 6228-6233.https://doi.org/10.1039/C3AY41275J

[18] D. Zhang, Y. Wang, J. Xie, W. Geng, H. Liu, Ionic-liquid-stabilized fluorescent probe based on S-doped carbon dot-embedded covalent-organic frameworks for

determination of histamine, Mikrochim. Acta187 (2020) 28.
https://doi.org/10.1007/s00604-019-3833-7

[19] F. Abazar, A. Noorbakhsh, Chitosan-carbon quantum dots as a new platform for highly sensitive insulin impedimetric aptasensor, Sens. Actuators, B 304 (2020) 127281. https://doi.org/10.1016/j.snb.2019.127281

[20] A. Ghafarloo, R. Emamali Sabzi, N. Samadi, H. Hamishehkar, Sensitive and selective spectrofluorimetric determination of clonazepam using nitrogen-doped carbon dots, J. Photochem. Photobiol., A 388 (2020) 112197.
https://doi.org/10.1016/j.jphotochem.2019.112197

[21] E. Topuz, C.A.M. van Gestel, The effect of soil properties on the toxicity and bioaccumulation of Ag nanoparticles and Ag ions in Enchytraeus crypticus, Ecotoxicol. Environ. Saf. 144 (2017) 330-337.
https://doi.org/10.1016/j.ecoenv.2017.06.037

[22] A. Cayuela, M.L. Soriano, S.R. Kennedy, J.W. Steed, M. Valcárcel, Fluorescent carbon quantum dot hydrogels for direct determination of silver ions, Talanta 151 (2016) 100-105.https://doi.org/10.1016/j.talanta.2016.01.029

[23] Z. Wang, D. Chen, B. Gu, B. Gao, T. Wang, Q. Guo, G. Wang, Biomass-derived nitrogen doped graphene quantum dots with color-tunable emission for sensing, fluorescence ink and multicolor cell imaging, Spectrochim. Acta, Part A 227 (2020) 117671.https://doi.org/10.1016/j.saa.2019.117671

[24] C. Wang, T. Hu, Z. Wen, J. Zhou, X. Wang, Q. Wu, C. Wang, Concentration-dependent color tunability of nitrogen-doped carbon dots and their application for iron (III) detection and multicolor bioimaging, J. Colloid Interface Sci. 521 (2018) 33-41. https://doi.org/10.1016/j.jcis.2018.03.021

[25] C. Cheng, M. Xing, Q. Wu, Preparation of carbon dots with long-wavelength and photoluminescence-tunable emission to achieve multicolor imaging in cells, Opt. Mater. 88 (2019) 353-358. https://doi.org/10.1016/j.optmat.2018.12.007

[26] A. Pathak, P. Suneesh, J. Stanley, T.S. Babu, Multicolor emitting N/S-doped carbon dots as a fluorescent probe for imaging pathogenic bacteria and human buccal epithelial cells, Microchim. Acta 186 (2019) 157. https://doi.org/10.1007/s00604-019-3270-7

Materials Research Forum LLC
https://doi.org/10.21741/9781644901250-4

Chapter 4

Antibacterial Quantum Dots

Jin-Chung Sin[1,2,3*], Ying-Hui Chin[3], Sze-Mun Lam[1,2,4], HongHu Zeng[1,2*], Hua Lin[1,2], Haixiang Li[1,2]

[1] College of Environmental Science and Engineering, Guilin University of Technology, Guilin 541004, China

[2] Guangxi Key Laboratory of Theory and Technology for Environmental Pollution Control, Guilin University of Technology, Guilin 541004, China

[3] Department of Petrochemical Engineering, Faculty of Engineering and Green Technology, Universiti Tunku Abdul Rahman, Jalan Universiti, Bandar Barat, 31900 Kampar, Perak, Malaysia

[4] Department of Environmental Engineering, Faculty of Engineering and Green Technology, Universiti Tunku Abdul Rahman, Jalan Universiti, Bandar Barat, 31900 Kampar, Perak, Malaysia

sinjc@utar.edu.my (Jin-Chung Sin), zenghonghu@glut.edu.cn (Honghu Zeng)

Abstract

The emergence and global spread of multi antibiotic-resistant bacteria underscored the need to find new alternative antimicrobial candidates. Graphene quantum dots have received tremendous attention as promising new microbicidal agents owing to their ease of production, excellent physicochemical properties and high biosafety. In this chapter, the synthesis and physicochemical characteristics of graphene quantum dots are reviewed. A recent research progress on their antibacterial activities and the reaction mechanisms are also discussed. Lastly, an outlook on future development of effective graphene quantum dots was suggested with the goal of addressing current limitation and motivating further research on this promising area.

Keywords

Graphene Quantum Dots, Synthetic Method, Photoluminescence, Antimicrobial Activity, Biomedical

Contents

1.   Introduction

Over the years, the continuous expansion in world population has increased the spread of pathogens and infectious diseases in our ecosystem. About 1.8 million deaths annually have been reported because of diarrhea, trachoma and cholera contraction, many of which have been linked to diseases caused by the pathogens like fungi, viruses, bacteria and parasites [1,2]. Antibiotics are usually applied in the treatment of bacterial and fungal diseases, nevertheless, nowadays these pathogens have started developing resistance towards the available antibiotics via de novo mutation or by attaining resistance genes from other organisms. For instance, *Staphylococcus aureus* (*S. aureus*) and *Klebsiella pneumoniae* (*K. pneumoniae*) bacteria can produce enzymes that are capable of altering and inactivating the antibiotics [3]. In addition, *Escherichia coli* (*E. coli*) can change the nature of outer membrane to protect the bacteria against the damage from antibiotics [4]. Therefore, the development of new and effective antimicrobial agents is urgently needed.

Quantum dots (QDs) materials have attracted increasing attention for the development of new analytical techniques for numerous applications due to their small nanometer-size (1-10 nm). As emerging member of QDs, graphene quantum dots (GQDs) have high attention for biomedical applications due to their easy synthesis, biocompatibility and low cytotoxicity. The zero-dimensional GQDs possessing sp^2 hybridized carbon atoms with abundant functional groups over the shell, due to which they demonstrated exceptional fluorescence properties. The GQDs in suspension can produce reactive oxygen species (ROS) upon light irradiation. Therefore, they displayed strong photodynamic effects, in which the light-excited compounds can trigger severe damages to the bacterial cells by oxidative decomposition of cellular components. In addition to these, they were low-cost and can be fabricated from organic wastes. Moreover, they were nondegradable in body fluids which made them to retain the antibacterial effects with time [5-7]. All these outstanding characteristics have resulted GQDs to be promising antibacterial agents. Herein, this chapter a review dealing with the recent research progress on antibacterial GQDs, including their synthetic methods, physicochemical characteristics and antibacterial actions is provided. Finally, future development of effective GQDs was suggested with the goal of addressing current limitation and motivating further research into this promising area.

2. Synthesis of graphene quantum dots (GQDs)

In general, the synthesis of GQDs can be achieved through top-down and bottom-up approaches. In top-down route, the macromolecule is broken down into smaller fluorescent units (<10 nm) either by chemical or physical approach. In bottom-up way, a basic building block such as carbon source is taken and a series of small molecules from the basic building block underwent carbonization, condensation and polymerization to form fluorescent quantum dots under optimized synthesized condition [8-10].

2.1 Top-down approaches

2.1.1 Electrochemical oxidation

In this method, the electrode materials (e.g. graphite rod, carbon nanotube, carbon paste) were adopted as working electrodes where the alkaline solution or acid solution were

Materials Research Forum LLC
https://doi.org/10.21741/9781644901250-4

used as electrolyte to provide an electrochemical system which is economic and environment friendly to generate quantum dots. This method relied on the using of electrode materials and electrolyte solution to generate the quantum dot with different properties in terms of cytotoxicity, surface states and photoluminescence performance [11-13]. Ahirwar et al. [13] reported the preparation of GQDs via an electrochemical exfoliation technique. The synthesis method was as follows: the bare graphite rods were treated at 1050°C for 5 min in order to create the defect induced graphite rod. Then, these graphite rods were applied as anode and cathode and immersed into electrolyte (mixture of citric acid and sodium hydroxide) solution. The sodium citrate salt and water were produced. When potential was applying on the system, hydrolysis of water happened. The produced OH$^-$ and oxygen occupied the Van der Waals gaps resulting to exfoliation of these graphite rods. Finally, the GQDs were generated owing to C-C cleavage through electric field. Fig. 1 [13] shows the schematic illustration of electrochemical exfoliation process resulted in the generation of GQDs.

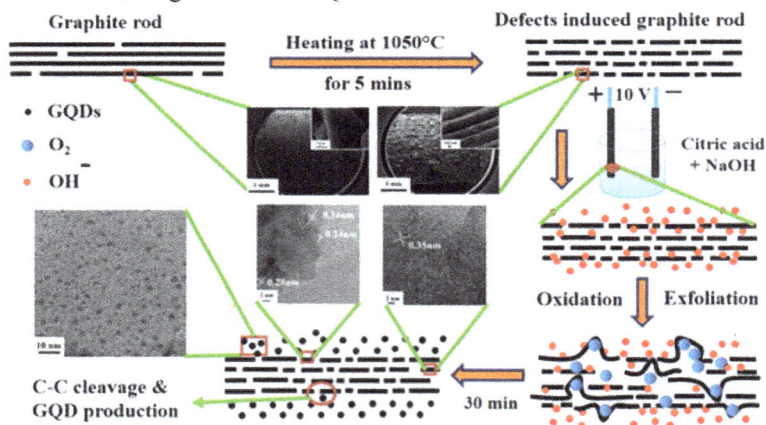

Fig. 1 Schematic diagram of the fabrication of GQDs by electrochemical exfoliation method [13].

2.2.2 Laser ablation

Laser ablation refers to a process where high-power laser pulse is directed onto a solid surface to a state of thermodynamic in which high temperature and pressure are created and heated up speedily and evaporated into plasma state and finally the crystallization of vapor to obtain nanoparticles [14-16]. Lin et al. [16] reported the production of GQDs via a pulsed laser ablation in graphene oxide (GO) solution. The graphene aqueous solution

Materials Research Forum LLC
https://doi.org/10.21741/9781644901250-4

(275 mg/L) was poured into a quartz cell on a rotational stage with an angular velocity of 80 rpm and exposed to the pulses generated from laser ablation machine under 415 nm, 10Hz, 10ns. The process of pulsed laser ablation of GO was performed under 2.58 J cm^{-2} for 5 to 40 min. The suspension products which contained GQDs were filtered. Fig. 2 [16] depicts the schematic diagram of the experiment setup for pulsed lase ablation.

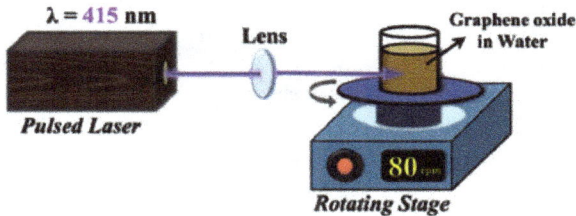

Fig. 2 Schematic diagram of setup for laser ablation in fabricating GQDs [16].

2.2.3 Arc discharge

Arc discharge is a process to induce the reorganization of quantum atoms that disintegrated from precursors in anodic electrode driven using gas plasma produced in a sealed reactor. A high-energy plasma was produced when the reactor temperature reached 4000K. The quantum dots were formed at the cathode. The synthesized quantum dots from this technique have exceptional water solubility but possessed a large particle size distribution [17-19]. Kim et al. [18] prepared the GQDs using arc discharge method. The synthesis step was as follows: high purity graphite was applied as cathode and anode. The seamless arc discharge in distilled water was conducted by applying a voltage (25 V) to initiate a discharge and moving the anode up and down to get in touch with the fixed cathode on the bottom. At the same time, the plasma zone generated between graphite electrodes heated the graphite and induced the graphene exfoliation with assistance of water pressure fluctuation. The obtained graphene were filtered and dispersed in organic solvent followed by ultra sonification. After centrifugation at 7000 rpm for 30 min, the graphene quantum dots were obtained. Fig. 3 [18] shows the schematic diagram of arc discharge in water to produce GQDs.

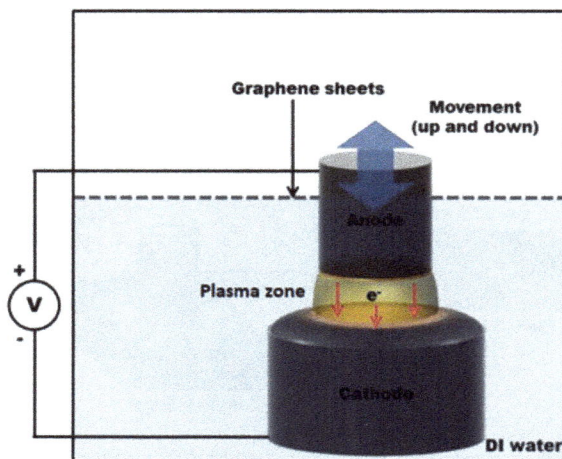

Fig. 3 Schematic diagram of arc discharge in water to fabricate GQDs [18].

2.2   Bottom-up approaches

2.2.1 Hydrothermal synthesis

It is an approach where the small organic molecules dissolved in solvent to give the reaction precursor. The reaction precursor was then heated at moderately high temperature and merged together to form the seeding cores and finally grew into quantum dots with particles size smaller than 10 nm [20]. A two-step microwave-assisted hydrothermal approach has been used to synthesize pyrrole-ring functionalized graphene quantum dots (p-GQDs) [21]. In brief, 5 mL of 50 mg/mL GO solution together with 20 mL of 65 % v/v HNO_3 were placed into a Teflon reactor and treated at 200°C for 5 min under 800 W of 30 Mpa microwave irradiation. After cooling to room temperature naturally, the GQDs solids were obtained by evaporating the solvents using a rotary evaporator. The obtained GQDs were re-dissolved into ammonium hydroxide solution and then treated exactly like the experimental conditions at the first step except the heating duration was 10 min. After evaporation and filtration, clear solution of p-GQDs with light-pink colour was obtained. This method is cost-effective, facile and easily scalable however it has a poor size distribution.

2.2.2 Microwave-assisted synthesis

Microwave-assisted method has been recognized as fast, effective and facile route to synthesize GQDs, in which GO/graphite as raw materials and fragmented them into smaller sheets. The greenish-yellow GQDs (gGQDs) were synthesized from GO nanosheets with concentrated HNO_3 and H_2SO_4 within 3 h via a one-pot microwave assisted approach. When the gGQDs were further reduced with $NaBH_4$ within 2 h, bright blue luminescent graphene quantum dots (bGQDs) were reported (Fig. 4) [22]. The quantum yield of gGQDs and bGQDs were found to be 11.7% and 22.9%, respectively [22].One-pot synthesis of GQDs in mass quantity from graphite powder via high-powered microwave irradiation under acidic and oxidative conditions have also been reported by Shin et al. [23]. Upon microwave irradiation, a large piece of oxidized graphite was observed instantaneously at lower power (300 W) whereas small-sized single- or bi-layer GQDs was found at higher power (600 W). They went further to explain the formation of small-sized GQDs was related to the multiply broken of large oxidized graphite through repeated redox reactions of graphite at higher microwave power. This facile and straightforward route also allowed a tremendously high yield (~70 wt%) of GQDs, which showed great potential for practical applications.

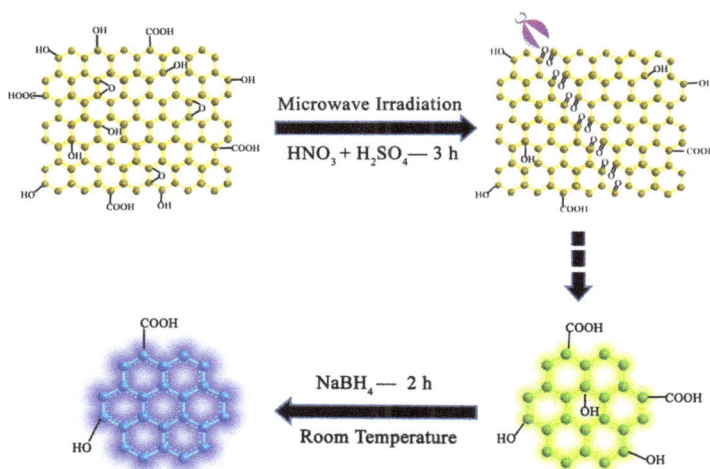

Fig. 4 Schematic diagram of the synthetic route for gGQDs and bGQDs [22].

2.2.3 Combustion method

In general, this is the method that involved combustion and carbonization process in synthesis of quantum dots. It was reported that the quantum dots can be synthesized through combustion of citric acid followed by carbonization process under high temperature [24,25]. Ye et al. [24] reported the synthesis of GQDs from bituminous coal. The synthesis path was as follows: 300 mg of bituminous coal was sonicated in concentrated H_2SO_4 (60 ml) and HNO_3 (20ml) for 2 h. It was followed by the heat treatment or combustion at 100 or 120°C for 24 h. The mixture was cooled to room temperature and then adjusted to pH 7 using 3 M NaOH. The mixture was filtered and concentrated using rotary evaporation to obtain solid GQDs. Fig. 5 [24] shows the structure of GQDs synthesized from bituminous coal.

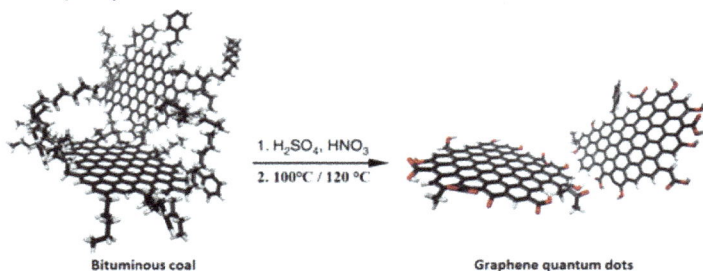

1. H_2SO_4, HNO_3

2. 100°C / 120 °C

Bituminous coal

Graphene quantum dots

Fig. 5 Schematic diagram for structure of GQDs synthesized from bituminous coal [24].

3. Physical and chemical properties

3.1 Absorbance

Generally, the quantum dots have broad absorption spectra which allowed the excitation by a wide range of wavelengths and the light absorption range of quantum dots varied based on their size [26]. Moreover, it was reported that the existence of surface functional groups imparted hydrophilicity of quantum dots and increased their capability for functionalization with biological, organic and polymeric molecules. This meant the introducing of different functional groups on the surface of quantum dots affected their absorption wavelength [27]. In addition, the preparation method of quantum dots can affect the absorption wavelengths [28]. The infrared spectrum of GQDs ranged from 3400 cm^{-1} to 1000 cm^{-1} which were attributed to -OH, -CH, C=O, aromatic and C-O-C stretching vibration bands, respectively. It was also reported that the GQDs displayed strong absorption in the UV region (230-320 nm), assigning to π-π* transition of C=C bonds. Besides, another shoulder peak in the range of 260-390 nm was caused by the n=π* transition of C=O bonds [29].

3.2   Photoluminescene

Photoluminescence (PL) is the light emission form of any matter after the absorption of photons that initiated by photoexcitation. The PL properties of graphene can be affected by the π energy states of the sp^2-hybridized carbon. In general, the photoluminescence emission can be divided into intrinsic state emission (electron-hole pair recombination, zigzag edge sites) and defect state emission (defect derived PL)[29,30]. These phenomena can modify the π-π electron of sp^2 domain energy gaps and affected the PL properties. Zhang et al. [29] reported that the PL properties of the GQDs can be affected by the surface status, pH, solvent, synthesis method and thickness of graphene layers. Fu and Lin [31] reported that the PL of GQDs have broad emission spectrum ranged from 450 to 750 nm where the green and red emission were excited by laser at 530 nm and 720 nm, respectively. Fig. 6 [30] shows the emission spectra excited by different wavelengths and their corresponding photographs of GQDs solution. Another study from Temerov et al. [32] discovered that the GQDs with different functional groups (hydroxyl, carboxyl, carbonyl) have various energy levels and resulting in emissive traps which affected the PL emission.

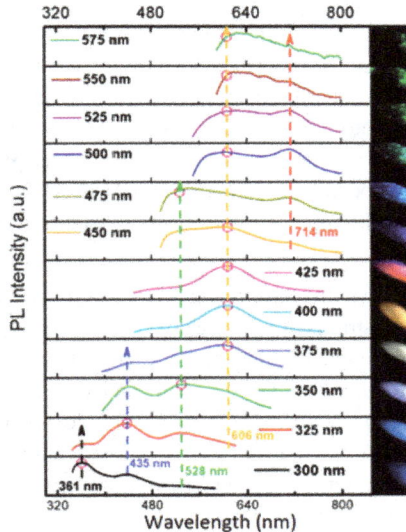

Fig. 6 The emission spectra excited by wavelength in the range of 320-800 nm and their corresponding photograph of GQDs solution [30].

3.3   Electroluminescence

Electroluminescence is the phenomena where photons/light generated through electrical excitation. Liu et al. [33] reported the enhancement of electroluminescence intensity 4 times greater than the bare vertical light-emitting diodes (VLEDs) when the VLEDs coated with N-doped graphene quantum dots. It was attributed to the photons recovered by extracting the light from waveguide modes. Wang et al. [34] synthesized the chlorine-doped GQDs based white-light-emitting diodes (WLEDs). This type of WLEDs showed good quality emitting and gratifying stability as well as excellent photostability and heat resistance. Generally, the thermal stress occurred and reduced the luminescence when the working time increased. Nonetheless, the GQDs based WLED still maintained 94% of the initial emission intensity over 180 min of operation, revealing good LED stability.

4.   Antibacterial activity of GQDs

4.1   Light-assisted antibacterial activity of GQDs

The emerging GQDs have received enormous attraction for their excellent biomedical applications due to their distinctive and tunable PL properties, nontoxic *in vitro* and *in vivo*, excellent stability in water, environmentally benign and small size. With their efficient light absorption over a broad range of light spectrum, GQDs are known as potential candidates for photodynamic therapy in treating cancer [6,35]. Correspondingly, light-excited GQDs can generate reactive oxygen species (ROS), which are recognized to inactivate the bacteria. Under light irradiation, GQDs can produce a large quantity of free electrons and holes. The photo generated electrons and holes can react with O_2 in air or water to generate excessive ROS including hydroxyl radicals (•OH) and singlet oxygen ($^1O_2$) [36,37]. It was well documented that the contact of ROS with the bacteria caused damage of cell membrane, leakage of intracellular substances and ultimately cell death [37-39]

Ristic et al. [6] reported that the electrochemically produced GQDs can act as new antibacterial agents for two strains of pathogenic bacteria, namely methicillin-resistant *S. aureus* and *E. coli*. They found that the photo excited GQDs had better antibacterial performance compared to those of GQDs alone and direct photolysis. The bacterial killing was revealed through the decrease in bacterial colonies as measured by standard plate count technique, the increment in propidium iodide intake verifying the disintegration of cell membrane, and the change of bacterial cell morphology as evidenced by the atomic force microscopy. Their reported results also suggested that the generated ROS disrupted the bacterial cell membrane and caused the cell death of microbes.

In another study, Kuo et al. [40] reported that the amino-functionalized N-doped GQDs presented superior ROS generation ability compared to unmodified GQDs, leading to an enhanced antibacterial performance. They went further to suggest that the oxidative stress induced by ROS caused the cell failed to maintain normal physiological redox-regulated functions, which then resulted in DNA damage and eventually led to bacterial death. More recently, Wang et al. [41] synthesized Cl-doped GQDs and analyzed their antibacterial activity under simulated sunlight irradiation. Compared to undoped GQDs, the Cl-doped GQDs exhibited impressive bactericidal capability towards *E. coli*. The enhanced antibacterial performance was attributed to their high ROS producing ability, which were identified in electron spin resonance (ESR) spectroscopy.

4.2   Shape and size of GQDs

Hui et al. [42] studied the effect of source material and bacterial shape matter on the antibacterial property of GQDs. The GQDs fabricated by rupturing C_{60} cages (C_{60}-GQD) with nonzero Gaussian curvature possessed utmost antibacterial activity towards *S. aureus*, while remained unreactive against *Bacillus subtilis* (*B. subtilis*), *Pseudomonas aeruginosa* (*P. aeruginosa*) and *E. coli*. The observed activity of C_{60}-GQD was reported to be interrelated with its disruption ability on the bacterial cell membrane. They also further explained that the match in the surface curvature of GQD with the bacterium can play vital role for the interaction of the GQD with bacterial cell surface. On the contrary, the GQD synthesized from GO (i.e., GO-GQD) with zero Gaussian curvature barely impacted the viability ratios of all bacterial strains studied, although both GQDs showed similar surface charges, sizes and compositions. Nevertheless, they have differences in the oxygen/carbon (O/C) ratio, functional group and zeta potential. Therefore, the detailed structural features affecting the antibacterial performance of GQDs was deserved to be examined further.

Bare GQDs were also studied for the size-dependent antibacterial activity towards *E. coli* [43]. Compared to the larger size GQDs at 50 nm, the smaller size GQDs with 15 nm penetrated the cell membrane more easily, leading to oxidative damage and membrane rupture. The oxidative stress and membrane disruption were attributed to the ROS generation as detected by the 3-(4,5-Dimethylthiazol-2-yl)-2,5-Diphenyltetrazolium Bromide (MTT) method

4.3 H_2O_2-assisted antibacterial activity of GQDs

Over the past decades, H_2O_2 has been widely used for disinfection and sterilization. Nevertheless, some bacterial species exhibited enhanced resistance to peroxide and therefore required higher concentrations of H_2O_2, which can cause harmful to biological system like human tissue [44,45]. To address this lapse, a combination of H_2O_2 with nontoxic photo excited GQDs can be an effective method for achieving high antibacterial performance at low dosage of H_2O_2. Sun et al. [45] combined GQDs with low H_2O_2concentrations and investigated their antimicrobial ability against bacterial cell. The results in their investigation showed that the combination of GQDs with H_2O_2 significantly inhibited the growth of both Gram-negative *E. coli* and Gram-positive *S. aureus* bacteria. More specifically, the combination of GQDs and H_2O_2 (1 and 10 mM) resulted the similar inactivation performance on *E. coli* and *S. aureus* as compared to the higher concentrations of individual H_2O_2 (100 mM and 1 M). Such enhancement effect was attributed to the peroxidase-like activity of GQDs to catalyze the decomposition of H_2O_2 to generate more •OH radicals. Further scanning electron microscopy investigation also showed that the GQDs/H_2O_2 treatment caused the bacterial surface became rough as the ROS can oxidize the cell membrane and resulted to membrane rupture. In contrast, the bacterial cell viability was also relied on the dose of GQDs.

4.4   Biomedical application of GQDs

Sun et al. [45] also assessed the antibacterial effect of GQD-Band-Aid in wound disinfection via *in vivo* studies with mice. It was found that the GQDs derived from GO yielded no antibacterial performance *in vitro*. Nevertheless, the antibacterial mechanism of GQDs and H_2O_2 enhanced the •OH radicals formation via high H_2O_2decomposition, which led to better antibacterial performance. They concluded that the GQD-Band-Aid has potential applications for injure healing/disinfection with the assistance of low concentration of H_2O_2

Conclusions

A series of methods has been used for synthesizing GQDs, in spite of that, there is a requirement for significant relationship between the synthetic technique and photo electrochemical parameters of GQDs. The understanding of these is essential in improving the production of quantum dots. Top-down approaches have been used to synthesize uniform structure of GQDs but usually with slightly lower yield and required harsh conditions which restricting their practical applications. Some bottom-up methods have also been widely employed to synthesize GQDs because of their simplicity; nevertheless, irregular nanostructures and wide size distributions were generated through bottom up approaches. With rapid advancement on this area, a timely account regarding the applications of GQDs in antibacterial is extremely essential. The biocompatibility of GQDs for different cell lines should be broadly investigated for safe biomedical applications. Moreover, the investigations of acute and chronic toxicity of GQDs on animals should be carried out. We confidently believed that the GQDs have extensive application prospect especially antibacterial and will be discovered more comprehensively. We also anticipated that many scientists will spend their efforts in this field and brought substantial scientific developments.

Acknowledgments

Supports from Universiti Tunku Abdul Rahman (UTARRF/2019-C1/L03), Ministry of Higher Education of Malaysia (FRGS/1/2016/TK02/UTAR/02/1 and FRGS/1/2019/TK02/UTAR/02/4), Guangxi Key Laboratory of Theory and Technology for Environmental Pollution Control (1801K012 and 1801K013) and Special Funding for Guangxi "Bagui Scholars" Construction Project are greatly acknowledged.

References

[1]   B. Das, S. Moumita, S. Ghosh, M.I. Khan, D. Indira, R. Jayabalan, S.K. Tripathy, A. Mishra, P. Balasubramanian, Biosynthesis of magnesium oxide (MgO) nanoflakes by using leaf extract of Bauhinia purpurea and evaluation of its antibacterial property against Staphylococcus aureus, Mater. Sci. Eng. C 91 (2018) 436-444. https://doi.org/10.1016/j.msec.2018.05.059

[2]   D. Sethi, R. Sakthivel, ZnO/TiO_2 composites for photocatalytic inactivation of Escherichia coli, J. Photochem. Photobiol. B 168 (2017) 117-123. https://doi.org/10.1016/j.jphotobiol.2017.02.005

[3]   S. Santajit, N. Indrawattana, Mechanisms of antimicrobial resistance in ESKAPE pathogens, Biomed. Res. Int. 2016 (2016) 2475067. https://doi.org/10.1155/2016/2475067

[4]   S.I. Miller, Antibiotic resistance and regulation of the gram-negative bacterial outer membrane barrier by host innate immune molecules, mBio 7 (2016) e01541-16. https://doi: 10.1128/mBio.01541-16

[5]   A. Abbas, L.T. Mariana, A.N. Phan, Biomass-waste derived graphene quantum dots and their applications, Carbon 140 (2018) 77-99. https://doi.org/10.1016/j.carbon.2018.08.016

[6]   B.Z. Ristic, M.M. Milenkovic, I.R. Dakic, B.M. Todorovic-Markovic, M.S. Milosavljevic, M.D. Budimir, V.G. Paunovic, M.D. Dramicanin, Z.M. Markovic, V.S. Trajkovic, Photodynamic antibacterial effect of graphene quantum dots, Biomaterials 35 (2014) 4428-4435. https://doi.org/10.1016/j.biomaterials.2014.02.014

[7] N.A.A. Anas, Y.W. Fen, N.A.S. Omar, W.M.E.M.M. Daniyal, N.S.M. Ramdzan, S. Saleviter, Development of graphene quantum dots-based optical sensor for toxic metal ion detection, Sensors 19 (2019) 3850. https://doi.org/10.3390/s19183850

[8] P. Devi, S. Saini, K.H. Kim, The advanced role of carbon quantum dots in nanomedical applications, Biosens. Bioelectron. 141 (2019) 111158. https://doi.org/10.1016/j.bios.2019.02.059

[9] A. Mehta, A. Mishra, S. Basu, N.P. Shetti, K.R. Reddy, T. Saleh, Band gap tuning and surface modification of carbon dots for sustainable environmental remediation and photocatalytic hydrogen production-A review, J. Environ. Manag. 250 (2019) 109486. https://doi.org/10.1016/j.jenvman.2019. 109486

[10] A. Valizadeh, H. Mikaeili, M. Samiei, S.M. Farkhani, N. Zarghami, M. Kouhi, A. Akbarzadeh, S. Dayaran, Quantum dots: synthesis, bioapplications, and toxicity, Nanoscale Research Lett. 7 (2012) 480. https://doi.org/10.1186/ 1556-276X-7-480

[11] M. Liu, Y. Xu, F. Niu, J. Gooding, J. Liu, Carbon quantum dots directly generated from electrochemical oxidation of graphite electrode in alkaline alcohols and the applications for specific ferric ion detection and cell imaging, Analyst 141 (2016) 2657-2664. https://doi.org/10.1039/C5AN02231B

[12] N.P. Shetti, D.S. Nayak, S.J. Malode, R.R. Kakarla, S.S. Shukla, T.M. Aminabhavi, Sensors based on ruthenium-doped TiO_2 nanoparticles loaded into multi-walled carbon nanotubes for the detection of flufenamic acid and mefenamic acid, Anal. Chim. Acta 1051 (2019) 58-72. https://doi.org/10.1016/j.aca.2018.11.041

[13] S. Ahirwar, S. Mallick, D. Bahadur, Electrochemical method to prepare graphene quantum dots and graphene oxide quantum dots, ACS Omega 2 (2017) 8343-8353. https://doi.org/10.1021/acsomega.7b01539

[14] C. Donate-Buendia, R. Torres-Mendieta, A. Pyatenko, E. Falomir, M.F. Alonso, G. Minguez-Vega, Fabrication by laser irradiation in a continuous flow jet of carbon quantum dots for fluorescence imaging, ACS Omega 3 (2018) 2735-2742.https://doi.org/10.1021/acsomega.7b02082

[15] P.G. Kuzmin, G.A. Shafeev, V.V. Bukin, S.V. Garnov, C. Farcau, R. Carles, B. Warot-Fontrose, V. Guieu, G. Viau, Silicon nanoparticles produced by femtosecond laser ablation in ethanol: size control, structural characterization, and optical properties, J. Phys. Chem. C 114 (2010) 15266-15273.https://doi.org/10.1021/jp102174y

Quantum Dots – Properties and Applications
Materials Research Foundations **96** (2021) 95-112

Materials Research Forum LLC
https://doi.org/10.21741/9781644901250-4

[16]   T.N. Lin, K.H. Chih, C.T. Yuan, J.L. Shen, C.A.J. Lin, W.R. Liu, Laser-ablation production of graphene oxide nanostructures: from ribbons to quantum dots, Nanoscale 7 (2015) 2708-2715. https://doi.org/10.1039/C4NR05737F

[17]   N. Arora, N.N. Sharma, Arc discharge synthesis of carbon nanotubes: comprehensive review, Diam. Relat. Mater. 50 (2014) 135-150.https://doi.org/10.1016/j.diamond.2014.10.001

[18]   S. Kim, J.K. Seo, J.H. Park, Y. Song, Y.S. Meng, White-light emission of blue-luminescent graphene quantum dots by europium (III) complex incorporation, Carbon 124 (2017) 479-485.https://doi.org/10.1016/j.carbon.2017.08.021

[19]   S. Yatom, J. Bak, A. Khrabryi, Y. Raitses, Detection of nanoparticles in carbon arc discharge with laser-induced incandescence, Carbon 117 (2017) 154-162.https://doi.org/10.1016/j.carbon.2017.02.055

[20]   S. Anwar, H.Z. Ding, M.S. Xu, X.L. Hu, Z.Z. Li, J.M Wang, L. Liu, L. Jiang, D. Wang, C. Dong, M.Q. Yan, Q.Y. Wang, H. Bi, Recent advances in synthesis, optical properties, and biomedical applications of carbon dots, ACS Appl. Bio. Mater. 2 (2019) 2317-2338. https://doi.org/10.1021/acsabm.9b00112

[21]   S. Chen, X. Hai, C. Xia, X.W. Chen, J.H. Wang, Preparation of excitation-independent photoluminescent graphene quantum dots with visible-light excitation/emission for cell imaging, Chem. Eur. J. 19 (2013) 15918-15923.https://doi.org/10.1002/chem.201302207

[22]   L.L. Li, J. Ji, R. Fei, C.Z. Wang, Q. Lu, J.R. Zhang, L.P. Jiang, J.J. Zhu, A facile microwave avenue to electrochemiluminescent two-color graphene quantum dots, Adv. Funct. Mater. 22 (2012) 2971-2979. https://doi.org/10.1002/adfm.201200166

[23]   Y.H. Shin, J.H. Lee, J.H. Yang, J.T. Park, K. Lee, S.J. Kim, Y.H. Park, H.Y. Lee, Mass production of graphene quantum dots by one-pot synthesis directly from graphite in high yield, Small 10 (2014) 866-870.https://doi.org/10.1002/smll.201302286

[24]   R. Ye, C. Xiang, J. Lin, Z. Peng, K. Huang, Z. Yan, N.P. Cook, E.L.G. Samuel, C. Hwang, G. Ruan, G. Ceriotti, A.O. Raji, A.A. Martí, J.M. Tour, Coal as an abundant source of graphene quantum dots, Nat. Commun. 4 (2013) 2943. https://doi.org/10.1038/ncomms3943

[25]   S. Li, S. Zhou, Y. Li, X. Li, J. Zhu, L. Fan, S. Yang, Exceptionally high payload of the IR780 iodide on folic acid-functionalized graphene quantum dots for targeted photothermal therapy, ACS Appl. Mater. Interfaces 9 (2017) 22332-22341.https://doi.org/10.1021/acsami.7b07267

[26] B. Bajorowicz, M.P. Kobylański, A. Gołąbiewska, J. Nadolna, A. Zaleska-Medynska, A. Malankowska, Quantum dot-decorated semiconductor micro- and nanoparticles: A review of their synthesis, characterization and application in photocatalysis, Adv. Coll. Interfac. 256 (2018) 352-372. https://doi.org/10.1016/j.cis.2018.02.003

[27] G. Muthusankar, C. Rajkumar, S. Chen, R. Karkuzhali, G. Gopu, A. Sangili, N. Sengottuvelan, R. Sankar, Sonochemical driven simple preparation of nitrogen-doped carbon quantum dots/SnO_2 nanocomposite: a novel electrocatalyst for sensitive voltammetric determination of riboflavin, Sensor. Actuator. B: Chem. 281 (2019) 602-612. https://doi.org/10.1016/j.snb.2018.10.145

[28] D. Huang, H. Zhou, Y. Wu, T. Wang, L. Sun, P. Gao, Y. Sun, H. Huang, G. Zhou, J. Hu, Bottom-up synthesis and structural design strategy for graphene quantum dots with tunable emission to the near infrared region, Carbon 142 (2019) 673-684. https://doi.org/10.1016/j.carbon.2018.10.047

[29] X. Zhang, C. Wei, Y. Li, D. Yu, Shining luminescent graphene quantum dots: Synthesis, physicochemical properties, and biomedical applications, Anal. Chem. 116 (2019) 109-121.https://doi.org/10.1016/j.trac.2019.03.011

[30] X. Li, S.P. Lau, L. Tang, R. Ji, P. Yang, Multicolour light emission from chorine-doped graphene quantum dots, J. Mater. Chem. C 1 (2013) 7308-7313.https://doi.org/10.1039/C3TC31473A

[31] W.X. Fu, J.F. Lin, Electrical and optical properties of the specimens with graphene quantum dots prepared by different number of wet transfer, Diam. Relat. Mater. 99 (2019) 107527. https://doi.org/10.1016/j.diamond.2019.107527

[32] F. Temerov, A. Belyaev, B. Ankudze, T.T. Pakkanen, Preparation and photoluminescence properties of graphene quantum dots by decomposition of graphene-encapsulated metal nanoparticles derived from Kraft lignin and transition metal salts, J. Lumin. 206 (2019) 403-411. https://doi.org/10.1016/j.jlumin.2018.10.093

[33] D. Liu, H. Li, B. Lyu, S. Cheng, Y. Zhu, P. Wang, D. Wang, X. Wang, J.Yang, Efficient performance enhancement of GaN-based vertical light-emitting diodes coated with N-doped graphene quantum dots, Opt. Mater. 89 (2019) 468-472. https://doi.org/10.1016/j.optmat.2019.01.026

[34] X. Wang, G. Wang, J. Li, Z. Liu, Y. Chen, L. Liu, J. Han, Direct white emissive Cl-doped graphene quantum dots-based flexible film as a single luminophore for

remote tunable UV-WLEDs, Chem. Eng. J. 361 (2019) 773-782.
https://doi.org/10.1016/j.cej.2018.12.131

[35] S. Ahirwar, S. Mallick, D. Bahadur, Photodynamic therapy using graphene
quantum dot derivatives, J. Solid State Chem. 282 (2020) 121107.
https://doi.org/10.1016/j.jssc.2019.121107

[36] W.S. Kuo, H.H. Chen, S.Y. Chen, C.Y. Chang, P.C. Chen, Y.I. Hou, Y.T. Shao,
H.F. Kao, C.L.L. Hsu, Y.C. Chen, S.J. Chen, S.R. Wu, J.Y. Wang, Graphene quantum
dots with nitrogen-doped content dependence for highly efficient dual-modality
photodynamic antimicrobial therapy and bioimaging, Biomaterials 120 (2017) 185-
194. https://doi: 10.1016/j.biomaterials.2016.12.022

[37] J. Ruan, Y. Wang, F. Li, R.B. Jia, G.M. Zhou, C.L. Shao, L.Q. Zhu, M. Cui, D.P.
Yang, S.F. Ge, Graphene quantum dots for radiotherapy, ACS Appl. Mater. Interfaces
10 (2018) 14342-14355. https://doi.org/10.1021/acsami.7b18975

[38] J.A. Quek, S.M. Lam, J.C. Sin, A.R. Mohamed, Visible light responsive flower-
like ZnO in photocatalytic antibacterial mechanism towards *Enterococcus faecalis* and
Micrococcus luteus, J. Photochem. Photobiol. B: Biol. 187 (2018) 66-75.
https://doi.org/10.1016/j.jphotobiol.2018.07.030

[39] F.D. Zhao, W. Gu, J. Zhou, Q. Liu, Y. Chong, Solar-excited graphene quantum
dots for bacterial inactivation via generation of reactive oxygen species, J. Environ.
Sci. Health C 37 (2019) 67-80. https://doi.org/10.1080/10590501.2019.1591701.

[40] W.S. Kuo, Y.T. Shao, K.S. Huang, T.M. Chou, C.H. Yang, Antimicrobial amino-
functionalized nitrogen-doped graphene quantum dots for eliminating multidrug-
resistant species in dual-modality photodynamic therapy and bioimaging under two-
photon excitation, ACS Appl. Mater. Interfaces 10 (2018) 14438-14446.
https://doi.org/10.1021/acsami.8b01429

[41] L.F. Wang, Y. Li, Y.M. Wang, W.H. Kong, Q.P. Lu, X.G. Liu, D.W. Zhang, L.T.
Qu, Chlorine-doped graphene quantum dots with enhanced anti- and pro-oxidant
properties, ACS Appl. Mater. Interfaces 11 (2019) 21822-21829.
https://doi.org/10.1021/acsami.9b03194

[42] L.W. Hui, J.L. Huang, G.X. Chen, Y.W. Zhu, L.H. Yang, Antibacterial property of
graphene quantum dots (both source material and bacterial shape matter), ACS Appl.
Mater. Interfaces 8 (2016) 20-25. https://doi.org/10.1021/acsami.5b10132

[43] S.H. Su, C.B. Shelton, J.J. Qiu, Size-dependent antibacterial behavior of graphene quantum dots, IMECE2013 15 (2014) 64872. https://doi.org/10.1115/IMECE2013-64872

[44] X.L. Dong, M.A. Awak, N. Tomlinson, Y.G. Tang, Y.P. Sun, L.J. Yang, Antibacterial effects of carbon dots in combination with other antimicrobial reagents, Plos One 12 (2017) 0185324. https://doi.org/10.1371/journal.pone.0185324

[45] H.J. Sun, N. Gao, K. Dong, J.S. Ren, X.G. Qu, Graphene quantum dots-band-aids used for wound disinfection, ACS Nano 8 (2014) 6202-6210. https://doi.org/10.1021/nn501640q

Quantum Dots – Properties and Applications   Materials Research Forum LLC
Materials Research Foundations **96** (2021) 113-144   https://doi.org/10.21741/9781644901250-5

Chapter 5

Computational Theories Used in the Study of Quantum Dots

N. Mhlanga[1,2]*, T.A. Ntho[2]

[1]DSI/Mintek Nanotechnology Innovation Centre, Randburg, South Africa

[2]Advanced Materials Division, Mintek, Randburg, South Africa

Nikiwem@mintek.co.za

Abstract

Quantum dots (QDs) are intriguing semiconductors with remarkable quantum confinement, optical and electrical properties which avails for various industrial and commercial applications to revolutionize our world. However, their optimal utilization hinges on the understanding of their properties and computational theories are imperative to explore both existing and new QDs properties. This chapter gives a comprehensive analysis of molecular mechanics and quantum mechanics computational approaches used in the study of the QDs properties.

Keywords

Quantum Dots, Density Functional Theory, Effective Mass Approximation, Schrodinger, Time-Dependent Density Functional Theory

Contents

Materials Research Forum LLC
https://doi.org/10.21741/9781644901250-5

1.   Introduction

Computational chemistry entails the use of theoretical ideas and computers to gain insight into the properties of molecules such as their physical and chemical properties [1]. These theoretical ideas are coupled with a set of techniques [2]. Quantum dots (QDs) are one such molecule studied using the computation theories. QDs are light-absorbing, fluorescing (luminescence) materials surging in numerous applications such as sensing due to their remarkable optical, electrical and high quantum yield properties [2]. However, their application is dependent on their structural design and material which significantly governs their application. Therefore, to maximize their potential in the various applications and gain insight into their mechanisms, theoretical studies are implemented. Quantum based computational theories (density functional theory (DFT), effective approximation (EMA) and time-dependent DFT) together with both classical and *ab-Initio* molecular mechanics methods are used for the computational prediction of QDs physico-chemical properties [2]. Table 1 [3-70] summarizes the application of the theories in various properties of the QDs. The application of the QDs is classified according to their most studied properties: electrical, and optical; a few examples are given on the less explored magnetic and thermal properties.

The optical property of the QDs is dependent on quantum confinement which is determined by their shape, size and chemical composition [3]. A size decrease of the QDs results in quantum confinement which widens the bandgap and the bandgap is described as the energy necessary to generate an electron and a hole at rest. As the QDs are illuminated with a photon characterized by higher energy than the bandgap's energy, an exciton is produced (electron-hole pair) [4, 71]. The distance between the two entities (hole and electron) is called the exciton Bohr radius (r_B) and is expressed by equation 1:

Materials Research Forum LLC

https://doi.org/10.21741/9781644901250-5

$$r_B = \frac{\hbar^2 \varepsilon}{e^2}\left(\frac{1}{m_e} + \frac{1}{m_h}\right)\ldots\ldots\ldots$$ (1)

Where $\hbar$ is the reduced Plank's constant, ε is the optical dielectric constant, e an electron charge and m_e, m_h are the effective masses of the electron and hole, correspondingly [4]. The quantum confinement effect is achieved when the radius of the QDs has a smaller diameter than its exciton hence results in the confinement of the electron and hole [71]. A gap exists between synthesis of the QDs e.g., core-shell QDs and detailed understanding of their optical properties. The experiments assume a lot of chemical reactions based on chemistry rules and use instrumentation to confirm the realities of the experiments which have shortfalls in visualization. Computational theories solve the issue of visualization by providing the necessary tools which will benefit applications of the QDs. For example, gaining insight into the optical properties of the QDs can improve their application in solar energy [3].

QDs are also called artificial atoms because the electron and the hole are carriers of distinct energies [71], hence they have an electrical property. During the excitation of the QDs from a ground state to an excited one, the electron and hole acquire higher energies which are determined by their electric structures. The empty hole serves as a mobile positive charge. After excitation, the excited bodies will lose their energy and jump back to the hole in a process called electron-hole recombination or relaxation. The released energy is the sum of the confinement, electron-hole, electron, bandgap and bound energy of exciton energies [4, 71]. The output energies from the excitation and relaxation processes are either radiative (emits photons) or non-radiative (emits phonons) [4]. Bera et al. [4] in their review on QDs and their multimodal application gives a comprehensive insight of these energies. The size and surface of the QDs also determine their electrical properties [3].

Materials Research Forum LLC
https://doi.org/10.21741/9781644901250-5

Table 1 A summary of the computational approaches (with references) used to study the properties of QDs and possible applications [3-70].

QDs properties	Application	Computational theories	References
Optical property	• Emitters for luminescent solar concentrations (in photovoltaics) [3]. • Solar cells (sensitized, dispersed solar cells) • Biological markers/imaging • Light-emitting devices • Optical amplifiers • Lasers [4]	• Effective Mass Approximation (EMA) [4] • Linear combination of atomic orbitals (LCAO) [4] • Density functional theory (DFT) • Time dependent DFT (TD-DFT) • Tight binding method (TB) • Monte Carlo • Time-dependent DF TB • Finite difference method (FDM) • k.p. model	[3-11] [12-15] [16-24] [19, 25-30] [31, 32] [33, 34] [35] [36-40] [41, 42]
Electrical property		• TD-DFT • EMA • Nondiabatic molecular dynamic • TB • DFT • FDM	[16, 43-49] [50, 51] [44] [52-54] [29, 54-65] [36, 66]
Thermal property	Thermoelectric devices [67]	• DFT	[68]
Magnetic properties	Bioimaging, nanomedicine [69]	• Monte Carlo	[70]

2. Computational approaches

2.1 Molecular mechanics

The classical molecular mechanics (MM) model is based on the coarse grain atom which clumps the nucleus and electrons into one particle and treats the nucleus as inert spheres evolving according to Newton's law of motion [72]. The MM calculations are based on empirical potential and they are used in the investigation of the properties of systems. MM offers the subsequent advantages; feasibility to run larger systems, calculations at low computational cost and is good for calculation of thermal properties. However, they are handicapped in the study of QDs due to the lack of electron-electron interactions.

They overlook the description of electronic/structural degrees of freedom, therefore, lacks accuracy in studying the optical properties of the QDs [3]. Although the lack of accuracy negates their application in QDs, they can be used in the study of complex systems such as QDs conjugated to gigantic biological moieties. In such cases, MM can be used solely or combined with other approaches such as quantum mechanics (QM).

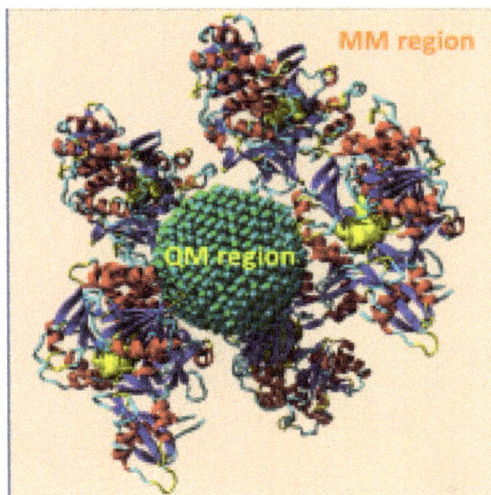

Fig. 1 ONIOM partitions of the CdSe in green and firefly luciferase protein (ribbon representation) system. Reprinted from optical signature of formation of protein corona in the firefly luciferase-CdSe quantum dot complex, 10/12, J.M. Elward, F.J.Irudayanathan, S.Nangia, A. Chakraborty, J. Chem. Theory Comput., 5224-5228, Copyright (2014)with permission from American chemical society [73].

Elward et al. [73] combined MM with QM to study the interaction of CdSe QDs with firefly luciferase protein. The complex CdSe protein complex was treated with different levels of theory to facilitate convergence. The CdSe was treated at the QM level of theory while MM was used for the protein. The interface of the CdSeQd and protein is dominated with electrostatic interactions produced by the partial charges on the protein hence the calculation had to consider the presence of the electrostatic field. Fig. 1 [74] shows the QM/MM partitioned QDs/protein complex.

Materials Research Forum LLC
https://doi.org/10.21741/9781644901250-5

2.2   Quantum mechanics

Quantum mechanical (QM) theory encompasses systems of interacting ions and electrons and is founded on the solution of the many bodied Schrodinger equation. Equation 2 depicts the compressed version of the Schrodinger equation [74].

$$H\Psi - E\Psi \ldots\ldots\ldots\ldots \tag{2}$$

H is the Hamiltonian operator and **E** is the energy of the system [77]. The Hamiltonian operator contains kinetic and potential energy produced as the ions and electrons are interacting as shown in equation 3 [74].

$$\underbrace{\frac{-\hbar^2}{2m}\nabla^2\,\Psi(r)}_{\text{Kinetic energy}} + \underbrace{V(r)\Psi(r)}_{\text{Potential energy}} = E\Psi(r)\ldots\ldots\ldots \tag{3}$$

Kinetic energy   Potential energy

$\hbar$ is Planck's constant divided by $2\,\pi$, m is the electron mass, ∇ is the Laplacian operator, Ψ is the wave function, V is the potential energy, E is the energy system, **(r)** signifies the functions of spherical polar coordinates (r, θ, φ) [74].

Born-Oppenheimer is one of the simplest approaches used to describe the quantum state of molecules. It approximates that the ions are moving slowly in space while the electrons respond instantaneously to an ionic motion and render a wavefunction dependent on the electronic degrees of freedom alone. Hence it neglects the movement of the atomic nuclei and basis the calculations on the motion of the electrons. This assumption is based on the masses and charges of the atomic nuclei and electrons [74, 75]. Firstly, there is a huge mass difference between the ions and atomic nuclei which renders the ions classical particles (MM) i.e., the ions kinetic quantum mechanical term is omitted and taken as a classical contribution. The electrons, which are smaller in size compared to the heavy nuclei move faster and possess a quicker response to forces [74, 75]. Secondly, the opposite charges carried by the electrons and atomic nuclei results in a mutually attractive force which accelerates both bodies. The magnitude of the acceleration is oppositely symmetrical to the mass hence the electrons will possess a higher acceleration magnitude compared to the atomic nuclei [75]. Therefore, the motion of the atomic nuclei is ignored by assuming they are stationery and the quantum state of solid molecules is described by the movement of the faster electrons. However, the stationary atomic nuclei position still determine the electronic energy and the resulting wavefunction [74]. Hence, even with the simplified Hamiltonian approximation, solving the many bodied Schrodinger remains a challenge as a result of the nature of the electrons. To solve the Schrodinger equation, the Hamiltonian term should take into account the electron-electron repulsions, yet it complicates it. The electron-electron interactions suggest a single function which is dependent on the coordinates of the electrons as an ideal wavefunction [75]. Hartree [74,75] separated the variables of the Hamiltonian to make it easy to solve the Schrodinger equation. Hartree-approximates the Hamiltonian from one electron eigenfunction i.e., non-interacting by introducing electrons [74, 75]. Then, Fock [75] modified Hatree's approximation into Hartree-Fock by introducing anti symmetrized wavefunctions instead of the one-electron wavefunctions.

Several theorems and approximations have been developed to solve the many bodied Schrodinger equation and the next sections discuss some of the popular ones with reference to their application in QDs simulations.

2.2.1 Density functional theory

Hohenberg, Kohn and Sham [76] solved the many bodied Schrodinger equation using DFT hence it is based on their theorems. The application of DFT in realistic calculations of solids and molecular properties such as energies, geometries and charge densities is growing [74]. DFT does not specify the wavefunction and hence single-particle equations are developed and approximations are introduced as required [74]. The DFT theorems utilize single-particle equations and an effective potential to calculate many-body electronic ground state [76]. The effective potential takes into account the following potentials; ionic, Hartree and the exchange-correlation potentials from the atomic cores, electron-electron interactions and many-body effects, respectively [76].

The application of DFT using pseudopotentials, plane-wave basis set and supercell geometry is becoming a standard for first-principle defect studies on semiconductors such as QDs [76]. The DFT tool has been used to study the QDs size effect, density of state of the surface topology because it controls the electronic and optical properties [3]. The DFT approach solves the Kohn-Shawn equation by using the self-consistent 1st principle local density approximation (LDA) calculations. Although DFT is becoming a standard tool for QDs calculations, it is not devoid of challenges. DFT is a slow tool which includes high computational cost and is not suitable for larger systems even though it has high accuracy. Also, studying complex surface passivation of the QDs using the DFT tools is challenging and hinders the understanding of the electronic structure of the QDs [3].

The DFT calculations are computationally challenging because of their self-consistent nature and the calculation of each one of the orbitals ψ_i per iteration [77]. However, calculation of semiconductors such as QDs does not require all the orbitals but concentrates only on a few states in the area of the gap that determines the optical and transparent properties [77]. The DFT/LDA calculations are known to underestimate the QDs bandgap [80]. The DFT self-consistent full calculation is alleviated through modifications such as the charge patching method (CPM). CPM, without compromising the accuracy of the results assumes that the charge density in the region of an atom is reliant on the local atomic surroundings of the atom in the absence of long-range external electric field [80]. A charge density of small prototypes is decomposed into contributions from individual atoms which are used to patch the charge density of a bulky system [80]. The accuracy of the CPM approach is comparable to the DFT calculations but it is limited in calculations of asymmetric QDs total dipole moments [77]. Table 2 adapted from Bodroski et al. [78] shows DFT and CPM calculated eigenenergies of C10H22, C20H42 and C40H82 and proved the comparability of the CPM and DFT methods.

Table 2   CPM and DFT eignenergengies (Ha). Reprinted from Gaussian basis implementation of the charge patching method, 368, Z. Bodroski, N. Vukmirović, S. Skrbic, J. Comput. Phys., 196-209, copyright (2018), with permission from Elsevier [78].

$C_{10}H_{22}$		$C_{20}H_{42}$		$C_{40}H_{82}$	
CPM	DFT	CPM	DFT	CPM	DFT
−0.309603	−0.309024	−0.278819	−0.278651	−0.275738	−0.276441
−0.290277	−0.289713	−0.278804	−0.278263	−0.275696	−0.276399
−0.290144	−0.289644	−0.278742	−0.277949	−0.275376	−0.275761
−0.281836	−0.281425	−0.278170	−0.277590	−0.275333	−0.275743
−0.279675	−0.279398	−0.276962	−0.276825	−0.267867	−0.268255
−0.279312	−0.278751	−0.276230	−0.275420	−0.252784	−0.253185
−0.278589	−0.277850	−0.275932	−0.275314	−0.239152	−0.239529
−0.276855	−0.276513	−0.261474	−0.260857	−0.227590	−0.227898
−0.275733	−0.275079	−0.235925	−0.235267	−0.218769	−0.218922
−0.230841	−0.229995	−0.218384	−0.217236	−0.213411	−0.213216

Chu et al. [79] explored both the DFT and CPM models to calculate the total charge density of two CdSe QDs linked by a Sn_2S_6 molecule, shown in Fig. 2 [79]. Using a diving and conquer method, the molecule was segmented into 3 parts and a small part of the 2-CdSe QDs shown by the red solid lines was cast out. Then using DFT the cutoff bonds were coated by pseudo hydrogen atoms. The CPM was used to make the charge density of the 2 CdSe QDs excluding the linker molecule. Subsequently, the 3 entities, 2

QDs and linker molecules were seamlessly connected to yield the total charge density of the whole system [82].

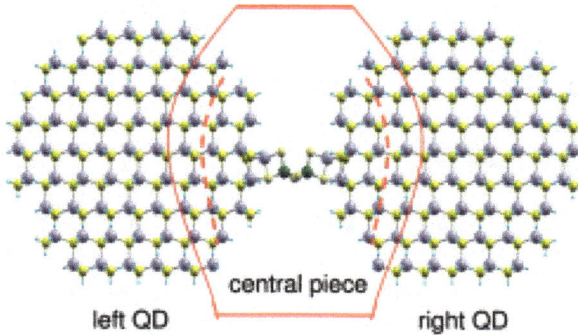

Fig. 2 Representation of two CdSe QDs conjugated by a Sn_2S_6 molecule used to calculate the total energies. Reprinted from charge transport in a QDs supercrystal, 115/43, I.-H. Chu, M. Radulaski, N. Vukmirovic, H.-P. Cheng, L.-W. Wang, J. Phys. Chem. C, 21409-21415, Copyright (2011), with permission from American chemical society [79].

2.2.2 Effective mass approximation

EMA models are analytical tools that study the macroscopic properties of mediums basing them on the properties and relation fractions of its components [80]. The inception of EMA models started with Briggeman and has been modified by Lavendor, Efros and Bruss, respectively [4, 80, 81]. The EMA approximation has been applied in solving numerous problems, they are the basis for macroscopically inhomogeneous media studies and are cemented for future applications as materials are shrinking to smaller sizes [80, 81]. EMA is a particle in a box model that assumes an infinite barrier of the boundary of the particle in a potential well [4] and it uses Schrodinger resembling equations based on effective parameters of the QDs' holes and electron to describe their exciton[41]. Equation 4 describes the energy of a particle that can assume any spot of choice in the box and holds for hole and electron in the QDs semiconductor [4].

$$E = \frac{\hbar^2 k^2}{2m^*} \ldots \ldots \ldots \tag{4}$$

Where $\hbar$ is the reduced Planck's constant, $\mathbf{k}$ is the wave vector and $\boldsymbol{m}^*$ is the effective mass [4].

The bandgap energy ($\boldsymbol{\Delta E_g}$) of the QDs with a diameter $\mathbf{R}$ is given by relation 5 and $\boldsymbol{\mu}$ is the reduced mass of an electron-hole pair, $\boldsymbol{\varepsilon}$ is the optical dielectric constant, $\boldsymbol{e}$ an electron charge, $\boldsymbol{m_e}$, $\boldsymbol{m_h}$ are effective masses for the electron and holes and $\boldsymbol{E_{Ry}}$ is Rydberg energy [4, 82].

$$\Delta E_g = \frac{\hbar^2}{2\mu R^2} - \frac{1.8e^2}{\varepsilon R} = \frac{\hbar^2\pi^2}{2R^2}\left(\frac{1}{m_e} + \frac{1}{m_h}\right) - \frac{1.786e^2}{4R} - 0.248E_{Ry}{}^* \tag{5}$$

The first part of equation 5 connects a particle in a box (quantum confinement) and the R of the QDs and the next part depicts the columbic interaction energy with a R^{-1} reliance [4].

Chukwocha et al. [83] simulated the quantum confinement effect of ZnS, CdSe and GaAs QDs using Brus' equation based on the EMA approximation. They explored equation 5 to calculate the quantum confinement effect and their results shown in Fig. 3 [83] confirmed a size dependent effect.

However, equation 5 fails to calculate the band gap of QDs with smaller diameters due to the overlay of the wavefunction with crystal boundaries [82]. To mitigate the size limitation Harrison (2005) used a corrected effective mass to consider the non-parabolicity of the energy bandgap (equation 6) [82].

$$m_{corr}{}^x(r) - m^x(r)\left[1 + \left(1 - \frac{m^y(r)}{m_0}\right)^2 \frac{E - V(r)}{Eg(r)}\right]\ldots\ldots\ldots \tag{6}$$

E_n is the energy of the considered level $\mathbf{n}$ and $\boldsymbol{m_0}$ is the electron mass.

The $E_g(r)$ term is size-dependent and the large values are utilized to compute the conduction valence bands' 1st level energies as they present the smallest deviations stimulated by nonporabolicity [82]. The bandgap values obtained from this energy level calculation are used to rectify the nonparabolicity on higher levels. This ballistic method is applied for QDs of different sizes, including the smaller ones [82].

Materials Research Forum LLC
https://doi.org/10.21741/9781644901250-5

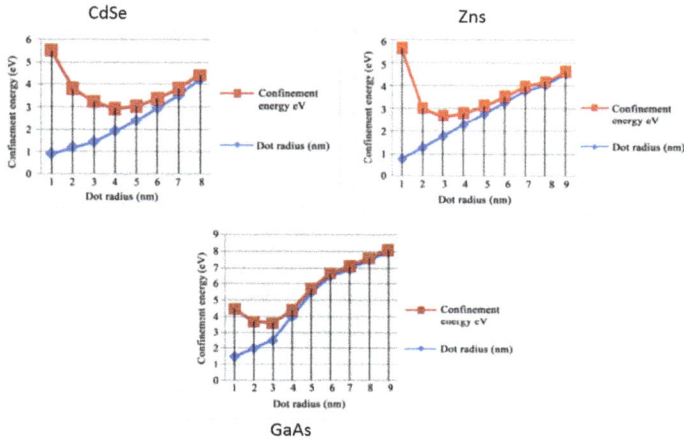

Fig. 3 CdSe, ZnS and GaAs QDs calculated radius and ground state confinement energy. Reprinted from theoretical studies on the effect of confinement on quantum dots using the Brusequation, 2, E.O. Chukwuocha, E.O., M.C. Onyeaju, T.S. Harry, WJCMP, 96-100, Copyright (2012) [83].

Although EMA is the simplest and most appropriate approach suitable for predicting with accuracy the size dependence of optical gaps i.e., optical and electronic properties of mostly large QDs systems, it treats the QDs surface as an infinite potential barrier around the core and neglects the surface effects [84]. This challenge is resolved by the use of semi-empirical modeling approaches which represent the QDs core using TB or pseudopotential approaches. The TB approach will be described in the subsequent paragraphs. More EMA and TB challenges still linger, such as the study of the QDs surface layer and QDs-ligand coupling due to the non-periodic and complicated potential at interfaces [84].

2.2.3 Linear combination of atomic orbitals

LCAO is a simple method that qualitatively gives a picture of the molecule's molecular orbitals (MOs). The LCAO approximation combines two s-atomic orbitals to form an s-bonding and σ^*antibonding MOs [85]. LCAO-MO predicts the evolution of electronic structures and their size dependence to the bandgap of clusters transition from atoms-molecules-QDs-bulk systems [4]. It is suitable for smaller sized QDs and cannot be used for large ones due to mathematical complexity [4].

2.2.4 Tight binding

TB is a simple semi-empirical approach similar to the LCAO and is used to calculate atomic structures of QDs. It is a very fast approach that suites the calculation of a larger system with a few thousand atoms in a unit cell. However, it lacks explicit basis functions [77]. Arunragsa et al. [86] successfully fabricated a room temperature gas sensor by drop-casting hydroxyl-functionalized graphene QDs (H-GQDs) onto a nickel electrode. The sensor showed high selectivity and sensitivity towards ammonia at room temperature with a linear relationship for 10 to 500 ppm concentrations. A synergy of the experimental and theoretical approaches was displayed in this paper as they used self-consistent charge density functional TB approach to study the mechanism of the sensor which was attributed to the OH-functional.

2.2.5 Time dependent DFT

TDDFT is a popular linear response approach that is appropriate to study the excitation of molecules since it allows robust electron-hole interactions at a reasonable computational cost. Its suitability, speed and cost elucidate it popularity for photoexcitations of nanomaterials and solids such as ligated QDs [84]. TDDFT solves a time-dependent Kohn-Shawn equation as shown in equations 7 and 8.

$$i\frac{\partial}{\partial t}\psi_i(r,t) = \left[\frac{1}{2}\nabla^2 + V(r,t)\right]\psi_i(r,t)\ldots\ldots \tag{7}$$

$$\text{And } \rho = \sum_{i=1}^{M} |\psi_i(r,t)^2 \ldots\ldots \tag{8}$$

The TDDFT potential $V(r,t)$ should constantly be determined by the charge density before t, however, in reality, the approach utilizes the adiabatic LDA. The LDA assumes that the potential is dependent on $\rho(r,t)$ and this dependence results in a functional form that is the same as LDA in TDDFT. Hence the approximation in TDDFT is termed as time-dependent LDA (TDLDA) [77]. To enable the calculation of the QDs optical properties using the TDDFT approach, an external electromagnetic field is added into equation 7 and it is solved by explicit integration in time [77].

The TDDFT is well suited for simulating optical properties of small molecules and clusters and it agrees with experimental measurements [77]. However, for large molecules, the accuracy of the TDDFT is compromised and it gives the same bandgap as LDA calculations. To alleviate this issue hybrid functional such as B3LYP can be used in the TDDFT calculations. The B3LYP models the total energy as a linear combination of the exact Hartree-Fock exchange with local and gradient-corrected exchange and

Materials Research Forum LLC

https://doi.org/10.21741/9781644901250-5

correlation terms [90]. And it can be used to attain accuracy in the bandgap calculations of some bulk molecules. Therefore, the TDDFT-B3LYP solves the TDLDA since it contains exchange integral which produces long-range Coulomb interactions [77]. Niaz et al. [87] used the TDDFT-B3LYP to study electro-optical properties of silicon QDs of diverse morphologies (stable spheres, elongated and reconstructed) endowed with different sizes. The TDDFT-B3LYP calculations showed high accuracy and comparability to both experimental and proven theoretical results (Fig. 4) [87]. The results could be further used to interpolate and extrapolate formulas for Si properties e.g., cohesive energy band gap, dielectric constant and index of refraction of silicon nanocrystals [87].

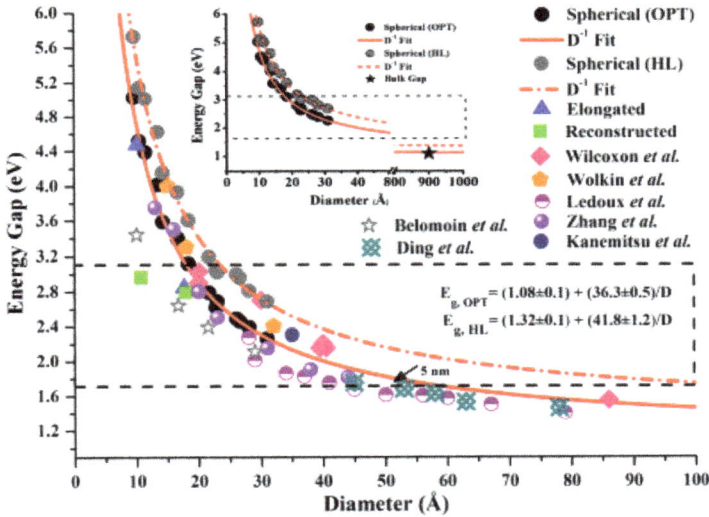

Fig. 4 Graphical plots showing calculated energy gap dependence on the diameter of Si QDs of various morphologies (spherical in red, elongated in blue and reconstructed in green). They are also compared to experimental data sourced from various researchers. A visible light range is depicted by the dotted rectangles and a silicon experimental band gap is shown by the black star in the insert. Reprinted from comprehensive ab-initio study of electronic, optical, and cohesive properties of silicon quantum dots of various morphologies and sizes up to infinity, 120/20, S.Niaz , A.D. Zdetsis, J. Phys. Chem. C, 11288-11298, Copyright (2016), with permission from American chemical society [87].

2.2.4 Other approaches

The configuration interaction (CI) method forms many-body excitations via slater determinants creation from single-particle states. Restricted sets of determinants form a Hilbert space which is used to translate the many-body Hamiltonian approximation. The CI wavefunction is assumed as equation 9 and is used to calculate QDs excitons.

$$\Psi = \sum_{v=1}^{Nv} \sum_{c=1}^{Nc} C_{v,c} \Phi_{v,c} \ldots\ldots \tag{9}$$

$\Phi_{v,c}$ is the slater determinant when the electron is excited from the valence band v to the conduction band c. The Hamiltonian eigenvalue is the represented by equation 10, where E_v and E_c are the single-particle eigenenergies, E is the exciton energy, $k_{vc,v'c'}$ and $\mathfrak{F}_{vc,v'c'}$ terms represent the exchange and Coulomb interactions, respectively.

$$\sum_{v'c'} H_{vc,v'c'} C_{v'c'} = \sum \left[(E_v - E_c)\delta_{v,v'}\delta_{c,c'} + k_{vc,v'c'} - \mathfrak{F}_{vc,v'c'} \right] C_{v'c'} = E C_{vc} \tag{10}$$

The eigenvalue problem is applicable in large molecules calculations e.g., pyramidal QDs made up of million atoms, many and few body excitations calculations [77].

Quantum Monte Carlo (QMC) is a method that uses variational (VMC) and diffusion monte Carlo (DMC) to solve the many bodied wave function and Schrodinger equation. Both the VMC and DMC with the use of pseudopotential are used in the calculation of systems of up to a dozen atoms [77].

The **k.p.** method describes the bulk band structure around a definite individual point structure in the Brillouin zone which has extended to heterostructures. The **k.p.** describes the Hamiltonian of an electron in a semiconductor using equation 11 [77]. The **k.p.** model shows accuracy in the calculation of the QDs exciton state and is comparable to experimental data [41].

$$\hat{H} = \frac{\hat{P}^2}{2} + V_0(r) + \hat{H}_{so} \ldots\ldots\ldots\ldots\ldots \tag{11}$$

$\hat{P}$momentum operator, $V_0(r)$ the crystal potential, $\hat{H}_{so}$ spin-orbit interaction Hamiltonian [77].

2.3   Evolution of the theoretical approaches in the band gap QDs calculations

In DFT the use of the Kohn-Shawn eigenvalue LDA and GGA approximations is not reliable for band gap calculations of semiconductors because it underestimates the band gap [88, 89]. Zhuravlev et al. [90] utilized the full-potential linear muffin-tin orbital to simulate electronic band structures of PbSe and CdSe QDs. Both the LDA and GGA approximations were used for the simulation and resulted in an underestimation and overestimation, respectively. To attain accuracy for the bandgap calculations of QDs several methods have been proposed and these methods include; self-interaction Perdew and Zunger correction method, the costly Green's functional and screened (GW) Coulomb method, LDA + U (coulomb energy parameter), exploitation of the grounded approximations in DFT e.g., TB, CPM, [78], time-dependent DFT or correction of the DFT-LDA bandgap via empirical approaches. The empirical pseudopotential method (EPM) was first initiated in the 1960s to calculate band structures for bulk semiconductors.   The EPM uses the plane-wave to fit the band structure of semiconductors. Semi-empirical (SEPM) is another approach that is used to calculate the band structure of semiconductors and it fits the pseudopotential directly to the LDA-calculated potential and alters them to correct the bandgap. Both the EPM and SEPM pseudopotentials construct a single particle Hamiltonian and the EPM can also be solved by the bulk Bloch bands methods [77].

Clark et al. [89] re-visited the HartreeFock approximation and utilized the screened exchange (SX) function derived before the local DFT functionals. The SX was used with the plane wave set and pseudopotentials in the CASTEP code to simulate band gap energies of various semiconductors. Fig. 5 [89] compares the DFT SX computed band gap (in purple) with GGA (PBE) DFT calculation (green) and the experimental band gap (red line) [89]. The CdSe had an SX calculated band gap of 1.88 eV comparable to the experimental 1.84 eV [89].

Thierry et al. [91] proposed a better and efficient technique to calculate the optical properties of semiconductors by solving the Schrodinger equation using a new formulation of a shooting method from EMA. The shooting method originates from Harrison's book and its speed and stability are improved via segmentation into two coupled equations with less floating points operations for application in 0,1 and 2-dimensional calculations. Relation 12 and 13 show the two shooting equations [91].

$$\psi_{n,l}(r + dr) = \frac{r\,dr\,m*(r)}{r+(N-1)m*(r)} \tilde{\psi}_{n,l}(r) + \psi_{n,l}(r)\ldots\ldots \tag{12}$$

$$\tilde{\psi}_{n,l}(r + dr) = \left(V(r) + \frac{h^2}{2r}\frac{l(l+N-2)}{2m*(r)} \right) - E\,\psi_{n,l}(r) + (\widetilde{\psi})_{n,l}(r)\ldots \tag{13}$$

Materials Research Forum LLC
https://doi.org/10.21741/9781644901250-5

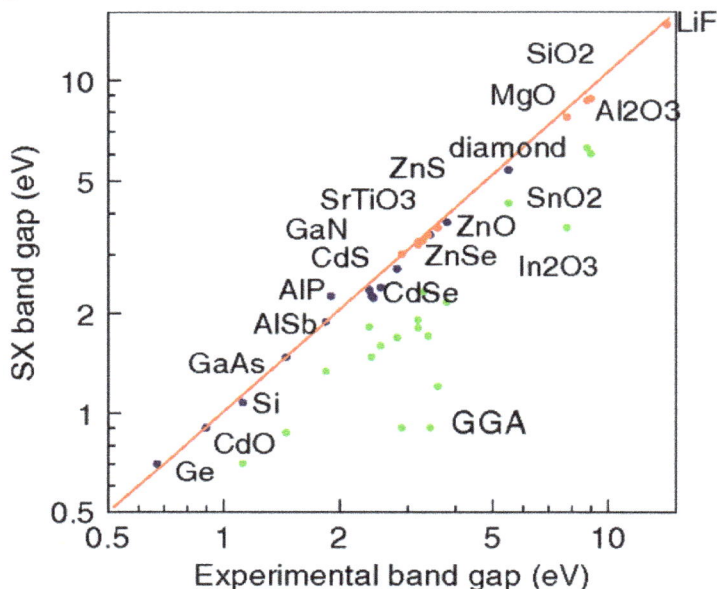

Fig. 5 SX and GGA (PBE) calculated band gaps for various compounds compared with experimental values. Reprinted from screened exchange density functional applied to solids, 82/8, S.J. Clark, J. Robertson, Phys. Rev. B, 085208, Copyright (2010), With permission from American Physical Society [89].

The equations were evaluated on CdSe and PbS QDs nanospheres and the results were corroborated with literature. Fig. 6 shows the calculated band gap of CdSe and Pbs (solid line and dashed line (correction) compared with experimental, SEPM, Brus (EMA) and TB. The calculated results correlated with SEPM, TB and experimental band gap values with exception of the Brus equation (EMA) for smaller QDs because it is not suitable for QDs with smaller band gap and increased quantum confinement as depicted in the Fig. [91].

Materials Research Forum LLC
https://doi.org/10.21741/9781644901250-5

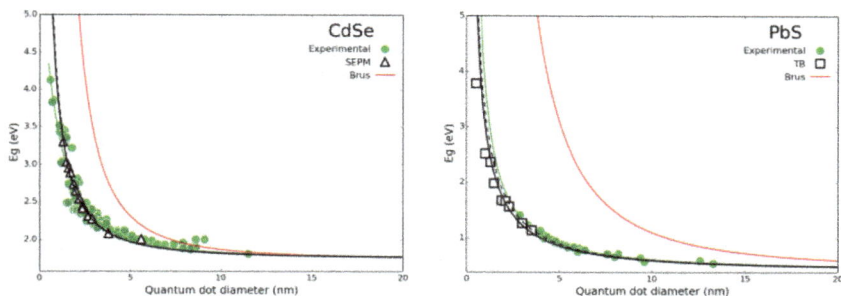

Fig.6 CdSe and PbS nanospheres calculated band gaps using EMA shooting method, SEPM, Brus (EMA) and TB compared to experimental data. Reprinted from the optimization of the optical properties of nanostructures through fast numerical approaches, 9161, S.J. Thierry, François, Judikaël Le Rouzo, François Flory, Gérard Berginc, LudovicEscoubas, SPIE, 916102, Copyright (2014) [91].

The accuracy of the computational approaches discussed in preceding paragraphs was evaluated by comparing calculated band gaps of hydrogen passivated Si QDs. Fig. 7 [63, 87, 93-95] compare the DFT and TB approaches used to calculate band gaps of SiH clusters endowed with various diameters. Different functionals and basis sets were explored for the DFT calculations. Fig. 7 shows corresponding data from Laref et al. DFT/PBE [92],Niaz et al. TDDFT/B3LYP [87] and Wang et al. (TD-DFTB) [93] while Anas et al. (DFT/LDA) [94] is slightly lower than the 3. The DFT/B3LYP Niaz et al. [87] plot aligns with bondwal et al. [63] CAM-B3LYP/cc-pVDZ. The band gap decreased with increasing SiH cluster diameter. Table 3 adapted from reference [96] list experimental band gap values for the clusters. The SiH_4 cluster with a diameter of less than 0.6 nm has an experimental band gap of approximately 9 eV. Bondwal et al. [63] CAM-B3LYP/cc-pVDZ and B3LYP/6-31G (d) theories overestimated SiH_4 band gap to about 11 eV and Laref et al. [92] DFT/PBE had 7.4 eV for a SiH cluster of 0.3 nm diameter. For the Si_5H_{12} cluster with an experimental band gap of 6.5 eV it aligned with Wang et al. [93] 6.4 eV TD-DFTB calculated band gap and was overestimated by Bondwal et al. [63]. $Si_{17}H_{36}$ band gap calculated by Bondwal et al. [63] DFT; CAM-B3LYP/cc-pVDZ (5.8 eV) and Niaz et al. [87] DFT/B3LYP (5.7 eV) and TDDFT/B3LYP (5.03 eV) were slightly closer to the experimental 5.47 eV (Table 2). $Si_{29}H_{36}$ experimental value of 4.2 is closer to Michael et al. [95] not shown in Fig. 7 who reported a theoretical value of 4.39 eV using TDDFT/LANL2D2 level of theory. $Si_{35}H_{36}$ 3.7 eV aligned with Anas et al. [94] DFT/LDA calculated 3.8 eV. The TD-DFTB and

TDDFT/B3LYP approaches showed a higher level of alignment to experimental data when compared with the other approaches shown in Fig. 7 [63, 87, 93-95]. Several approaches can be used to calculate the optical properties of QDs and each offers its limitations. However, with appropriate fine-tuning of functionals and basis sets accuracy of the approaches can be attained.

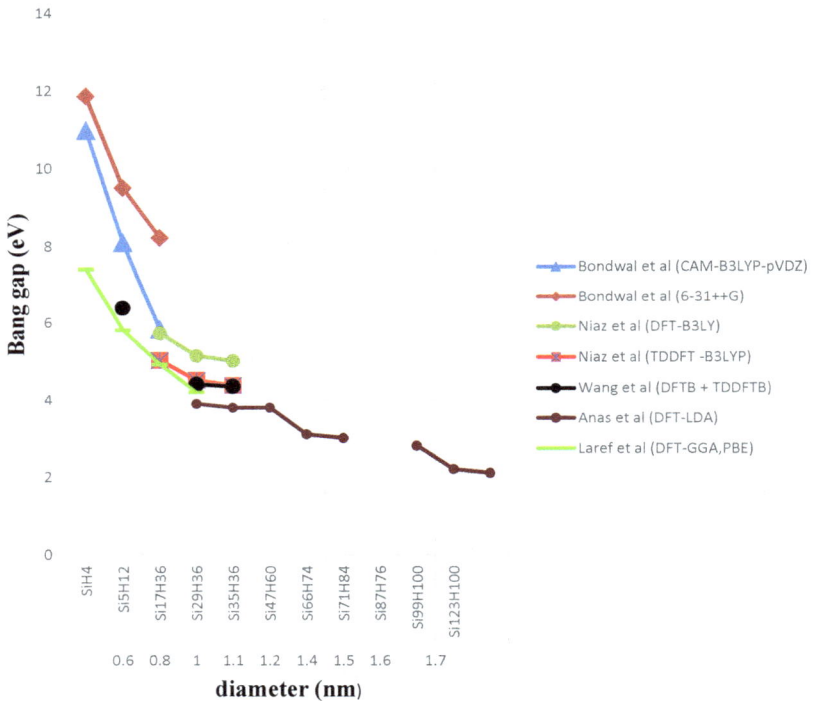

Fig.7 Comparison of hydrogen passivated Si QDs (SiH₄, Si₅H₁₂, Si₁₇H₃₆, Si₂₉H₃₆, Si₃₅H₃₆, Si₄₇H₆₀, Si₆₆H₇₄, Si₇₁H₈₄, Si₈₇H₇₆, Si₉₉H₁₀₀, Si₁₂₃H₁₀₀) theoretical band gap calculated using various theories. Bondwal et al. [63] calculated the HOMO-LUMO gap using DFT; CAM-B3LYP/cc-pVDZ and B3LYP/6-31G (d) level of theories. Niaz et al. [87] compared DFT/B3LYP and TDDFT/B3LYP. Wang et al. [93] used a computationally capable density functional-based TB approach (DFTB) and its time-dependent linear response(TD-DFTB). Anas et al. [94] calculated the band gap using DFT/LDA. Laref et al. [92] used DFT/PBE.

Table 3 Optical properties of hydrogen passivated Si QDs. Reprinted from electronic and optical properties of small hydrogenated silicon quantum dots using time-dependent density functional theory, 2015, M.M. Anas, G. Gopir, J. Nanomater., Copyright (2015), With permission from creative commons attribution license [96]

Quantum dots			Optical properties	
Diameter (nm)	Empirical formula	Optical threshold (eV)	First noticeable peak (eV)	Second noticeable peak (eV)
—	SiH_4	8.1, 8.0 [9], 8.0 [11]	8.54, 8.7 [9], 8.2 [11]	9.45, 9.6 [9], 9.1 [11]
0.6	Si_5H_{12}	5.5, 5.7 [10], 5.6 [11]	6.7, 6.5 [10], 6.4 [11]	8.12
0.8	$Si_{17}H_{36}$	4.0	5.47	6.72
1.0	$Si_{29}H_{36}$	3.5	4.2	4.8
1.1	$Si_{35}H_{36}$	3.43	3.7	4.4
1.2	$Si_{47}H_{60}$	3.4	4.1	5.05
1.4	$Si_{66}H_{74}$	3.3	3.72	4.93
1.5	$Si_{71}H_{84}$	3.2	3.81	5.98
1.6	$Si_{87}H_{76}$	3.0	4.78	6.72

Zerveas et al. [97] computed band gaps of quantum wells (InAs, GaAs and $In_{0.5}Ga_{0.47}As$) using DFT, empirical TB. K.p and EMA. Fig. 8 [97] shows a comparison of their calculated band gaps using TB, K.p (closed bounds and open bounds) and DFT with experimental. Quantum wells theoretical band gaps with a diameter of 10 nm had good agreement with experimental data while smaller than 10 nm overestimated and bigger than 10 nm underestimated the band gap. Besides the accuracy of the theoretical approach with the experimental data, simulation time is also critical in choosing a theoretical approach. The duration of the simulation is determined by the thickness of the QDs which governs the number of atoms in the DFT and TB codes, and the k.p code's grid size. The EMA code has negligible computational burden because it depends on the analytical band, hence it was excluded by ref [97] in the comparison of simulation duration. They compared the average computational of TB, k.p (4 bands), k.p (8 bands) and DFT and Fig. 9 [97] shows the computational times of the codes as a function of the thickness of the quantum well. For quantum wells with diameters of 5-7 nm the k.p (4 bands) is the cheapest while k.p (8 bands) is expensive. However, as the diameter increased, DFT gets more expensive (time spent solving eigenvalue problems), followed by k.p (8 bands). Increasing the k.p. bands improve the accuracy of the calculation but it comes at a cost whereas the TB approach is expected to be more accurate compared to k.p (8 bands). Hence the TB code is a better option because its saves on computational time [97].

Materials Research Forum LLC

https://doi.org/10.21741/9781644901250-5

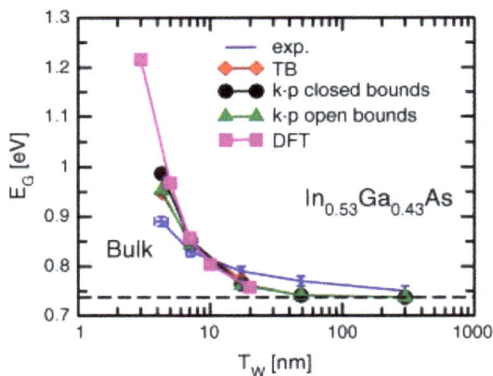

Fig. 8 Comparison of calculated In$_{0.5}$Ga$_{0.47}$As band gap with experimental. Reprinted from comprehensive comparison and experimental validation of band-structure calculation methods in III–V semiconductor quantum wells, 115, Zerveas, George, et al., 92-102, Copyright (2016), with permission from Elsevier [97].

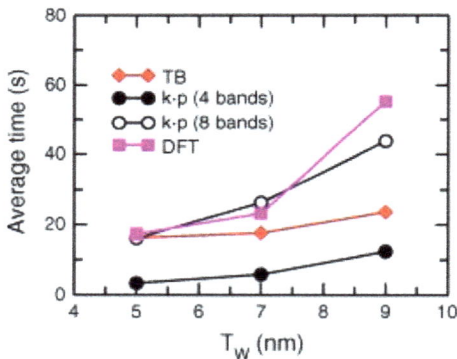

Fig.9 Computational time as a function of the thickness of the quantum wells. Reprinted from comprehensive comparison and experimental validation of band-structure calculation methods in III–V semiconductor quantum wells, 115, Zerveas, George, et al., 92-102, Copyright (2016), with permission from Elsevier [97]

Materials Research Forum LLC
https://doi.org/10.21741/9781644901250-5

Conclusions and future perspectives

The application of QDs in various fields such as the biological field is growing. With such growth, QDs-based gadgets are expected to be significant contributors to the economy. The popularity of the QDs is due to their consummate photostability, size tunability and photoluminescence, potential to produce multiple excited electron holes pairs from a single phonon [84]. Hence, they are key players in the fabrication of gadgets such as biological biosensors and solar cells. The fabrication of these gadgets hinges in the synthesis of reliable, reproducible QDs with outstanding properties. The high-quality gadgets depend on the understanding of the QDs properties to enable their engineering for optimum usage. For example, the application of QDs in photophysics demands an understanding of the QDs surface chemistry and their sensitivity to passivation and defeat [84]. Experimental approaches are failing to resolve this issue although recently a breakthrough was attained using scanning tunneling microscopy yet several issues remain unresolved. The pending issues include charge transfer, photoluminescence achieved by passivation and manipulation of carrier multiplications [84]. Both computational and experimental studies are significant in the understanding of the QDs properties. Gaining insight into the quantum confinement property of the QDs which control the optical, electronic, thermal and magnetic properties is essential for both current and future applications. Future applications include even new QDs materials e.g. carbon QDs which are a fairly new material with potential in electrocatalysis yet understanding of their novel design and fabrication is still missing [98]. Both computational and experimental approaches are an imperative insight into these issues. MM and QM based theoretical models described in the preceding sections can be used in the study of the QDs. The MM theory, however, lacks accuracy in studying the QDs optical properties because it disregards the description of electronic/structural degrees of freedom. But it can still be used as a screening tool before using the accurate QM model. The QM theory has several single electron and many bodied models that are used in QDs calculation such as the simple and popular EMA, simple K.p model, TB, DFT, TDDFT, LCAO, CI and QMC. Their advantages and disadvantages have been explained broadly in the chapter. The choice of the computation theory is determined by the type of the QDs (size), properties in question and envisaged applications and available theoretical resources and their capabilities. For example, QDs semiconductors can be classified into freestanding and embedded QDs and each has unique differences which determine the theoretical models that can be used for each although they are overlaps. The Freestanding QDs suit molecular electronic structure calculations, while embedded QDs are suitable for EMA [2].

First-principle quantum-chemical methods DFT and TDDFT show reasonable accuracy for QDs calculations hence remain a preference for QDs structural, optical properties in molecular and solid calculations [84]. However, they are limited by non-linear scaling with system size and uncertainties related to the chemical composition and morphology of the QDs surface [84]. The magnetic and thermal properties of QDs are not well studied as compared to their optical and electrical; hence, more theoretical work still has to explore these properties for future applications.

Acknowledgements

The authors acknowledge the financial support provided by the South African Department of Science and Innovation (DSI) and Mintek.

References

[1]   C. Dykstra, G. Frenking, K. Kim, G. Scuseria, Theory and applications of computational chemistry, the first forty years. Elsevier, 2011.

[2]   O. Lehtonen, D. Sundholm, T. Vänskä, Computational studies of semiconductor quantum dots, Phys. Chem. Chem. Phys. 10 (228) 4535-4550. https://doi.org/10.1039/B804212H

[3]   Y. Hong, Y. Wu, S. Wu, X. Wang, J. Zhang, Overview of computational simulations in quantum dots,Isr. J. Chem. 59 (2019) 661-672. https://doi.org/10.1002/ijch.201900026

[4]   D. Bera, L. Qian, T.-K. Tseng, P.H. Holloway, Quantum dots and their multimodal applications: a review, Materials 3 (2010) 2260-2345. https://doi.org/10.3390/ma3042260

[5]   A.L. Kaledin, D. Kong, K. Wu, T. Lian, D.G. Musaev, Quantum confinement theory of auger-assisted biexciton recombination dynamics in type-I and quasi type-II quantum dots,J. Phys. Chem. C. 122 (2018) 18742-18750. https://doi.org/10.1021/acs.jpcc.8b04874

[6]   N.S. Makarov, P.C. Lau, C. Olson, K.A. Velizhanin, K.M. Solntsev, K. Kieu, S. Kilina, S. Tretiak, R.A. Norwood, N. Peyghambarian, Two-photon absorption in CdSe colloidal quantum dots compared to organic molecules, ACS nano 8 (2014) 12572-12586. https://doi.org/10.1021/nn505428x

[7]   N. Zeiri, A. Naifar, S.A.-B. Nasrallah, M. Said, Theoretical studies on third nonlinear optical susceptibility in CdTe–CdS–ZnS core–shell–shell quantum dots,

Photonic. Nanostruct. 36 (2019) 100725.
https://doi.org/10.1016/j.photonics.2019.100725

[8]   S. Nasa, S. Purohit, Linear and third order nonlinear optical properties of GaAs quantum dot in terahertz region, Physica E Low Dimens. Syst. Nanostruct. 118 (2019) 113913. https://doi.org/10.1016/j.physe.2019.113913

[9]   J. Vinasco, A. Radu, A. Tiutiunnyk, R. Restrepo, D. Laroze, E. Feddi, M. Mora-Ramos, A. Morales, and C. Duque, Revisiting the adiabatic approximation for bound states calculation in axisymmetric and asymmetrical quantum structures,Superlattice Microst. 138 (2019) 106384. https://doi.org/10.1016/j.spmi.2019.106384

[10]   F. Aydin, H. Sari, E. Kasapoglu, S. Sakiroglu, I. Sokmen, Anisotropy dependence of the optical response in an impurity doped quantum dot under intense laser field,Physica E Low Dimens. Syst. Nanostruct. 114 (2019) 113566. https://doi.org/10.1016/j.physe.2019.113566

[11]   P. Borah, D. Siboh, P. Kalita, J. Sarma, and N. Nath, Quantum confinement induced shift in energy band edges and band gap of a spherical quantum dot, Physica B Condens. 530 (2018) 208-214. https://doi.org/10.1016/j.physb.2017.11.046

[12]   P. Ganesan, L. Senthilkumar, The influence of interfaces and intra-band transitions on the band gap of CdS/HgS and GaN/X (X= InN, In0. 33Ga0. 67N) core/shell/shell quantum dot quantum well–A theoretical study, Physica. E Low. Dimens. Syst. Nanostruct. 74 (2015) 204-212. https://doi.org/10.1016/j.physe.2015.07.002

[13]   F. Qu, F.V. Moura, F.M. Alves, R. Gargano, Optical tunability of magnetic polaron stability in single-Mn doped bulk GaAs and GaAs/AlGaAs quantum dots, Chem. Phys. Lett. 561 (2013) 107-114. https://doi.org/10.1016/j.cplett.2013.01.042

[14]   C. Pemmaraju, Valence and core excitons in solids from velocity-gauge real-time TDDFT with range-separated hybrid functionals: An LCAO approach, Comput. Condens. Matter. 18 (2019) e00348. https://doi.org/10.1016/j.cocom.2018.e00348

[15]   S. Saha, P. Sarkar, Tuning the HOMO–LUMO gap of SiC quantum dots by surface functionalization, Chem. Phys. Lett. 536 (2012) 118-122. https://doi.org/10.1016/j.cplett.2012.03.107

[16]   R. Bertel, M. Mora-Ramos, J. Correa, Electronic properties and optical response of triangular and hexagonal MoS2 quantum dots. A DFT approach, Physica ELow Dimens. Syst. Nanostruct. 109 (2019) 201-208. https://doi.org/10.1016/j.physe.2019.01.021

[17]   V. Sharma, H.L. Kagdada, J. Wang, P.K. Jha, Hydrogen adsorption on pristine and platinum decorated graphene quantum dot: A first principle study,Int. J. Hydrogen. Energ. (2019). https://doi.org/10.1016/j.ijhydene.2019.09.021

[18]   D. Lee, J.L. DuBois, Y. Kanai, Importance of Excitonic Effect in Charge Separation at Quantum-Dot/Organic Interface: First-Principles Many-Body Calculations, Nano Lett. 14 (2014) 6884-6888. https://doi.org/10.1021/nl502894b

[19]   J. Feng, Q. Guo, H. Liu, D. Chen, Z. Tian, F. Xia, S. Ma, L. Yu, L. Dong, Theoretical insights into tunable optical and electronic properties of graphene quantum dots through phosphorization, Carbon 155 (2019) 491-498. https://doi.org/10.1016/j.carbon.2019.09.009

[20]   R. Das, N. Dhar, A. Bandyopadhyay, D. Jana, Size dependent magnetic and optical properties in diamond shaped graphene quantum dots: A DFT study, J. Phys. Chem. Solids. 99 (2016) 34-42. https://doi.org/10.1016/j.jpcs.2016.08.004

[21]   I. Bryndal, J. Lorenc, L. Macalik, J. Michalski, W. Sąsiadek, T. Lis, J. Hanuza, Crystal structure, vibrational and optic properties of 2-N-methylamino-3-methylpyridine N-oxide–Its X-ray and spectroscopic studies as well as DFT quantum chemical calculations, J. Mol. 1195 (2019) 208-219. https://doi.org/10.1016/j.molstruc.2019.05.064

[22]   M. V. Mukhina, V.G. Maslov, A.V. Baranov, A.V. Fedorov, A.O. Orlova, F. Purcell-Milton, J. Govan, Y.K. Gun'ko, Intrinsic chirality of CdSe/ZnS quantum dots and quantum rods, Nano lett. 15 (2015) 2844-2851. https://doi.org/10.1021/nl504439w

[23]   J. Nagakubo, T. Nishihashi, K. Mishima, K. Yamashita, First-principles approach to the first step of metal–phosphine bond formation to synthesize alloyed quantum dots using dissimilar metal precursors, Chem. Phys. 528 (2020) 110512. https://doi.org/10.1016/j.chemphys.2019.110512

[24]   M. Algarra, V. Moreno, J.M. Lázaro-Martínez, E. Rodríguez-Castellón, J. Soto, J. Morales, A. Benítez, Insights into the formation of N doped 3D-graphene quantum dots. Spectroscopic and computational approach,J. Colloid. Interf. Sci. 561 (2020) 678-686. https://doi.org/10.1016/j.jcis.2019.11.044

[25]   A. D. Laurent, D. Jacquemin, TDDFT benchmarks: a review, Int. J. Quantum Chem. 113 (2013) 2019-2039. https://doi.org/10.1002/qua.24438.

[26]   D. Raeyani, S. Shojaei, S. Ahmadi-Kandjani, Optical graphene quantum dots gas sensors: Theoretical study, SuperlatticeMicrost. 114 (2018) 321-330. https://doi.org/10.1016/j.spmi.2017.12.050

[27] S.S. Yamijala, M. Mukhopadhyay, S.K. Pati, Linear and nonlinear optical properties of graphene quantum dots: A computational study, J. Phys. Chem. 119 (2015) 12079-12087. https://doi.org/10.1021/acs.jpcc.5b03531

[28] D. Mombrú, M. Romero, R. Faccio, Á.W. Mombrú, Electronic and optical properties of sulfur and nitrogen doped graphene quantum dots: A theoretical study, Physica E Low Dimens. Syst. Nanostruct. 113 (2019) 130-136. https://doi.org/10.1016/j.physe.2019.05.004

[29] S. Gopalakrishnan, P. Kolandaivel, Electronic, optical and magnetic properties of Co, Fe and Ni doped (ZnX) 6;(X= O, S & Se) quantum dots–A DFT study, Comput. Theor. Chem. 1111 (2017) 56-68. https://doi.org/10.1016/j.comptc.2017.04.005

[30] F. Gao, C.-L. Yang, M.-S. Wang, X.-G. Ma, W.-W. Liu, Theoretical studies on the feasibility of the hybrid nanocomposites of graphene quantum dot and phenoxazine-based dyes as an efficient sensitizer for dye-sensitized solar cells, Spectrochim. Acta A 206 (2019) 216-223. https://doi.org/10.1016/j.saa.2018.08.012

[31] Q.R. Dong, Y. Li, C. Jia, F.-L. Wang, Y.-T. Zhang, C.-X. Liu, Electrically-induced polarization selection rules of a graphene quantum dot, Solid State Commun. 273 (2018) 55-59. https://doi.org/10.1016/j.ssc.2018.02.009

[32] H. Ryu, D. Nam, B.-Y. Ahn, J.R. Lee, K. Cho, S. Lee, G. Klimeck, M. Shin, Optical TCAD on the Net: A tight-binding study of inter-band light transitions in self-assembled InAs/GaAs quantum dot photodetectors, Math. Comput. Model. 58 (2013) 288-299. https://doi.org/10.1016/j.mcm.2012.11.024

[33] K.A. Nguyen, P.N. Day, R. Pachter, Understanding structural and optical properties of nanoscale CdSe magic-size quantum dots: insight from computational prediction, J. Phys. Chem. 114 (2010) 16197-16209. https://doi.org/10.1021/jp103763d

[34] J. Shu, X. Zhang, P. Wang, R. Chen, H. Zhang, D. Li, P. Zhang, J. Xu, Monte-Carlo simulations of optical efficiency in luminescent solar concentrators based on all-inorganic perovskite quantum dots, Physica B Condens. 548 (2018) 53-57. https://doi.org/10.1016/j.physb.2018.08.021

[35] F. Trani, G. Scalmani, G. Zheng, I. Carnimeo, M.J. Frisch, V. Barone, Time-dependent density functional tight binding: new formulation and benchmark of excited states, J. Chem. Theory Comput. 7 (2011) 3304-3313. https://doi.org/10.1021/ct200461y

Materials Research Forum LLC
https://doi.org/10.21741/9781644901250-5

[36]   F. Zaouali, A. Bouazra, M. Said, Numerical modelling of electronic and optical properties of isolated and self-assembled InAs/InP quantum dots,Optik 182 (2019) 731-738. https://doi.org/10.1016/j.ijleo.2019.01.075

[37]   S.N. Mohajer, A. Ibral, J. El Khamkhami, E.M. Assaid, Quantum confined Stark effects of single dopant in polarized hemispherical quantum dot: Two-dimensional finite difference approach and Ritz-Hassé variation method, Physica B Condens. 537 (2018) 40-50. https://doi.org/10.1016/j.physb.2018.01.061

[38]   S.N. Mohajer, A. Ibral, J. El Khamkhami,E.M. Assaid, Energies and wave functions of an off-centre donor in hemispherical quantum dot: Two-dimensional finite difference approach and ritzvariational principle, Physica B Condens. 497 (2016) 51-58. https://doi.org/10.1016/j.physb.2016.05.028

[39]   M. Choubani, H. Maaref, F. Saidi, Nonlinear optical properties of lens-shaped core/shell quantum dots coupled with a wetting layer: effects of transverse electric field, pressure, and temperature, J. Phys. Chem. Solids. 139 (2020) 109226. https://doi.org/10.1016/j.jpcs.2019.109226

[40]   J.S. Ahn, Finite difference method for the arbitrary potential in two dimensions: Application to double/triple quantum dots, Superlattice. Microst. 65 (2014) 113-123. https://doi.org/10.1016/j.spmi.2013.10.044

[41]   M. Garagiola, O. Osenda, Excitonic states in spherical layered quantum dots. Physica E Low Dimens. Syst. Nanostruct. 116 (2020) 113755. https://doi.org/10.1016/j.physe.2019.113755

[42]   Y. Liu, S. Bose, W. Fan, Effect of size and shape on electronic and optical properties of CdSe quantum dots,Optik. 155(2018) 242-250. https://doi.org/10.1016/j.ijleo.2017.10.165

[43]   S. Kilina, K.A. Velizhanin, S. Ivanov, O.V. Prezhdo, S. Tretiak, Surface ligands increase photoexcitation relaxation rates in CdSe quantum dots, ACS nano. 6 (2012) 6515-6524. https://doi.org/10.1021/nn302371q

[44]   K. Hyeon-Deuk, O.V. Prezhdo, Multiple exciton generation and recombination dynamics in small si and cdse quantum dots: An ab initio time-domain study, ACS nano. 6 (2012) 1239-1250. https://doi.org/10.1021/nn2038884

[45]   C. Dong, X. Li, P. Jin, W. Zhao, J. Chu, J. Qi, Intersubunit electron transfer (IET) in quantum dots/graphene complex: what features does IET endow the complex with? J. Phys. Chem. C, 116 (2012) 15833-15838. https://doi.org/10.1021/jp304624y

Materials Research Forum LLC
https://doi.org/10.21741/9781644901250-5

[46]   Y. Li, H. Shu, X. Niu,J. Wang, Electronic and optical properties of edge-functionalized graphene quantum dots and the underlying mechanism. J. Phys. Chem. C, 119 (2015) 24950-24957. https://doi.org/10.1021/acs.jpcc.5b05935

[47]   S. Zhai, P. Guo, J. Zheng, P. Zhao, B. Suo, Y. Wan, Density functional theory study on the stability, electronic structure and absorption spectrum of small size g-C3N4 quantum dots, Comput. Mater. Sci. 148 (2018) 149-156. https://doi.org/10.1016/j.commatsci.2018.02.023

[48]   H-P. Li, Z.-T. Bi, R.-F. Xu, K. Han, M.-X. Li, X.-P. Shen, Y.-X. Wu, Theoretical study on electronic polarizability and second hyperpolarizability of hexagonal graphene quantum dots: Effects of size, substituent, and frequency, Carbon, 122 (2017) 756-760. https://doi.org/10.1016/j.carbon.2017.07.033

[49]   C. Wang, Y. Ding, X. Bi, J. Luo, G. Wang, Y. Lin, Carbon quantum dots-Ag nanoparticle complex as a highly sensitive "turn-on" fluorescent probe for hydrogen sulfide: a DFT/TD-DFT study of electronic transitions and mechanism of sensing, Sensor. Actuat. B-Chem. 264 (2018) 404-409. https://doi.org/10.1016/j.snb.2018.02.186

[50]   A.L. Kaledin, T. Lian, C.L. Hill, D.G. Musaev, A hybrid quantum mechanical approach: Intimate details of electron transfer between type-I CdSe/ZnS quantum dots and an anthraquinone molecule, J. Phys. Chem. B, 119 (2015) 7651-7658. https://doi.org/10.1021/jp511935z

[51]   A. El Aouami, E. Feddi, N. El-Yadri, N. Aghoutane, F. Dujardin, C. Duque, H.V. Phuc, Electronic states and optical properties of single donor in GaN conical quantum dot with spherical edge. Superlattice Microst. 114 (2018) 214-224. https://doi.org/10.1016/j.spmi.2017.12.043

[52]   Q. Dong, Electrical spin switch in a two-electron triangular graphene quantum dot. Physica E Low Dimens, Syst. Nanostruct. 116 (2020) 113779. https://doi.org/10.1016/j.physe.2019.113779

[53]   A. Tiutiunnyk, C. Duque, F. Caro-Lopera, M. Mora-Ramos, J. Correa, Opto-electronic properties of twisted bilayer graphene quantum dots. Physica E Low Dimens, Syst. Nanostruct. 112 (2019) 36-48. https://doi.org/10.1016/j.physe.2019.03.028

[54]   H. Abdelsalam, H. Elhaes, M.A. Ibrahim, First principles study of edge carboxylated graphene quantum dots, Physica B Condens. 537 (2018) 77-86. https://doi.org/10.1016/j.physb.2018.02.001

Materials Research Forum LLC
https://doi.org/10.21741/9781644901250-5

[55] H. Abdelsalam, V.A. Saroka, W.O. Younis, Phosphorene quantum dot electronic properties and gas sensing, Physica E Low Dimens. Syst. Nanostruct. 107 (2019) 105-109. https://doi.org/10.1016/j.physe.2018.11.012

[56] D. Gabay, X. Wang, V. Lomakin, A. Boag, M. Jain, A. Natan, Size dependent electronic properties of silicon quantum dots—An analysis with hybrid, screened hybrid and local density functional theory,Comput. Phys. Commun. 221 (2017) 95-101. https://doi.org/10.1016/j.cpc.2017.08.005

[57] N. Thongsai, P. Jaiyong, S. Kladsomboon, I. In, P. Paoprasert, Utilization of carbon dots from jackfruit for real-time sensing of acetone vapor and understanding the electronic and interfacial interactions using density functional theory, Appl. Surf. 487 (2019) 1233-1244. https://doi.org/10.1016/j.apsusc.2019.04.269

[58] N. Pattarapongdilok, V. Parasuk, Theoretical study on electronic properties of curved graphene quantum dots, Comput. Theor. Chem. 1140 (2018)86-97. https://doi.org/10.1039/C8CP01403E

[59] T.D. Rodríguez, J. Reyes-Nava, M. Pacio, H. Juárez, and J. Muñiz, Theoretical study on the electronic structure properties of a PbS quantum dot adsorbed on TiO$_2$ substrates and their role on solid-state devices, Comput. Theor. Chem. 1100 (2017) 83-90. https://doi.org/10.1021/jp952869n

[60] A. Samia, E. Feddi, C. Duque, M. Mora-Ramos, V. Akimov, J. Correa, Optoelectronic properties of phosphorene quantum dots functionalized with free base porphyrins, Comput. Mater. Sci. 171 (2020) 109278. https://doi.org/10.1016/j.commatsci.2019.109278

[61] Y. Liu, L. Du, K. Gu, M. Zhang, Effect of Tm dopant on luminescence, photoelectric properties and electronic structure of In2S3 quantum dots,J. Lumin. 217 (2020) 116775. https://doi.org/10.1016/j.jlumin.2019.116775

[62] H. Abdelsalam, V.A. Saroka, M. Ali, N.H. Teleb, H. Elhaes, M.A. Ibrahim, Stability and electronic properties of edge functionalized silicene quantum dots: A first principles study,Physica E Low Dimens. Syst. Nanostruct. 108 (2019) 339-346. https://doi.org/10.1016/j.physe.2018.07.022

[63] S. Bondwal, P. Debnath, P.P. Thankachan, Structural, electronic and optical properties of model silicon quantum dots: A computational study, Physica E Low Dimens. Syst. Nanostruct. 103 (2018) 194-200. https://doi.org/10.1016/j.physe.2018.05.037

[64]   N. Li, Z. Liu, S. Hu, Q. Chang, C. Xue, H. Wang, Electronic and photocatalytic properties of modified MoS2/graphene quantum dots heterostructures: A computational study,Appl. Surf. 473 (2019) 70-76. https://doi.org/10.1016/j.apsusc.2018.12.122

[65]   H. Abdelsalam, H. Elhaes, M.A. Ibrahim, Tuning electronic properties in graphene quantum dots by chemical functionalization: Density functional theory calculations,Chem. Phys. Lett. 695 (2018) 138-148. https://doi.org/10.1016/j.cplett.2018.02.015

[66]   F. Dujardin, E. Assaid, E. Feddi, New way for determining electron energy levels in quantum dots arrays using finite difference method, Superlattice. Microst. 118 (2018) 256-265. https://doi.org/10.1016/j.spmi.2018.04.027

[67]   D. Kennes, D. Schuricht, V. Meden, Efficiency and power of a thermoelectric quantum dot device, EPL. 102 (2013) 57003. https://doi.org/10.1209/0295-5075/102/57003

[68]   J.D. Castaño-Yepes, D. Amor-Quiroz, Super-statistical description of thermo-magnetic properties of a system of 2D GaAs quantum dots with gaussian confinement and Rashba spin–orbit interaction, Physica. A, (2019) 123871. https://doi.org/10.1016/j.physa.2019.123871

[69]   A. Armaşelu, Quantum Dots and Fluorescent and Magnetic Nanocomposites: Recent Investigations and Applications in Biology and Medicine. Nonmagnetic and Magnetic Quantum Dots, Stavrou, Vasilios N., ed. BoD–Books on Demand, 2017. https://doi: 10.5772/intechopen.70614

[70]   R. Masrour, A. Jabar, Size and diluted magnetic properties of diamond shaped graphene quantum dots: Monte Carlo study,Physica A. 497 (2018) 211-217. https://doi:10.1016/j.physa.2017.12.141

[71]   T. Pisanic Ii, Y. Zhang, T. Wang, Quantum dots in diagnostics and detection: principles and paradigms,Analyst. 139(2014) 2968-2981. https://doi.org/10.1039/C4AN00294F

[72]   S. Hug, Classical molecular dynamics in a nutshell, in biomolecular simulations, (2012) 127–152. https://doi:10.1007/978-1-62703-017-5_6

[73]   J.M. Elward, F.J. Irudayanathan, S. Nangia, A. Chakraborty, Optical signature of formation of protein corona in the firefly luciferase-CdSe quantum dot complex, J. Chem. Theory Comput. 10 (2014) 5224-5228. https://doi.org/10.1021/ct500681m

[74]   E. Kaxiras, Atomic and electronic structure of solids. Cambridge University Press. 2003. https://doi.org/10.1017/CBO9780511755545

[75]   T.J. Zielinski, E. Harvey, R. Sweeney, D.M. Hanson, Quantum states of atoms and molecules, J. Chem. Educ. 82 (2005) 1880. https://doi.org/10.1021/ed082p1880.2.

[76]   J. Neugebauer, C.G. Van de Walle, Theory of hydrogen in GaN, in N.H Nickel (ed) R.K Wallardson, E.R. Weber. Hydrogen in Semicondictros IISemiconduct. Semimet. 61 (1999) 479-502.Academic Press, boston.

[77]   N Vukmirovic´, L.-W.W., Quantum Dots: Theory.Elsevier B.V, 2011.

[78]   Z. Bodroski, N. Vukmirović, S. Skrbic, Gaussian basis implementation of the charge patching method, J. Comput. Phys., 368 (2018) 196-209. https://doi.org/10.1016/j.jcp.2018.04.032

[78]   I.-H. Chu, M. Radulaski, N. Vukmirovic, H.-P. Cheng, L.-W. Wang, Charge transport in a quantum dot supercrystal, J. Phys. Chem. C, 115 (2011) 21409-21415. https://doi.org/10.1021/jp206526s

[80]   D. Stroud, The effective medium approximations: Some recent developments, Superlattice. Microst., 23 (1998) 567-573. https://doi.org/10.1006/spmi.1997.0524

[81]   M.A.F. Richard A. Dudley, Engineered Materials and Metamaterials: Design and Fabrication, SPIE, 2017.

[82]   F. Flory, Y.-J. Chen, C.-C. Lee, L. Escoubas, J.-J. Simon, P. Torchio, J. Le Rouzo, S. Vedraine, H. Derbal-Habak, I. Shupyk, Optical properties of dielectric thin films including quantum dots, Appl. opt., 50 (2011) C129-C134. https://doi.org/10.1364/AO.50.00C129

[83]   E.O. Chukwuocha, M.C. Onyeaju, T.S. Harry, Theoretical studies on the effect of confinement on quantum dots using the brus equation, 2012. https;//doi.org: 10.4236/wjcmp.2012.22017

[84]   S.V. Kilina, P.K. Tamukong, D.S. Kilin, Surface chemistry of semiconducting quantum dots: theoretical perspectives, Acc. Chem. Res. 49 (2016) 2127-2135. https://doi.org: 10.1021/acs.accounts.6b00196

[85]   J. Pipek, P.G. Mezey, A fast intrinsic localization procedure applicable for abinitio and semiempirical linear combination of atomic orbital wave functions, J. Chem. Phys., 90 (1989) 4916-4926. https://doi.org/10.1063/1.456588

[86]   Y.S. Sarun Arunragsa, W. Pon-On, C. Wongchoosuk,Hydroxyl edge-functionalized graphene quantum dots for gas-sensing applications, Diam. Relat. Mater., (2020) 107790. https://doi.org/10.1016/j.diamond.2020.107790

[87]   S. Niaz, A.D. Zdetsis, Comprehensive ab initio study of electronic, optical, and cohesive properties of silicon quantum dots of various morphologies and sizes up to infinity, J. Phys. Chem. C., 120 (2016) 11288-11298. https://doi.org/10.1021/acs.jpcc.6b02955

[88]   P.J. Hasnip, K. Refson, M.I. Probert, J.R. Yates, S.J. Clark, C.J. Pickard, Density functional theory in the solid state, Philos. T. R. Soc. A. 372 (2014) 20130270. https://doi.org/10.1098/rsta.2013.0270

[89]   S.J Clark, J. Robertson, Screened exchange density functional applied to solids, Phys. Rev. B. 82 (2010) 085208. https://doi.org/10.1103/PhysRevB.82.085208

[90]   K. Zhuravlev, PbSe vs. CdSe: Thermodynamic properties and pressure dependence of the band gap. Phys. B: Condens. Matter, 394 (2007) 1-7. https://doi.org/10.1016/j.physb.2007.01.030

[91]   F. Thierry, J. Le Rouzo, F. Flory, G. Berginc, L. Escoubas. Optimization of the optical properties of nanostructures through fast numerical approaches. in Nanophotonic Materials XI. International Society for Optics and Photonics. 2014. https://doi.org/ 10.1117/12.2061042

[92]   A. Laref, N. Alshammari, S. Laref, S. Luo, Surface passivation effects on the electronic and optical properties of silicon quantum dots, Sol. Energy Mater. Sol. Cells, 120 (2014) 622-630. https://doi.org/10.1016/j.solmat.2013.10.005

[93]   X. Wang, R. Zhang, T.A. Niehaus, T. Frauenheim, Excited state properties of allylamine-capped silicon quantum dots, J. Phys. Chem. C., 111 (2007) 2394-2400. https://doi.org/10.1021/jp065704v

[94]   M. Anas, A. Othman, G. Gopir. First-principle study of quantum confinement effect on small sized silicon quantum dots using density-functional theory. in AIP Conference Proceedings, American Institute of Physics, 2014. https://doi.org/10.1063/1.4895180

[95].   M.G. Mavros, D.A. Micha, D.S. Kilin, Optical properties of doped silicon quantum dots with crystalline and amorphous structures, J. Phys. Chem. C. 115 (2011) 19529-19537. https://doi.org/10.1021/jp2055798

[96]   M.M.-A Anas,. G. Gopir, Electronic and optical properties of small hydrogenated silicon quantum dots using time-dependent density functional theory,J. Nanomat., 2015 (2015).https://doi.org/10.1155/2015/481087

[97]   G. Zerveas, E. Caruso, G. Baccarani, L. Czornomaz, N. Daix, D. Esseni, E. Gnani, A. Gnudi, R. Grassi, M. Luisier, Comprehensive comparison and experimental validation of band-structure calculation methods in III–V semiconductor quantum wells, Solid State Electron. 115 (2016) 92-102.https://doi.org/10.1016/j.sse.2015.09.005

[98]   X. Wang, Y. Feng, P. Dong, J. Huang, A mini review on carbon quantum dots: preparation, properties and electrocatalytic application, Front. Chem. 7 (2019) 671.https://doi.org/10.3389/fchem.2019.00671

Quantum Dots – Properties and Applications
Materials Research Foundations **96** (2021) 145-168

Materials Research Forum LLC
https://doi.org/10.21741/9781644901250-6

Chapter 6

Application of Quantum Dots in Sensors

N. Mhlanga*, P. Tetyana

DSI/Mintek Nanotechnology Innovation Centre, Randburg, South Africa

Advanced Materials Division, Mintek, Randburg, South Africa

Nikiwem@mintek.co.za

Abstract

Nanotechnology has presented the science community with a platform to fabricate nanomaterials suitable for myriad applications. Quantum dots (QDs) are one such nanomaterial which surged in the last decade in applications such as sensing. Sensing uses sensors defined as an ensemble of bio/chemical materials that can recognize a corresponding molecule of interest. Sensors are of paramount importance in safeguarding the ecosystem from both natural and synthetic pollutants. The chapter reports on the application of the QDs in the development of sensors.

Keywords:

Quantum Dots, Sensing, Chemosensors, Biosensors, Luminescence

Contents

1.   Introduction

The search for improved diagnostic and sensing systems for detection and monitoring purposes has been the focus of research in various fields. This has birthed an interface amongst various study areas within the science arena, with nanotechnology at the centre of it all [1]. The use of nanomaterials to improve the performance of existing products and assays has increased immensely in the last decade. This has been prompted by the discovery of essential inherent properties that these nanomaterials contain. Quantum dots (QDs) have been the extensively explored nanomaterials lately, due to their exceptional properties compared to other nanomaterials. These are small fluorescent nanoparticles (NPs) dominating in research because of their enthralling properties. These semiconducting nanomaterials have been used successfully in innumerable applications and have established themselves as nanomaterials of the future. They are endowed with a quantum confinement phenomenon which offers unique optical, electrical, photochemical, semiconductor, catalytic, and magnetic properties [2-4]. The quantum confinement effects instigate an enclosure of electronic charge carriers within these QDs which enables the modification or tuning of the energy of distinct electronic energy states and optical transistors using their size and shape [5, 6]. In this chapter, we explore the applications of these nanomaterials in the fabrication of sensors.

2.   Properties of QDs used in sensing applications

QDs are nanocrystals comprised of semiconductor materials of group II – IV (e.g. CdSe, CdTe), III - V (e.g. GaAs, InAlAs) or IV - VI (e.g. Si QDs, graphene and carbon QDs) elements [7]. These nanocrystals often referred to as artificial atoms or zero-dimensional electron systems, are described as particles that exhibit physical dimensions lesser than the exciton Bohr radius of the bulk material [8]. The Bohr radius is the distance between an electron in the conduction band and its hole in the valence band, in an exciton. When a semiconductor is excited, an electron is transferred from the valence band to the conduction band, generating an electron-hole pair that is transformed into an exciton by weak Coulomb forces that exist between the electron and the hole [9].

According to literature, QDs range between 2 - 20 nm in diameter, and thus contain between hundred to a few thousand atoms. It has also been argued that the diameter of QDs should strictly be below 10 nm [10]. Drbohlavova et al. [11] believe that the dimensions of a QD are controlled or dependant on the material from which they are formed, and a system is considered to be a QD when quantum confinement effects occur.

This size range allows QDs to exhibit unique properties transitional between those of their bulk materials and individual atoms or discrete molecules. This is usually motivated by the high surface to volume ratio of these NPs which emanates from their reduced size [12]. QDs have been seen to contain an inherent ability to glow a particular color when exposed to light of a particular wavelength, with the resulting color dependent on their size, as shown in Fig. 1[13, 14]. Smaller particles appear bluer while larger particles are redder in appearance. This is because small QDs have a larger bandgap and therefore emit higher energy photons whereas larger QDs have a smaller bandgap and emit low energy photons [15,16].

Fig. 1 Size-dependent color emission observed in QD solutions of various sizes. The QDs size increases from left to right. Image obtained from [14].

The bandgap, therefore, depends on the size of the QDs and increases with a decrease in the size of the QDs. This is one of the most interesting properties that have made QDs superior over most nanostructures and is known as the quantum confinement effect whereby the size of the nanostructure determines its bandgap, as the dimensions are decreased to levels below the Bohr exciton radius [17]. Typical QD NPs inherit a core/shell structure which is characterized by an inorganic core and a shell, and an aqueous ligand serving as a coating material for further functionalization as depicted in Fig. 2 [18,19].

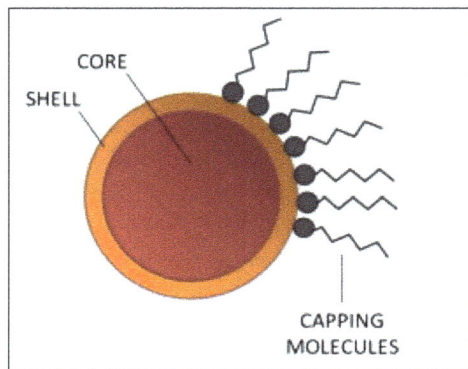

Fig. 2 An illustration of the structure of QDs comprising a core, shell and capping ligand [19].

3.   Quantum dot-based sensors

A sensor is an integrated device that uses either a chemical or biological recognition element to report analytical information [20]. This chapter classifies sensors into two groups, i.e., chemosensors and biosensors, based on the nature of the recognition element. Semiconducting nanoparticles are valuable components in both sensor systems. QDs have been applied in sensors as augments or replacements of conventional materials such as organic dyes that have been used extensively in past years. The performance of these nanomaterials in sensing devices has been seen to surpass that of conventional materials due to several properties they posses. These include; their large absorption cross-section, excellent stability, resistance to bleaching [20], sharp and adjustable emission spectra, blinking capability, high emission quantum yield, sensitivity, multiplexing potential [21] increased luminescence lifetime, narrow emission bands and wider adsorption bands [6]. Sensors are crucial for day-to-day life to monitor small compounds such as drugs, pesticides, pollutants, explosives, food additives, heavy metals and other materials which threaten the ecosystem [20].

The use of QDs in sensing applications is one of the main contributors that have increased the market of QDs enabled products, which is estimated to reach $5.69 billion by the end of 2020 [22]. As stated previously, QDs are used in a myriad of applications including camera sensors, biosensors, image sensors, and chemical sensors. The camera sensors research has been pioneered by market leaders Samsung and LG [22]. QDs have also been used tremendously in the pharmaceutical industry, in which the University of Shanghai has filed 118 patents [22]. Other companies that have used the QDs in improving their products include the TCL and BOE Technology Groups which have filed patents for use of QDs in biomedical devices such as probes and sensors [22]. Fig.3 [22] shows the increase in the number of patents which incorporates QDs, from 2007 till 2018.

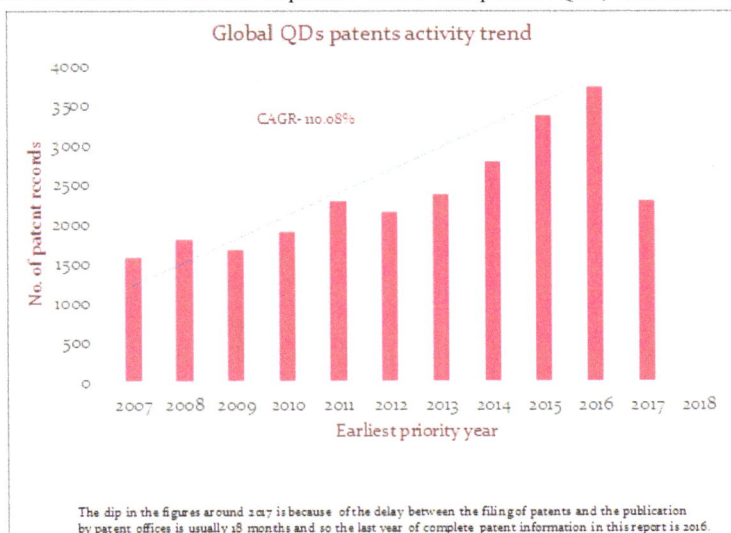

Fig. 3 Global QDs patents trend between 2007 and mid-2017 [22].

The application of QDs in sensors is governed by their optoelectrical properties. Excitation of the QDs valence band electrons to the conduction band results in an exciton i.e., an electron-hole pair. After excitation, relaxation occurs whereby the electron and hole recombine which generates luminescence. The generated luminescence is used in the sensing applications [23] and it is described as the release of light by a molecule catalyzed by various phenomena such as chemical reactions, subatomic motions, electrical energy or stress [24]. The optical process of the excitement of electrons from the conduction band to the valence band creates charge carriers in the form of excited electron and empty hole which is a mobile positive charge. Ensuing the excitement the electrons begin to lose the energy and jump to levels below the conduction band and recombine with the hole (electron-hole-recombination) [25]. The generated luminescence is determined by the trigger or source of the excitation. Hence, there are various types of luminescence-based on the trigger; photoluminescence (PL), chemiluminescence (CL), bioluminescence (BL), electroluminescence (EL), cathodoluminescence, radioluminescence, sonoluminescence and triboluminescence [24].

3.1 Classification of sensors

As previously illustrated, we have classified sensors into two groups, namely chemosensors and biosensors, based on the sensing element used. In subsequent sections, we will discuss the application of QDs in chemosensors and biosensors.

3.1.1 Chemosensors

A chemosensor is defined as a combination of a molecule and an analyte which forms a chemical system to convey a palpable response such as a fluorescence signal or color change. Chemosensors are mainly used in the detection of harmful environmental toxins such as ions, Au, Cu and small molecules [1, 26, 27]. The architecture of a chemosensor includes two components; a site with a tunable molecular property (receptor), and a transduction mechanism to convey the recognition into a signal. Fig. 4 [28] below illustrates the procedure through which a chemosensor results in the formation of a signal.

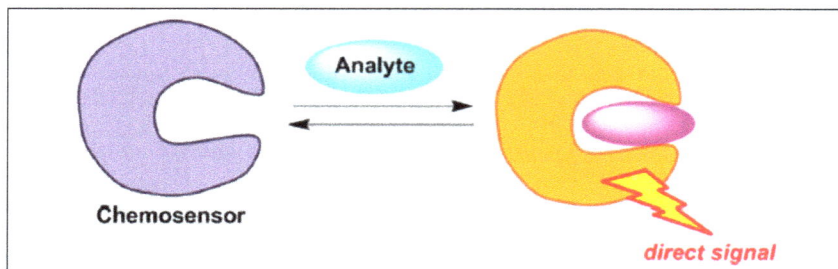

Fig. 4 Interaction of the sensor with an analyte to produce a signal [28]. Reprinted with permission from Elsevier.

When a chemosensor interacts with an analyte a signal can be observed through several forms, i.e., enhancement, quenching and shifting of the fluorescence maxima [27]. Conventionally, chemosensors have relied on the use of several organic dyes such as coumarin, rhodamine and many more as signal reporters [27]. Sensitive, efficient, and simple chemosensors are of paramount importance in the detection of toxic ions and small molecules which could impose adverse effects on the ecosystem. The environment is polluted by heavy metals produced by mines, foundry factories and they are a concern for the ecosystem [29]. QDs have been key players in the sensing of ions ever since 2002 and continue to be the backbone for chemosensors owing to their inimitable photophysical properties, ligand binding ability, flexible surface chemistry and functionalization which maintains their pristine luminescence property [30].

One key attribute of chemosensors is increased selectivity to analytes of interest. The selectivity of QDs based chemosensors can be enhanced using various strategies including; (1) Application of chelating ligands with high affinity for analytes such as ions. The chelating ligands include small molecules such as glutathione (GSH), mercaptoacetic acid, N-acetyl-L-cysteine, thioglycolic acid, L-cysteine, and dithizone; (2) use of a host-guest system whereby the surface of QDs is functionalized with supramolecular host ligands and the host materials can be modified to enhance their selectivity toward a specific analyte; (3) using the ability of the analyte (such as Hg^{2+} heavy metal) to form sulfide or telluride particles due to their slow-solubility. An example is Hg^{2+} heavy metal ions that can form HgS or HgTe particles which modifies the surface of the QDs; (4) application of dyes such as rhodamine-based Hg^{2+} used to increase selectivity towards Hg^{2+} ions; (5) use of biological molecules such as oligonucleotides, metalloproteins and DNA enzymes that have shown selectively towards some metal ions; (6) Formation of specific and strong d^{10}–d^{10} metallophilic bonds between metal ions [30].

3.1.2 Biosensors

The discovery of QDs and their superior properties (both optical and fluorescence) dates back to the early 1980s, but it is until recently that these intriguing nanomaterials have found widespread use in the biosensors field [31]. A biosensor is defined as a device that detects chemical compounds (or any other material) through specific biochemical reactions that are mediated by various biological molecules, thereby generating a signal. The signal generated can either be electrical, thermal or optical [32]. In recent years QDs have been used in various modes of biosensing applications that can be classified either by the type of interface element, the labeling substance and the signal converter. Herein, we have grouped biosensors into fluorescent, bioluminescent, chemiluminescent and photoelectrochemical biosensors [33, 34], and are explained thoroughly in subsequent sections. As mentioned earlier, QDs have demonstrated exceptional optical and fluorescence properties which surpass those of organic dyes and genetically engineered proteins which have been the materials of choice in most biosensing applications [35]. Biosensors have been heralded for their ability to selectively detect analytes in solution, rapidly and with ease. As such, biosensors are currently in use in many scientific fields including medical diagnosis, drug development, pathological services, food safety testing and environmental monitoring [34].

3.2   Sensing signals in QD based sensors

3.2.1   Luminescence based sensors

QDs have been used to augment the performance of luminescence-based sensors. This group is composed of a number of sensors that emit light as a means of detection signal. Luminescence is defined as a phenomenon whereby an electronically excited compound emits light as it returns to its ground or non-excited state. Luminescence sensors are classified based on the source of excitation energy. These include PL, CL and EL based sensors [36].

3.2.2   Photoluminescence (PL) sensors

PL based sensors emit light following a process known as photo-excitation. In here, a compound emits light after it has absorbed photons. These sensors can be subdivided into fluorescence and phosphorescence based sensors [37] with fluorescence-based sensors the most widely used.

Fluorescence sensors exploit the phenomenon of fluorescence as a reporter signal during detection. Fluorescence is described as a phenomenon whereby a molecule or system absorbs high energy light (with a short wavelength) and subsequent emission of low energy light (with a higher wavelength) [38]. Fluorescence-based sensors are the most widely used form of sensors amongst QDs based sensors, owing to several reasons. These include their increased sensitivity, ease of use, their ability to be used without the need for expensive equipment and their ability to perform multiplex analysis. In this category of sensors, QDs have been used as fluorescent labels which can serve as either a donor or acceptor due to their high quantum yield, high photostability and increased lifetimes. They have replaced fluorescent proteins and conventional organic dyes which presented various limitations such as lack of photochemical stability, weak signals, short fluorescence lifetimes and the need for a sophisticated tunable excitation source [39, 40]. The most commonly used QDs in QD-FRET include cadmium telluride (CdTe), cadmium selenide (CdSe), Indium Gallium Phosphide (InGaP), indium phosphide (InP) [41].

Fluorescence sensors generate signals using techniques such as the fluorescence resonance energy transfer (FRET), where one material acts as a donor fluorophore and another as an acceptor fluorophore bound to a biomolecule for signal transduction. In here, energy is conveyed from a high energy fluorophore acting as a donor species to a low energy acceptor through long-range dipole-dipole interactions. QD based FRET-based sensors have been used in various applications within the biomedical field [42-44]. QDs show excellent performance in the development of QD based fluorescent probes but have come under fire due to their high toxicity and expensive synthesis routes. Alternatively, carbon dots (CQDs) have been used to circumvent these limitations. CQDs are known to have low toxicity [29] high solubility, good stability and low synthesis costs [45], absence of blinking, excellent biocompatibility [46], good thermal and electrical conductivity, tunable PL and surface and solubility in a wide range of solvents [47]. These attributes render CQDs suitable candidates in the development of QD based fluorescence probes. CQDs fluorescence probes have been used in the detection of heavy metals such as Fe^{3+}, Hg^{2+}, Cd^{2+} and Cu^2. Fluorescence sensors for the detection of pesticides and inorganic materials have also been developed. The fluorescent probes detect the analyte by quenching the luminescence of the CQDs [48].

Qian et al. [46] reported on heteroatom doping of CQDs using silicon (SiQDs). The SiQDs showed visible fluorescence, high quantum yield, lower cellular toxicity and photobleaching resistance. The probes displayed strong fluorescence quenching for Fe (III) and a H_2O_2 quenching via charge transfer. The SiCQDs could also be recycled by inclusion of melamine which will form a stable with the H_2O_2 and hence remove it from the surface [46]. Ren et al. [47] explored nitrogen-doped graphene QDs (N-GQDs) probes for the detection of ferric ions (Fe^{3+}). The N-GQDs showed rapidity in the detection of the Fe^{3+} ions via static quenching mechanism.

3.2.3 Chemiluminescence sensors

Chemiluminescence (CL) sensors emit light as a result of a chemical reaction. Additionally, these type of sensors can further be divided into bioluminescence and electrochemiluminescence sensors which emit light following a biochemical and an electrochemical reaction, respectively [48]. CL sensors boast of using simple instrumentation, wide linear range, enhanced sensitivity and lack of background scattering light interferences which has resulted in their popularity. CQDs based CL sensors have been applied in the sensing of heavy metals such as Cu^{2+} and organic molecules [49].

Ge et al. [50] fabricated a novel CL based sensor for the detection of deltamethrinpyrethroid insecticide using CdTe QDs. This sensor was fabricated following a layer-by-layer self-assembly approach of CdTe QDs and molecularly imprinted polymers modified on the bottom of a 96 well plate resembling a glass slide. The sensor was effectively used in the sensing of the pesticide on real samples and achieved low detection limits of 0.018 µg/mL. Also, Tang et al. [51] developed a novel chlorine sensor using carbon nitride QDs (g-CNQDs). They determined the mechanism of the sensor by injecting the g-CNQDs with sodium hypochlorite (NaClO) which prompted a strong CL reaction. The sensor showed sensitivity towards free chlorine radicals with a detection limit of 0.01 µm within a 0.02-10 µM detection range [51].

Electrochemiluminescence (ECL), also known as electrogeneratedchemiluminescence, is the emission of light following an electrochemical reaction. In here, a molecule is excited via an applied potential at an electrode surface and thereby emits light when it undergoes an energy relaxation process [29,52]. ECL is simply a cross-play between chemiluminescence and electrochemistry and is slowly gaining popularity and becoming a powerful tool in the fabrication of immunoassays and other sensors due to its high sensitivity, low background signal, label-free ability, simplified set up and ease of control [53]. The use of QDs in ECL sensors spans as early as 2002 when QDs were observed to have ECL capabilities through the study of silicon QDs [53]. Since this discovery, numerous attempts have been made to develop ECL sensors using QDs. Dong et al. [54] developed a new ECL sensor system made up of GQDs and l-cysteine (l-Cys) that produces a strong cathodic ECL signal. The GQDs/l-Cys sensor was tested using lead ions (Pb^{2+}) where it showed rapidity, reliability and selectivity with a detection limit of 70 nM within a 100 nM - 10 µM dynamic range.

Also, a novel ECL sensor based on the immobilization of CQDs on graphene (GR) for detection of chlorinated phenols (CPs) in water was proposed by Yang et al. [55]. The CPs are indirectly detected by the presence of pentachlorophenol (PCP) in the environment which is a hard task due to ultratrace concentrations and real –time analysis failure. The novel ECL produces the signal from the interaction of the analyte with excited QDs and it uses a co-reactant $S_2O_8^{2-}$ and successfully detected the PCP in real-time with a high sensitivity of 1.0×10^{-12} M concentration in a wide linear range from 1.0×10^{-12} to 1.0×10^{-8} M using water samples.

3.2.4 Electrochemical based sensors

Electrochemical based transducers are the most commonly used in the development of sensors. These involve a chemical reaction between an electrode and the analyte to produce a qualitative or quantitative signal [56]. Conversion of the chemical reaction into a readable signal is achieved by the use of a potentiometric, amperometric, impedimetric, and conductometric techniques [57]. The electrode can be photo-irradiated which expands electrochemical sensors into photoelectrochemical sensors. CQDs and GQDs as a result of their optical property can be used as a coating of an electrode which is photo irradiated to enable electron transfer [49]. The photoelectrochemical sensing mechanism offers higher sensitivity and low background when contrasted with the conventional electrochemical sensing as a result of its separate excitation and detection source [58]. The QDs based photoelectrochemical sensor entails immobilization of the QDs on an electrode surface using an organic linker. The illumination of the QDs generates a photocurrent that is specific or determined by the analyte at the vicinity of the electrode, i.e., its type and concentration [59]. The QDs grafted on the electrode surface are illuminated with an appropriate potential which triggers the movement of electrons to produce the photocurrent. Hence, the concentration of the analyte determines the electron direction and amplitude of the photocurrent analyte [59].

Zhang et al. [60] fabricated a mercury photoelectrochemical sensor by coating a combination of low and wider band materials (Ag$_2$S and ZnS QDs respectively) on an indium tin oxide (ITO) electrode surface. The modified electrode was successfully evaluated on a solution containing Hg^{2+} ions (0.010-1000 nM concentration) showing a detection limit of 1.0 pM. Li et al. [61] developed an electrochemical sensor for detecting nitrile using nitrogen-doped GQDs engineered with nitrogen-doped carbon nanofibers (NGQDs@NCNFs). This sensor was tested on water (lake and tap water) and food samples using cyclic voltammetry (CV) and electrochemical impedance spectroscopy (EIS) where it showed high sensitivity, selectivity, reproducibility, a detection limit of 3 μM and a wide linear range 5–300 μM, R^2 = 0.999; 400–3000 μM, R^2 = 0.997). Also, Fu et al. [62] modified an indium tin oxide (ITO) electrode with N-GQDs to prepare an electrochemical sensor for the detection of mercury ions in aqueous solution. The electrode showed a concentration-dependent response against Hg^{2+} thereby showing great potential for use in the development of a sensor. Using CV and EIS, the Hg^{2+} N-GQDs modified electrode registered a detection limit of 10 ppb with an accumulation time of 32 s.

4.   Signal amplification strategies in quantum dot-based sensors

4.1   Forster (Fluorescence) resonance energy transfer

Forster (Fluorescence) resonance energy transfer (FRET) involves the non-radiative movement of the excitation energy from a donor at an excited state to an acceptor at a ground state [63]. FRET does not rely on the life span of the excited state but forms a

Materials Research Forum LLC
https://doi.org/10.21741/9781644901250-6

ground-state complex between the luminescent material and the quencher. The complex absorbs lights and returns to the ground state without emitting photons and is measured by the absorption spectra [23]. FRET has been used since its discovery back in 1922, where conventional FRET dyes characterized by emission from UV to near infrared region were used. These included dyes such as cyanine dyes and texas red which were seen to be the best due to their minute size, high quantum yields, solubility and bio-conjugation. However, these suffered from pH sensitivity and photobleaching and many more other drawbacks. QDs have since been used to improve the sensitivity and efficiency of FRET [44]. In here, QDs can either be used as fluorescent donors or acceptors, depending on the type of fluorophore [64]. The rate at which the energy is transferred in FRET is reliant on the distance separating the donor and acceptor molecules, their orientation relatively to each other and also their spectral overlap. For optimum performance, a distance of $2 - 8$ nm is acceptable [65].

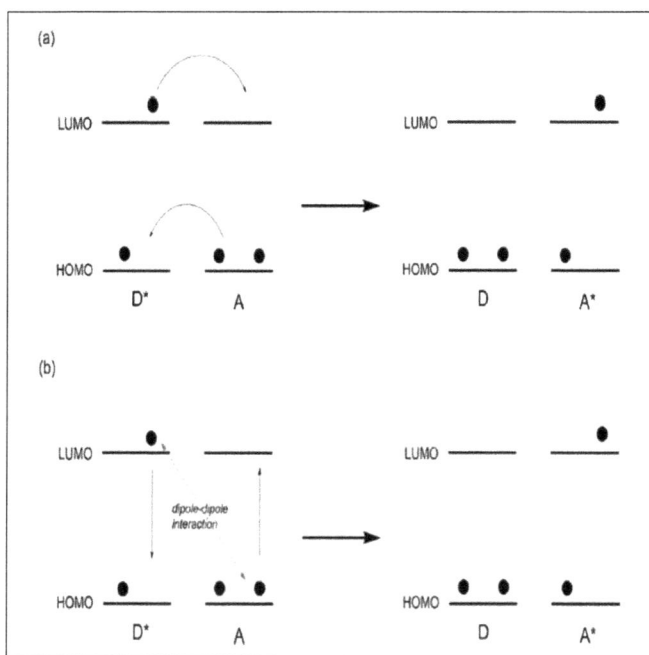

Fig.1 *Illustration of the Dexter and Forster mechanisms used in FRET-based sensors [28]. Image reprinted with permission from Elsevier.*

FRET transduction mechanism can be categorized into 2 types, based on their electronic interactions. These include the Dexter and Forster interactions, as illustrated in Fig. 5 [28]. Dexter is an orbital overlap induced electron transfer and has an exponential-distance reliance with a range of about 1 nm while Forster is facilitated by dipole-dipole interactions between an excited donor molecule and an acceptor molecule in the ground state [28, 30]. Almost close to FRET is inner filter effect (IFE) which also requires an overlap of the absorption spectra and the emission spectra of the QDs in the presence of the quencher [66].

4.2   Bioluminescence resonance energy transfer and Chemiluminescent resonance energy transfer

The bioluminescence (BRET) and chemiluminescence energy transfer (CRET) are phenomena that involve the non-radioactive transfer of energy from an excited state donor, generated through a chemical reaction, to a fluorescent acceptor, usually 10 nm apart [67]. BRET differs from CRET in that the reaction is biochemical. Therefore BRET is a non-radioactive phenomenon that generates an excited state donor via biochemical reactions [68]. BRET and CRET based sensors offer increased sensitivity and reliability, and they are also very easy to perform and are cost-effective. Proteins and small molecule fluorophores are used as donor molecules while organic dyes serve as acceptor molecules [67]. According to Dale et al. [69], three requirements must be met to perform a successful BRET assay. These include; (1) molecules of interest should be functionalized with the appropriate donor and acceptor molecules in a way that will not compromise their function; (2) tagged molecules should be localized properly in suitable experimental conditions; and (3) setup should be coupled to an instrument that is capable of monitoring energy transfer in real-time using live cells.

Just as in FRET, QDs have been used in BRET to enhance the performance of BRET sensors. Conventional energy acceptors in BRET are known to have small stokes shifts which result in poor spectral separation of donor and acceptor emissions. QDs, due to their large stoke shifts; have been used as suitable alternatives [58]. Moreover, QD based conjugates used herein can localize chemiluminescent and bioluminescent reactions to the surface of the QDs thereby providing the proximity required for CRET and BRET sensing [68]. Recently, numerous assays have been developed to demonstrate the feasibility of using QDs as acceptors in BRET systems. Fig. 6 demonstrates the use of QDs in developing a BRET sensor for the detection of protease activity. Luciferase protein was used as a donor and attached to the acceptor (QD) via a peptide substrate. In the absence of protease (A), the probe remains intact and as a result, luciferase donates energy to the QD thereby leading to bioluminescence-induced QD emission (red line).

Contrary, the presence of protease (B) results in the cleavage of luciferase from the QD thereby disrupting BRET, leading to a loss of bioluminescence-induced QD emission [70]. Fig. 6 [71] illustrates the application of QD based BRET in the analysis of protease activity.

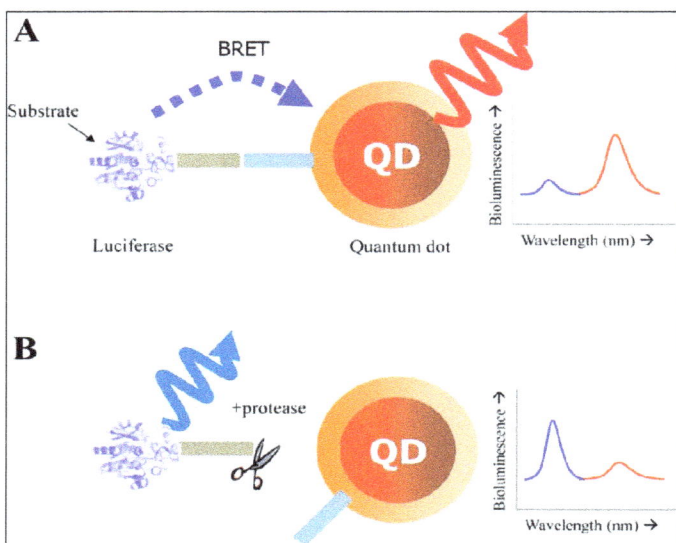

Fig. 6 Illustration of a QD-BRET sensor used for detecting protease activity [71].

4.3   Photoinducedelectron transfer (PET)

The photoinduced electron transfer (PET) mechanism involves the disruption of the electron-hole pair recombination which results in the quenching of the QDs luminescence due to the presence of the analyte. For electron transfer to occur, the QDs conduction band edge and valence band edge should be aligned. The movement of electrons can be in two directions as depicted by Fig. 7 [20]. This Figure illustrates the transfer of excited electrons from the LUMO of the donor molecule (D^*) to the LUMO of the accepting molecule (A^*) and forms radical ion pair ($\boldsymbol{D}^{\circ+}, \boldsymbol{A}^{\circ-}$) or transfer from the HOMO of the A^* to the HOMO of the D^* and forms a different radical ($\boldsymbol{D}^{\circ-}, \boldsymbol{A}^{\circ+}$) [23, 28].

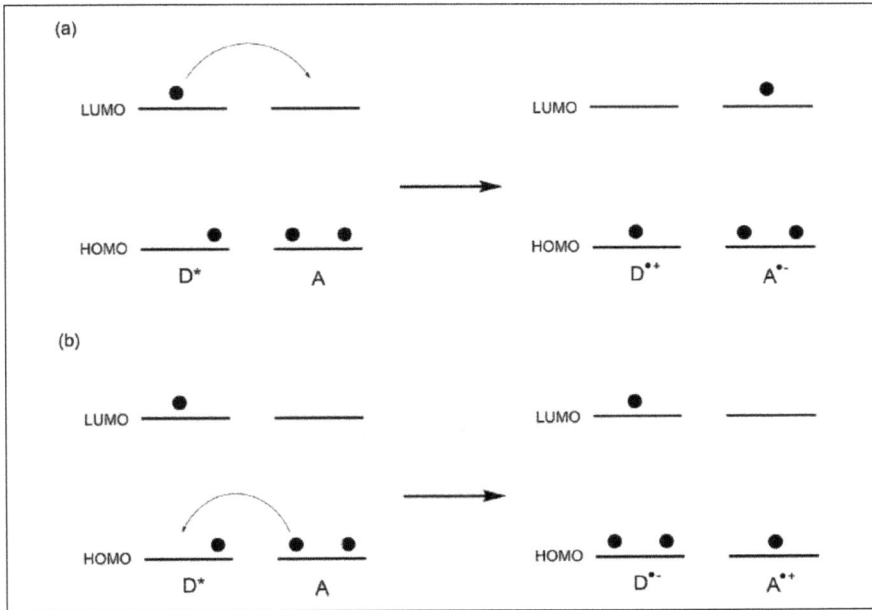

Fig. 7 Illustration of the PET process [20]. Reprinted from with permission from Elsevier.

An efficient QD-PET based sensor was developed and used to detect zinc ions (Zn^{2+}), as depicted in Fig. 8 [71]. Mercaptopropionic acid-capped CdSe/ZnS QDs were covalently attached to azamacrocycles to form a PET fluorophore-ligand. During the testing of the sensor, the absence of Zn^{2+} resulted in the transfer of the photogenerated hole on the QDs to an energy level on the ligand thus getting trapped and thereby blocking recombination with the QDs. This leads to charge separation which subsequently results in the emission of QDs being turned off. In the presence of Zn^{2+} ions, the emission is turned on since the energy level of the ligand is no longer available [71].

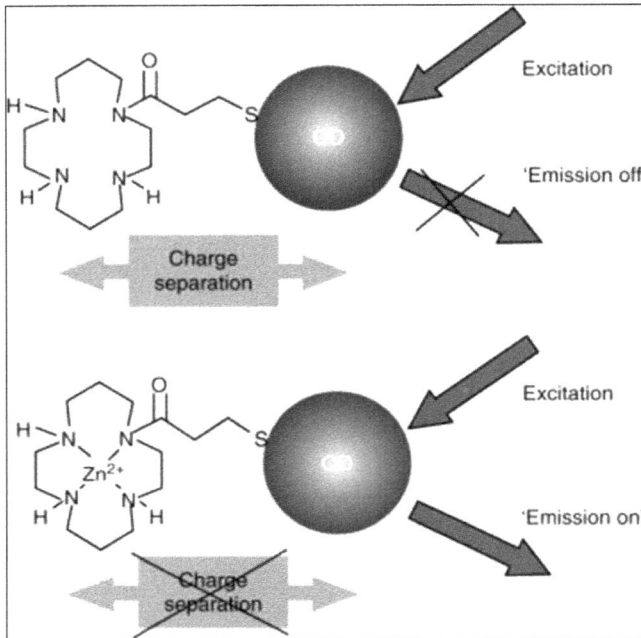

Fig. 8 Illustration of fluorescence switching on and off by the presence and absence of Zn^{2+}, through the photoinduced electron transfer mechanism [71].

4.4   Charge transfer

The optical excitation of QDs produces an exciton (electron-hole pair) which upon relaxation produces fluorescence or illumination. For the electron-hole recombination to generate efficient fluorescence, the rate of radiative recombination must be comparable with the non-radiative recombination rate. The exciton can be transferred to other acceptor molecules or emitted as photons. If charge transfer occurs between the donor QD and an acceptor molecule, it enables an "on" and "off" fluorescence switching mechanism, as shown in Fig. 9 [30]. This is a basic phenomenon for "turn on/off" sensors. These are distance-dependent and require that the molecules be in contact.

Materials Research Forum LLC
https://doi.org/10.21741/9781644901250-6

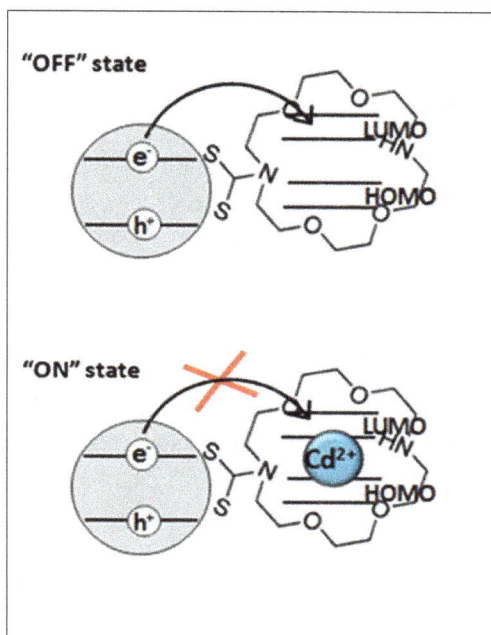

Fig. 9 Flow transfer process between QD donor molecule and an acceptor via charge transfer [30].

Conclusions

Sensors are useful analytical tools that have found widespread use in various fields such as environmental monitoring, clinical diagnostics, food testing and many others. Thus, the performance of these assays is of utmost importance in the sphere of mankind. Owing to their exceptional electrical and optical properties, QDs have revolutionized the field of chemosensors and biosensors. As highlighted earlier, these semiconducting nanomaterials offer numerous advantages that enhance the performance of these assays. These nanomaterials have been studied immensely in various fields of research and have emerged as the nanomaterials of choice where rapid and reliable analyses are concerned. Thus, the emergence of QDs based sensors with high sensitivity and good specificity will enhance the quality of life through improving clinical diagnosis and environmental monitoring in future.

Acknowledgements

The authors recognize the financial support provided by the South African Department of Science and Innovation (DSI) and Mintek.

References

[1]   A.M. Smith, S. Dave, S. Nie, L. True, X. Gao, Multicolor quantum dots for molecular diagnostics of cancer, Expert. Rev. Mol. Diagn. 6 (2006) 231-44. https://doi.org/10.1586/14737159.6.2.231

[2]   K. Krishnaswamy, V. Orsat, Sustainable delivery systems through green nanotechnology, in: A.M. Grumezescu (Ed.), Nano-and Microscale Drug Delivery Systems, Elsevier Inc., Amsterdam, 2017, pp. 17-32.

[3]   S. Suri, G. Ruan, J. Winter, C.E. Schmidt, Microparticles and Nanoparticles. In: B.D. Ratner, A.S. Hoffman, F.J. Schoen, J.E. Lemons (Eds.), Biomaterials Science: An Introduction to Materials in Medicine, Elsevier Inc., Amsterdam, 2013, pp. 360-388.

[4]   J. Njuguna, F. Ansari, S. Sachse, H. Zhu, V. Rodriguez, Nanomaterials, nanofillers, and nanocomposites: types and properties, in: J. Njunguna, K. Pielichowski, H. Zhu (Eds.), Health and Environmental Safety of Nanomaterials, Woodhead Publishing Limited, Cambridge, 2014, pp. 3-27.

[5]   A.M. Smith, S. Nie, Semiconductor nanocrystals: structure, properties, and band gap engineering, Acc. Chem. Res.43 (2010) 190-200. https://doi.org/10.1021/ar9001069

[6]   L. Cui, X.-P. He, G.-R. Chen, Recent progress in quantum dot based sensors, RSC. Adv.5 (2015) 26644-26653. https://doi.org/10.1039/C5RA01950H

[7]   A. Valizadeh, H. Mikaeili, M. Samiei, S.M. Farkhani, N. Zarghami, M. Kouhi, A. Akbarzadeh,S. Davaran, Quantum dots: synthesis, bioapplications, and toxicity, Nanoscale Res. Lett.7 (2012) 1 –14. https://doi.org/10.1186/1556-276X-7-480

[8]   P. Mishra, G. Vyas, M.S. Harsoliya, J.K. Pathan, D. Raghuvanshi, P. Sharma, A. Agrawal, Quantum dot probes in disease diagnosis, J. Pharm. Pharm. Sci. Rev. 1 (2011) 42–46. Corpus ID: 53340737.

[9]   F. Pinaud, X. Michalet, L.A. Bentolila, J.M. Tsay, S. Doose, J.J. Li, G. Iyer, S. Weiss,Advances in fluorescence imaging with quantum dot bio-probes, Biomaterials, 27 (2006) 1679-87. https://doi:10.1016/j.biomaterials.2005.11.018

Materials Research Forum LLC
https://doi.org/10.21741/9781644901250-6

[10]   W. Wei, Z. Jun-Jie, Optical applications of quantum dots in biological system,Sci. China. Chem. 54(2011) 1177 – 1184. https://doi.org/10.1007/s11426-011-4311-1

[11]   D. Aldakov, A. Lefrançois, P. Reiss, Ternary and quaternary metal chalcogenide nanocrystals: synthesis, properties and applications, J. Mater. Chem. C.1(2013) 3756 – 3776. https://doi.org/10.1039/C3TC30273C

[12]   W.R. Algar, M. Massey, U.J.Krull, The application of quantum dots, gold nanoparticles and molecular switches to optical nucleic-acid diagnostics, Trends Anal. Chem. 28(2009) 292–306. https://doi.org/10.1016/j.trac.2008.11.012

[13]   A.F.E. Hezinger, J.Teßmar, A. Gopferich, Polymer coating of quantum dots – A powerful tool toward diagnostics and sensorics, Eur. J. Pharm. Biopharm. 68(2008) 138–152. https://doi.org/10.1016/j.ejpb.2007.05.013

[14]   M.K. Wagner, F. Li, J. Li, X.F. Li, X.C. Le, Use of quantum dots in the development of assays for cancer biomarkers, Anal. Bioanal. Chem. 397 (2010) 3213–3224. https://doi.org/10.1007/s00216-010-3847-9

[15]   P. Zrazhevskiy, M. Sena, X. Gao, Designing multifunctional quantum dots for bioimaging, detection, and drug delivery, Chem. Soc. Rev. 39 (2010) 4326-4354. https://doi.org/10.1039/B915139G

[16]   F. Hetsch, N. Zhao, S.V. Kershaw, A.L. Rogach, Quantum dot field effect transistors, Mater. Today. 16 (2013) 312 – 325. https://doi.org/10.1016/j.mattod.2013.08.011

[17]   W.W. Yu, E. Chang, R.Drezek, V.L. Colvin, Water-soluble quantum dots for biomedical applications, Biochem.Biophys. Res. Commun. 348 (2006) 781–786. https://doi.org/10.1016/j.bbrc.2006.07.160

[18]   C. Zeng, A. Ramos-Ruiz, J.A. Field, R. Sierra-Alvarez, Cadmium telluride (CdTe) and cadmium selenide (CdSe) leaching behavior and surface chemistry in response to pH and O_2, J. Enviro.Manage. 154(2015) 78-85. https://doi.org/10.1016/j.jenvman.2015.02.033

[19]   D. Vasudevan, R.R. Gaddam, A. Trinchi, I. Cole, Core–shell quantum dots: Properties and applications, J. Alloys Compd. 636 (2015) 395-404. https://doi.org/10.1016/j.jallcom.2015.02.102

[20]   N. Chaniotakis, R. Buiculescu, Semiconductor quantum dots in chemical sensors and biosensors, in: K.C. Honeychurch (Eds.), Nanosensors for Chemical and Biological Applications, Woodhead Publishing Limited, Cambridge, 2014, pp. 267-294.

[21]   A. Lesiak, K. Drzozga, J. Cabaj, M. Bański, K. Malecha, A. Podhorodecki, Optical sensors based on II-VI quantum dots, Nanomaterials, 9 (2019) 192. https://doi.org/10.3390/nano9020192

[22]   A. intel, Quantum Dots Market and Patent Infographics. 2019.

[23]   C.M. Gonzalez, J.G. Veinot, Silicon nanocrystals for the development of sensing platforms, J. Mater. Chem. C.4 (2016) 4836-4846. https://doi.org/10.1039/C6TC01159D

[24]   B. Valeur, B.M.N. Berberan-Santos, A brief history of fluorescence and phosphorescence before the emergence of quantum theory, J Chem. Educ.88 (2011) 731-738. https://doi.org/10.1021/ed100182h

[25]   T. Pisanic, Y. Zhang, T. Wang, Quantum dots in diagnostics and detection: principles and paradigms, Analyst, 139 (2014) 2968-2981. https://doi.org/10.1039/c4an00294f

[26]   P. Rani, Chemosensor and its applications, IRJRR, 3 (2015) 1-10.

[27]   W. Sun, S. Guo, C. Hu, J. Fan, X. Peng, Recent development of chemosensors based on cyanine platforms, Chem.Rev.116 (2016) 7768-7817. https://doi.org/10.1021/acs.chemrev.6b00001

[28]   G. Fukuhara, Analytical supramolecular chemistry: colorimetric and fluorimetric chemosensors, J. Photochem. Photobiol. C: 42 (2020) 100340. https://doi.org/10.1016/j.jphotochemrev.2020.100340

[29]   M.L. Viger, L.S. Live, O.D. Therrien, D. Boudreau, Reduction of self-quenching in fluorescent silica-coated silver nanoparticles, Plasmonics, 3 (2008) 33-40. https://doi.org/10.1007/s11468-007-9051-x

[30]   M. Vázquez-González, C. Carrillo-Carrion, Analytical strategies based on quantum dots for heavy metal ions detection, J. Biomed.Opt. 19 (2014) 101503. https://doi.org/10.1117/1.JBO.19.10.101503

[31]   Z. Altintas, F. Davis, F.W.Scheller, Applications of Quantum Dots in Biosensors and Diagnostics, in: Z. Altintas (Eds.), Biosensors and Nanotechnology: Applications in Healthcare Diagnostics, John Wiley & Sons, Inc., New Jersey, 2018, pp. 185-200.

[32]   R. Monošik, M. Streďansky, E.Šturdik, Biosensors - classification, characterization and new trends, ActaChim. Slov.5 (2012) 109-120. https://doi.org/10.2478/v10188-012-0017-z

Materials Research Forum LLC
https://doi.org/10.21741/9781644901250-6

[33] F. Ma, C. Li, C. Zhang, Development of quantum dot-based biosensors: principles and applications, J. Mater. Chem. B. 6(2018)6173-6190. https://doi.org/10.1039/C8TB01869C

[34] S. Hong, C. Lee, The current status and future outlook of quantum dot-based biosensors for plant virus detection, Plant.Pathol. J. 34 (2018) 85–92. https://doi.org/10.5423/PPJ.RW.08.2017.0184

[35] P.D. Howes, R. Chandrawati, M.M. Stevens, Colloidal nanoparticles as advanced biological sensors, Science. 346 (2014)1247390-10. https://doi.org/10.1126/science.1247390

[36] C.A. Marquette, L.J. Blum, Electro-chemiluminescent biosensing, Anal. Bioanal. Chem. 390 (2008) 155-168. https://doi.org/10.1007/s00216-007-1631-2.

[37] B. Valeur, M.N. Berberan-Santos, A brief history of fluorescence and phosphorescence before the emergence of quantum theory, J. Chem. Educ. 88(2011) 731-738. https://doi.org/10.1021/ed100182h

[38] A.J. Sutherland, Quantum dots as luminescent probes in biological systems,Curr. Opin. Solid St. M. 6(2002) 365-370. https://doi.org/10.1016/S1359-0286(02)00081-5

[39] U. Resch-Genger, M.Grabolle, S. Cavaliere-Jaricot, R.Nitschke, T. Nann, Quantum dots versus organic dyes as fluorescent labels, Nat. Methods, 5 (2008) 763–775. https://doi.org/10.1038/nmeth.1248

[40] S.B. Rizvi, S.Ghaderi, M.Keshtgar, A.M. Seifalian, Semiconductor quantum dots as fluorescent probes for *in vitro* and *in vivo* bio-molecular and cellular imaging, Nano Rev. 1(2010) 5161-5176. https://doi.org/10.3402/nano.v1i0.5161

[41] A. Shamirian, A. Ghai, P.T. Snee, QD-based FRET probes at a glance, Sensors (Basel).15 (2015) 13028–13051. https://doi.org/10.3390/s150613028

[42] K. Boeneman, J.B. Delehanty, K. Susumu, M.H. Stewart, J.R. Deschamps, I.L. Medintz, Quantum Dots and Fluorescent Protein FRET-Based Biosensors. In: Zahavy E., Ordentlich A., Yitzhaki S., Shafferman A. (Eds.) Nano-Biotechnology for Biomedical and Diagnostic Research. Advances in Experimental Medicine and Biology, Springer, Dordrecht, 2012. 733: pp. 63-74.

[43] J. Shi, F. Tian, J.Lyu, M.Yang, Nanoparticle based fluorescence resonance energy transfer (FRET) for biosensing applications, J. Mater. Chem. B. 3(2015)6989-7005. https://doi.org/10.1039/C5TB00885A

[44] F. Ma, C. Li, C.Zhang, Development of quantum dot-based biosensors: principles and applications, J. Mater. Chem. B. 6 (2018) 6173-6190. https://doi.org/10.1039/C8TB01869C

[45] H. Feng, Z. Qian, Functional carbon quantum dots: a versatile platform for chemosensing and biosensing, Chem. Rec. 18(2018) 491-505. https://doi.org/10.1002/tcr.201700055

[46] Z. Qian, X. Shan, L. Chai, J. Ma, J. Chen, H. Feng, Si-doped carbon quantum dots: a facile and general preparation strategy, bioimaging application, and multifunctional sensor, ACS Appl. Mater. Interfaces.6 (2014) 6797-6805. https://doi.org/10.1021/am500403n

[47] Q. Ren, L. Ga, J. Ai, Rapid Synthesis of highly fluorescent nitrogen-doped graphene quantum dots for effective detection of ferric ions and as fluorescent ink, ACS Omega. 4(2019) 15842-15848. https://doi.org/10.1021/acsomega.9b01612

[48] A. Ravalli, D. Voccia, I. Palchetti, G. Marrazza, Electrochemical, electrochemiluminescence, and photoelectrochemical aptamer-based nanostructured sensors for biomarker analysis, Biosensors (Basel). 6(2016) 39. https://doi.org/10.3390/bios6030039

[49] M. Li, T. Chen, J.J. Gooding, J. Liu, Review of carbon and graphene quantum dots for sensing, ACS sens.4 (2019) 1732-1748. https://doi.org/10.1021/acssensors.9b00514

[50]. S. Ge, C. Zhang, F. Yu, M. Yan, J. Yu, Layer-by-layer self-assembly CdTe quantum dots and molecularly imprinted polymers modified chemiluminescence sensor for deltamethrin detection, Sens. Actuators B Chem. 156(2011) 222-227. https://doi.org/10.1016/j.snb.2011.04.024

[51] Y. Tang, Y. Su, N. Yang, L. Zhang, Y. Lv, Carbon nitride quantum dots: a novel chemiluminescence system for selective detection of free chlorine in water, Anal. Chem. 86 (2014) 4528-4535. https://doi.org/10.1021/ac5005162

[52] M.M. Richter, Electrochemiluminescence (ECL), Chem. Rev. 104 (2004) 3003–3036. https://doi: 10.1021/cr020373d

[53] X. Chen, Y. Liu, Q. Ma, Recent advances in quantum dot-based electrochemiluminescence sensors, J. Mater. Chem. C.6 (2018) 942-959. https://doi.org/10.1039/C7TC05474B

[54]. Y. Dong, W. Tian, S. Ren, R. Dai, Y. Chi, G. Chen, Graphene quantum dots/l-cysteine coreactant electrochemiluminescence system and its application in sensing

Materials Research Forum LLC
https://doi.org/10.21741/9781644901250-6

lead (II) ions, ACS Appl. Mater. Interfaces. 6(2014) 1646-1651.
https://doi.org/10.1021/am404552s

[55]   S. Yang, J. Liang, S. Luo, C. Liu, Y. Tang, Supersensitive detection of chlorinated phenols by multiple amplification electrochemiluminescence sensing based on carbon quantum dots/graphene, Anal.Chem. 85(2013) 7720-7725.
https://doi.org/10.1021/ac400874h

[56]   N.F. Carter, G.R. Chambers, G.J. Hughes, S. Scott, G.S. Sanghera, J.L. Watkin, Electrochemical sensor. 1997, Google Patents.

[57]   S.K. Mahadeva, J. Kim, Conductometric glucose biosensor made with cellulose and tin oxide hybrid nanocomposite, Sens. Actuators B: Chem. 157(2011) 177-182.
https://doi.org/10.1016/j.snb.2011.03.046

[58]   F. Jafari, A. Salimi, A. Navaee, electrochemical and photoelectrochemical sensing of dihydronicotinamide adenine dinucleotide and glucose based on noncovalently functionalized reduced graphene oxide-cadmium sulfide quantum dots/polyanile blue nanocomposite, Electroanalysis. 26 (2014) 1782-1793.
https://doi.org/10.1002/elan.201400164

[59]   Z. Yue, F. Lisdat, W.J. Parak, S.G. Hickey, L. Tu, N. Sabir, D. Dorfs, N.C. Bigall, Quantum-dot-based photoelectrochemical sensors for chemical and biological detection. ACS Appl. Mater. Interfaces. 5 (2013) 2800-2814.
https://doi.org/10.1021/am3028662

[60]   L. Zhang, P. Li, L. Feng, X. Chen, J. Jiang, S. Zhang, C. Zhang, A. Zhang, G. Chen, H. Wang, Synergetic Ag2S and ZnS quantum dots as the sensitizer and recognition probe: A visible light-driven photoelectrochemical sensor for the "signal-on" analysis of mercury (II), J. Hazard. Mater. (2019) 121715.
https://doi.org/10.1016/j.jhazmat.2019.121715

[61]   L. Li, D. Liu, K. Wang, H. Mao, T. You, Quantitative detection of nitrite with N-doped graphene quantum dots decorated N-doped carbon nanofibers composite-based electrochemical sensor, Sensor. Actuator. B Chem. 252(2017) 17-23.
https://doi.org/10.1016/j.snb.2017.05.155

[62]   C.-C. Fu, C.-T. Hsieh, R.-S. Juang, S. Gu, Y.A. Gandomi, R.E. Kelly, K.D. Kihm, Electrochemical sensing of mercury ions in electrolyte solutions by nitrogen-doped graphene quantum dot electrodes at ultralow concentrations, J. Mol. Liq.(2020) 112593. https://doi.org/10.1016/j.molliq.2020.112593

[63]   J. Saha, A.D. Roy, D.Dey, D. Bhattacharjee, S.A. Hussain, Role of quantum dot in designing FRET based sensors, Mater.5 (2018) 2306–2313. https://doi.org/10.1016/j.matpr.2017.09.234

[64]   K.F. Chou, A.M. Dennis, Förster resonance energy transfer between quantum dot donors and quantum dot acceptors, Sensors (Basel).15 (2015) 13288–13325. https://doi.org/0.3390/s150613288

[65]   M.F. Frasco, N.Chaniotakis, Semiconductor quantum dots in chemical sensors and biosensors, Sensors. 9(2009) 7266-7286. https://doi.org/10.1533/9780857096722.2.267

[66]   Y. Yang, T. Zou, Z. Wang, X. Xing, S. Peng, R. Zhao, X. Zhang, Y. Wang, The fluorescent quenching mechanism of N and S Co-doped graphene quantum dots with Fe^{3+} and Hg^{2+} ions and their application as a novel fluorescent sensor, Nanomaterials 9(2019) 738. https://doi.org/10.3390/nano9050738

[67]   E. Hwang, J. Song, J. Zhang, Integration of nanomaterials and bioluminescence resonance energy transfer techniques for sensing biomolecules, Biosensors. 9 (2019) 42 – 58. https://doi.org/10.3390/bios9010042

[68]   W.R. Algar, A.J.Tavares, U.J. Krull, Beyond labels: a review of the application of quantum dots as integrated components of assays, bioprobes, and biosensors utilizing optical transduction, Anal. Chim. Acta. 673 (2010)1-25. https://doi.org/10.1016/j.aca.2010.05.026

[69]   N.C. Dale, E.K.M. Johnstone, C.W. White, K.D.G. Pfleger, NanoBRET: The bright future of proximity-based assays, Front. Bioeng. Biotechnol. 7(2019). 56. https://doi: 10.3389/fbioe.2019.00056

[70]   G.B. Kim, Y-P. Kim, Analysis of protease activity using quantum dots and resonance energy transfer, Theranostics.2 (2012) 127-138. https://doi: 10.7150/thno.3476

[71]   M.J. Ruedas-Rama, E. A. H. Hall, Azamacrocycle activated quantum dot for zinc ion detection, Anal. Chem. 80 (2008) 8260–8268. https://doi.org/10.1021/ac801396y

Quantum Dots – Properties and Applications
Materials Research Foundations **96** (2021) 169-190

Materials Research Forum LLC
https://doi.org/10.21741/9781644901250-7

Chapter 7

Applications of Quantum Dots in Supercapacitors

Sanjeev Kumar Ujjain[1]*, Preety Ahuja[1]

[1]Research Initiative for Supra-Materials, Shinshu University, 4-17-1 Wakasato, Nagano-City 380-8553, Japan

drsanjeevkujjain@gmail.com

Abstract

Quantum dots (QDs) are a new class of zero-dimensional (0D) nanomaterials having unique electronic and optical properties along with biocompatibility, chemical inertness, dispersibility in water, and high specific surface area that gives them potential for biological, optoelectronic and energy related applications. Among them, charge storage supercapacitor (SC) devices have been intensively studied as the nano-sized QDs act as an excellent interface to stimulate an enhanced interaction between electrode and electrolyte resulting in superior charge storage properties of the SC. In this chapter, the latest research progress on the five representative types of QDs namely carbon nanodots (CNDs), graphene QDs (GQDs), polymer QDs (PQDs), transition metal oxide (TMO) and dichalcogenide (TMD) QDs are comprehensively introduced and their influence on the final charge storage properties of supercapacitor devices is emphatically discussed in detail. Finally, a brief outlook is given, pointing out the challenges which remain to be settled before adoption of QDs can be of widespread utility for near future energy-functional devices.

Keywords

Quantum Dots, Carbon and Graphene, Polymer Quantum Dots, Transition Metal Oxide, Supercapacitors

Contents

1.   Introduction

The supercapacitors (SCs) possess high potential among electrochemical energy storage devices owing to their elevated power density and high charge storage capacity along with rapid charge / discharge rate, long cycling stability, lightweight and compact size [1-5]. Based on their charge storage mechanism, SCs can be classified into two categories. The first is electrical double layer capacitors (EDLCs), where energy storage occurs through adsorption-desorption of ions. Carbonaceous materials like activated porous carbons, graphene, carbon nanotubes (CNTs) and carbon fibers having large specific surface area (SSA) along with high electrical conductivity are dominantly utilized for EDLC storage. As EDLC involves a non-faradic process which depends on the electrochemical accessibility of the available surface sites, which limits their charge storage resulting in inferior energy density. In order to enhance their energy density, EDLCs are mostly incorporated with pseudo-capacitive materials which involve Faradaic redox activities at the electrode-electrolyte interface. Such materials exhibit 10-100 times higher charge storage compared to EDLCs and hence delivers high energy density without declining the power density which is most crucial for high performance SCs [6]. The widely used pseudo-capacitive materials involve transition metal oxides (TMO), dichalcogenides (TMD), and conducting polymers (CPs) like polyaniline (PANI), polypyrrole (PPy) or poly(3,4-ethylenedioxythiophene) (PEDOT). Although the TMO, TMD and CPs possess high charge storage capacity, however their utilization is limited due to their meager cycling stability owing to structural variation due to swelling and shrinking during continuous charge/discharge cycles. So, hybrid materials resulting from the composites of TMO, TMD and CPs with high SSA carbonaceous materials, which can accommodate the structural deformations during charge/discharge process are used as SCs electrode to enhance their rate capability and cycling stability [7].

Since both the EDLC and pseudocapacitance are interfacial phenomena, so at present a lot of research work is devoted on design and synthesis of hybrid nanocomposite materials with high electrical conductivity, large SSA along with superior wettability to enhance the electrode/electrolyte interfacial charge transfer of SCs [8]. To this end, considerable attempts have been made to synthesize electrode materials using nanoparticulates of TMO, TMD, CPs etc. with nanocarbons such as graphene nanosheets, CNTs and carbon nanofibers [6,9], which enhance their performance. Downsizing electrode materials to nanometer size can enhance electrolytic accessibility and abbreviate the ionic diffusion resulting in improved electrochemical performances [10,11]. However, the cost involves with their production and processing increases the overall price of end products. In addition, the lower capacity contribution of such carbon materials towards the volumetric capacitance of the assembled device limits their miniaturization, which is non-desirable for future generation electronic applications.

To overcome the aforementioned problem, quantum dots (QDs) which are zero-dimensional (0-D) materials have attracted lot of attention since their discovery in 2004. QDs demonstrate several advantageous features desired for enhancing the electrochemical properties of electrode materials for charge storage: (a) the size can be tailored as small as 1-2 nm, such small size have the highest surface-to volume ratio which facilitate large capacity charge storage, (b) the diminished size also shorten the intercalating ions diffusion length in the solid phase which enhances the charge/mass transfer, (c) the high dispersion can deliver a huge space to buffer the volumetric changes occur during charge/discharge of the active materials [12] and (d) their surface can be easily functionalized to improve wettability which would realize rapid electron-transfer kinetics and charge-transport owing to intimate contact at the electrode/electrolyte interface, resulting in an enhanced specific capacitance (C_{sp}), rate capability and cycling stability [6,13].

In this chapter, various QDs which are being used as the electrode materials for SCs are described. Further, we briefly designate two different type of SCs based on their architectural assembly, i.e. Symmetric supercapacitors (SSCs) and Asymmetric supercapacitors (ASSCs) and summarize the utilization of different forms of QDs and their nanocomposites as electrode materials for such SCs. We primarily focus on their typical electrochemical charge storage performances using different techniques and principles. In the final section, further developments and perspectives for the application of QDs based nanomaterials towards realizing high performance SCs are discussed.

2.   Type of quantum dots (QDs) for supercapacitor electrode

Recent advances in the field of quantum dots (QDs) for electrochemical charge storage properties have offered mainly five different class of nanosized materials involving transition metal oxides (TMO) QDs, transition metal dichalcogenide (TMD) QDs, polymer quantum dots (PQDs), carbon nanodots (CNDs) and graphene quantum dots (GQDs) as represented in Fig. 1. The TMO-QDs offers multiple valency of metal ions for pseudo-capacitive reactions, while TMD-QDs possess 2D covalently bonded structures having 3-atom layer (X-M-X, where X = chalcogenides and M= metal) attached by interlayer van der Waals force. These 2D layered structure realize layered spaces for ion intercalation and active sites for charge storage [14-16]. Nickel oxide (NiO) [17], hematite Fe_2O_3 [18-20], niobium pentoxide (Nb_2O_5) [21], tin oxide (SnO_2) [22], tungsten oxide (WO_{3-x}) [13] along with vanadium nitride QDs [23] and Co_3O_4 nanosheets with embedded Ag QDs [24] have been regarded as promising active materials. In addition, tungsten disulfide (WS_2) QD sand molybdenum disulfide (MoS_2) dichalcogenides with excellent conductivities are considered for very fast pseudocapacitors [14,25].

PQDs which are highly dispersible in aqueous solutions consist of carbon core composed of cross-linked polymers synthesized from linear monomers or polymers. As a consequence, their properties can be tuned with proper choice of monomers or polymers with heteroatoms such as oxygen, nitrogen, sulfur or phosphorous for variety of applications [26-29]. Mostly, nanostructured CPs PANI, PPy and PEDOT are hybridized with carbon or graphene QDs or TMO to form binary or ternary nanocomposites for improving their electrochemical performance for SC application [33-35].

Carbon nanodots (CNDs) are nanosized spherical carbon materials having multiple carbon nanoparticles sizes < 10 nm and lack crystal lattice. They are composed of sp^2 hybridized carbon core covered with amorphous shell having several functional groups rich edges, as they are mostly derived from organic sources. These CNDs have fascinated several researchers over last decade owing to their high chemical stability, superior conductivity, broadband optical absorption, high SSA and act as an environmentally benign alternative for toxic QDs [36-40]. Furthermore, graphene quantum dots (GQDs) can be considered as a subset of CNDs which are derived from graphene oxide (GO) or graphene. GQDs constitute of one or very few layers of graphene with lateral dimensions less than 10 nm having 0.24 nm average crystal lattice size. Although GQDs demonstrate some of physical and chemical properties analogous to graphene, however several properties diverge due to the dominance of edge effects and quantum confinement [41-44].

Materials Research Forum LLC
https://doi.org/10.21741/9781644901250-7

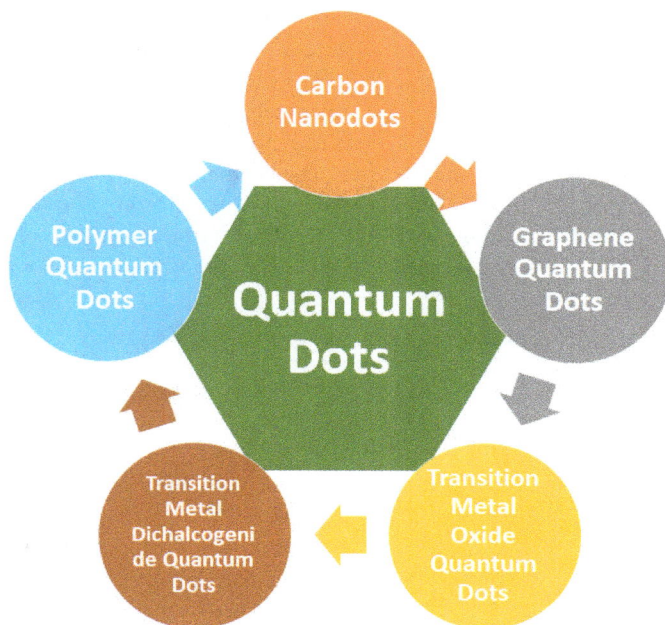

Fig. 1 Pictorial representation of types of quantum dots for supercapacitor applications.

Several reports suggest that QDs due to their small size and desired functional group perform as an accomplished interface which promotes the synergy among the electrode/electrolyte resulting in superior SC performances. In addition, nanocomposite materials derived using CNDs or GQDs having large SSA, with other forms of QDs involving TMO, TMD or PQDs possess extra interspaces and pathways for ion/mass transfer and volumetric changes during charge/discharge which provides stable cycling performance of the fabricated SC device [6,9,13,45].

3.   Types of supercapacitors

SCs can be classified into two categories: Symmetric supercapacitors (SSCs) and asymmetric supercapacitors (ASSCs), based on type of electrodes employed for charged storage in the assembled SC as shown in Fig. 2. The SSCs involve same electro-active material for both the positive and negative electrodes, while ASSCs integrates a pseudocapacitors (battery type) as positive with an EDLC type negative electrode. The performance of ASSCs can be benefitted from the electrochemical properties of both the materials. Their charge storage involves ionic and electronic charge parting at the interface along with Faradaic charge transfer. Before fabricating ASSCs, several factors should be optimized with regard to electrodes material and structure which correspondingly modify effective surface area, charge transfer processes leading to augmented performance. Generally, the working voltage window of an aqueous electrolyte based SSCs is limited up to 1.23 V [46]. In contrast, the ASSCs offer full-scale (sum of mode of positive and negative) utilization of the working voltage window which can be extended up to 2.0 V.

As the energy density is directly proportional to working voltage window,

$$E = \frac{1}{2}CV^2 \tag{1}$$

So, the energy performance of SSC is hampered due to limitation associated with the aqueous electrolyte electrochemical stability window (1.23 V). In contrast, ASSCs demonstrate enhanced charge storage with augmented energy density due to wide operational voltage window. Notably, in order to utilize the complete spectrum of working voltage window, the charges stored by both positive and negative electrodes should necessarily be balanced. Recent report [47] on ASSCs employing numerous electrode materials in aqueous electrolytes advocate high energy densities. SSCs with both positive and negative electrode material identical are primarily nanostructured carbon//carbon, CP//CP, TMO//TMO, TMD//TMD or their nanocomposites assembled in symmetric configuration [48-53]. While for ASSCs, activated carbon (AC), reduced graphene oxide (rGO) or other carbonaceous nanomaterials are commonly used as a characteristic negative electrode materials and CPs, TMOs, TMDs or their hybrid nanocomposites are employed as the positive electrode materials [54,55].

The later sections describe the role of QDs in improving SCs performances and their hybrid counterparts with pseudocapacitive constituents in SSCs and ASSCs.

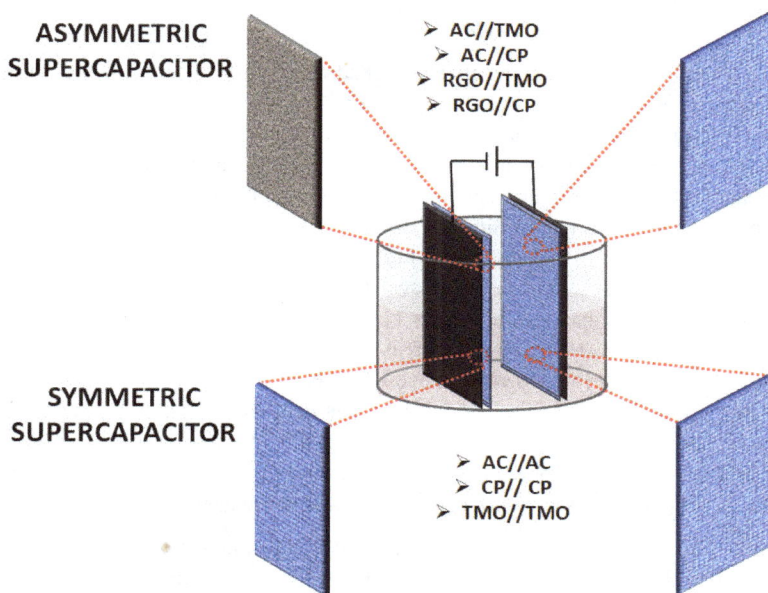

Fig. 2 Electrode assembly for symmetric and asymmetric supercapacitor.

3.1   QDs for symmetric supercapacitors

In the perspective of SC application, QDs showed much improved performance due to their intriguing properties which involve high SSA, mobility and superior dispersing ability in different solvents [56]. In addition, owing to their nanometric scale and consequent quantum effects, they possess higher intrinsic capacitance [57]. Carbon quantum dots (CQDs) aerogels were developed by initiating sol-gel polymerization of resorcinol and formaldehyde [58]. These CQDs were then pyrolyzed under argon to form 3D interconnected structure with ordered tunnels and mesopores. The fabricated SC displayed high C_{sp} of 294 Fg^{-1} at 0.5 Ag^{-1} (improved value compared to CQD free electrodes) and was able to retain 94% initial C_{sp} over 1000 cycles. Furthermore, highly porous composite from biomass derived organic CNDs were also prepared by carbonization process [59]. It involved protein rich biomass precursor, which underwent carbonization firstly at low temperature of 200°C, followed by high temperature treatment at 750°C. The resulting SC possesses high SSA owing to hierarchical porous structure, along with high C_{sp} (337.3 Fg^{-1} at $1Ag^{-1}$). They were also prepared by employing citric acid and L-cystein to produce heteroatom doped CQDs (N,S-CQDs), with highly improved properties [60].

CQDs composites with AC developed by Kumar et al. [61] showed three times better performance as compared to AC electrodes, owing to their placement in the pores of AC. Likewise, cauliflower leaf wastes was further utilized for producing biomass-derived N-doped CQDs and hybridized it with reduced graphene oxide (rGO) [62]. Here, CQDs as intercalators, prevent the aggregation rGO sheets and consequently, led to high SSA composite. Also, presence of nitrogen moiety offers richer pseudoactive sites, revealing C_{sp} of 278 Fg^{-1} at 0.2 Ag^{-1} along with high rate capability. Long-term cycling durability was evidenced by 96.1% C_{sp} retention over 10,000 galvanostatic charge/discharge (GCD) cycles. Later, Zucchini derived CQDs composite with rGO were further explored by deoxygenating in hydriodic acid vapor to introduce iodine heteroatoms in the resulting composite [63]. The specific capacitance was measured to be 374 Fg^{-1} at 2 mVs^{-1} with 268 Fg^{-1} at 1000 mVs^{-1}. The capacitance was decayed by only 28.3% demonstrating superior rate capability along with long term cycling stability (93.8% even after 10,000 cycles at 10 Ag^{-1}). Authors attributed this excellent performance to micro/mesoporous structure with enhanced electrical conductivity and consequently higher ionic diffusion.

Recently, GQDs SCs are also attracting considerable interest owing to the synergistic effects from the properties of graphene and QDs. Electrical method was employed by Hu et al. [64] to deposit GQDs on aligned single walled carbon nanotubes (SWCNTs). They obtained C_{sp} of 44 $mFcm^{-2}$, with approx. 200% improvement as compared to bare SWCNTs. The microporous structure resulted from GQDs and aligned SWCNTs scaffolds facilitated the ionic transportation leading to not only high capacitance but also the stability to the resulting composite [65]. Likewise, other porous carbon structures and three-dimensional (3-D) aerogel have been composited with GQDs and employed for SC electrodes with much improved performance [58,66].

In addition, SCs in micro form was also studied by employing GQDs as their electrode materials. Liu et al. [67] utilized GQDs based micro-supercapacitor prepared by electrodeposition method. It displayed excellent rate capability with short relaxation time. High cycling stability was also evidenced by 97.8% capacitive retention after 5000 cycles.

3.1.1 Carbon QDs and TMO nanocomposites

To further improve the energy density of the fabricated SSCs, the electrode material say CDs were hybridized with pseudocapacitive TMOs owing to their active redox sites and high dispersibility in solvents [68]. CQD was hybridized with RuO_2 and used as SSC electrode [69]. The composite with 41.9 wt% Ru content displayed highest capacitive characteristics i.e. C_{sp} of 594 Fg^{-1} at 1 Ag^{-1} with 77% retainment at 50 Ag^{-1}. With this high rate capability, the electrode also demonstrated improved cycling durability and coulombic efficiency as compared to pristine RuO_2. This augmented capability of the composite was ascribed to the preclusion of RuO_2 agglomeration with efficient dispersion in CQDs.

Wei et al. [70] decorated CDs by $Ni(OH)_2$ with improved electrical conductivity and employed them in SSC assembly. Results involved the uniform structure formation with high C_{sp} of 2900 Fg^{-1} at 1 Ag^{-1} alongwith good rate capability with 2051 Fg^{-1} at 10 Ag^{-1}. Also, one-dimensional (1-D) composites have been synthesized by coating NiO on CQDs, delivering high C_{sp} of 1858 Fg^{-1} at 1 Ag^{-1} with outstanding rate capability [71]. CNDs were also entrapped inside V_2O_5 nanobelts to augment the conductivity and ion propagation ability of the hybrid composite [72]. As obtained composite, $CND@V_2O_5$ proved a potential electrochemically active material with augmented three electrode electrochemical performance in 1 M Na_2SO_4 displaying ~4.5 times higher activity than bare V_2O_5 electrodes. It was also explored towards solid state SC assembly with Lithium based gel as the electrolyte, which provide energy density of 60 W h kg^{-1} with high power density of 4.1 kW kg^{-1} at 5 Ag^{-1} and excellent cycling stability.

In addition, another research group also synthesized $CDs/NiCo_2O_4$ electrodes by following hydrothermal treatment [73]. CDs were initially prepared by treating the solution of citric acid, ethylene diamine and phosphorous oxide in an autoclave to improve their properties by nitrogen or phosphorous functionalization. The resulted composites of $CDs/NiCo_2O_4$ possessed different morphologies based on CDs amount in the composites. Interestingly, their method is able to produce multi-structure morphology as compared to other methods involving template processing and chemical deposition. Moreover, Ni-Al layered double hydroxide (LDH) have also been composited with ultrathin nanoplate-like CQDs by a solvothermal method [74]. The synergistic effect of the constituents resulted in remarkable electrochemical performance with excellent specific capacitance of 1794 Fg^{-1} at 2 Ag^{-1}, along with remarkable cycling stability. Additionally, amorphous FeOOH QDs graphene hybrid was prepared and found to exhibit superior electrochemical performance, which overpowers bare iron oxides/hydroxides-based counterparts [75]. It exhibited large C_{sp} 365 Fg^{-1} in potential range of 0 to -0.8 V with outstanding cycling stability i.e. 90 % capacitance retention over 20,000 cycles. Although the C_{sp} increased to 403 and 1243 Fg^{-1} when voltage was extended to -1.0 V and -1.25 V but, compromises the rate capability and cycle performance. Also, Fe_2O_3 nanodots with nitrogen-doped graphene sheets ($Fe_2O_3NDs@NG$) was prepared via one-pot solvothermal method [20]. The resulting optimized composite has the potential of delivering high C_{sp} of 274 Fg^{-1} at 1 Ag^{-1} with brilliant rate capability. It further displayed outstanding long-term stability with capacitive retention of 75.3 % after 100, 000 cycles, which is the best reported stability till date for iron oxides in alkaline electrolytes. Such excellent characteristic was attributed to actively accessible redox sites along with facilitated electron and ionic diffusion throughout the composite.

3.1.2 Carbon QDs and TMD nanocomposites

Among QDs, TMDs have attracted considerable interest in past few years. Studies are rapidly advancing on GQDs with TMD heterostructures in the field of SSCs. They mollify the electrochemical requirement owing to their high electrical conductivity and specific capacity. To this end, WS_2 QDs having uniform size are developed and their pseudo-capacitance properties are examined without and with surface treatment with 1,2-ethanedithiol. The fabricated SSC attained specific capacitance of 457 Fg^{-1} and cycling stability (8000 cycles) at a coulombic efficiency of 81% [14]. In addition, Ghorai et al. [25] synthesized defect-rich QDs having uniform size from 1-3 nm by WS_2 nanosheets fragmentation and examined the charge storage performance of WS_2 QDs and nanosheets for assembled solid state SSCs. The assembled solid state SSC of WS_2 QD electrodes show a high areal C_{sp} (28 $mFcm^{-2}$) with high energy density (1.49 $\mu Whcm^{-2}$) at a current density of 0.1 $mAcm^{-2}$ with cycling stability 80% after 10,000 cycles. It is worth noting that their performance is limited by their low stability owing to destructions during electrochemical cycling. On this account, Huang et al. [76] integrated $NiCo_2S_4$ with GQDs prevent the structural variations by providing stable and strong conducting network, which also overpower their dissolution and aggregation. $GQDs/NiCo_2S_4$ have been prepared by hydrothermal method. It displayed larger C_{sp} as compared to their pristine counterparts which author attributes to GQDs inclusion and resulted unique structure.

3.1.3 Carbon QDs and polymer nanocomposites

Like pseudocapacitive TMOs, their polymeric hybridization also played vital role in improving the electrochemical performance. All solid state flexible SSC was fabricated by employing polymerization of pyrrole with CDs deposited on steel mesh. Jian et al. [31] observed the effect of including CDs on facilitation of polymerization process. Also, their fabricated electrodes illustrated high C_{sp} 308 Fg^{-1} in solid state assembly with high cycling stability. Their flexibility was also evidenced by their strain tolerating ability with no considerable change in performance when subjected to bending from 0 to 180°. CQDs were also incorporated into ferrous coordinated PPy which restrain volume alterations in PPy leading to improved cycling performance towards electrochemical evaluation [33].

In addition to PPy, CQDs were distributed on interior and surface of PANI nanowires and showed reasonably high C_{sp}, when utilized as SSC electrode [34]. Authors ascribed this improved performance to integration of CQDs which upgraded the conductivity and relieved the volumetric change in electrode during charging/discharging process.

3.1.3 Carbon QDs based ternary nanocomposites

CQDs owing to their unique properties can enhance the bonding strength between multiple materials by combining well with inorganic and organic materials. Earlier studies indicated that CQDs can be used as conducting bridge within composite, offering excellent electrochemical activeness towards high C_{sp} and cycle stability. Not only in binary composite, they showed even better performance in their ternary composites. Zhang et al. [30] used CDs to connect GO with PPy, and resulted in improved electron and ionic transportation leading to their potential application in SSC. The active material (GO/CDs/PPy) was synthesized by in-situ polymerization of pyrrole with GO/CDs composite. Their sandwich structure involved CDs in between GO and PPy promoting electron transportation in the composite with consequent declining internal resistance. In addition, it also augmented the interfacial interaction among the constituents which increase the dielectric constant of the resulting composite. Markedly, the composite offers C_{sp} of 576 Fg^{-1} at 0.5 Ag^{-1}, along with high energy density of 30.1 Wh kg^{-1} at power density of 250 W kg^{-1}. The fabricated SSC displayed superior cycling stability. Authors also demonstrated their practical application by powering 59 light–emitting diode indicators over 60s by assembling five such SSCs in series. Recently, they acted as the communicating bridge for linking MnO_2 with graphene, resulting in MnO_2/CQDs/graphene aerogel (MnO_2/CQDs/GA) with three-dimensional (3-D) net like structure [77]. Their unique structure offered large SSA with abundant electron transport pathways, indicating excellent electrochemical performance. The SSC fabricated with this electrode displayed high C_{sp} 721 Fg^{-1} at 1 Ag^{-1} and notable long cycling stability of 92.3% over 10,000 cycles.

In one work, a new architecture was developed by De et al. [78] constituting carbon hollow-structured nanospheres (CHNS) with core–shell copper sulfide@CQDs. Resulting composite, CuS@CQDs@CHNS was achieved by in situ hydrothermal processing and employed as SSC electrode. Excitingly, the electrochemical characteristics of high C_{sp} (618 Fg^{-1} at a current density of 1 Ag^{-1}) and an outstanding rate capability was accomplished with extended cycle life (95% capacitance retention after 4000 cycles). In another work [79], CDs act as efficient dispersing and reducing agent for graphene and $KMnO_4$ to synthesize (MnO_x)–graphene hybrid for SSC electrode fabrication. They are initially obtained from the pyrolysis of ammonium citrate with superior dispersibility in water owing to the presence of oxygen- and nitrogen-containing moieties. Additionally, it showed strong interaction via π–π stacking graphene and resulted in water-soluble CD/graphene nanocomposites (CDGs). Subsequently, MnO_x were grown on CDGs after reacting with $KMnO_4$ by sacrificial oxidation of CDs. The resulted composite was then employed in flexible solid-state SSC and exhibited good capacitive activity with very superior GCD cyclic stability (>10 000 cycles).

Further, GQDs have also been utilized to fabricate highly conductive nanofiber by utilizing PEDOT on polyvinyl alcohol-GQD-cobalt oxide $(PVA-GQD-Co_3O_4)$ composite for SSC [80]. The resulting cauliflower-like structure showed the uniform coating of PEDOT $PVA-GQD-Co_3O_4$ nanofibers. Its fabricated SSC displayed high charge storage capacity with low equivalent series resistance (ESR). It also possessed high specific energy and power with excellent rate capability.

3.2   QDs for asymmetric supercapacitors

3.2.1   Carbon QDs and TMO nanocomposites

MnO_2 membrane was prepared by synthesizing ultralong MnO_2 nanowires with CQDs [81]. CQDs encouraged the formation of 1-D MnO_2 with augmented wettability between electrode and electrolyte. It revealed C_{sp} of 340 Fg^{-1} at 1 Ag^{-1} with 76% capacitive retention. Furthermore, their ASSCs with AC displayed a high energy density of 33.6 Wh kg^{-1} at power density of 1.0 kW kg^{-1} with excellent durability under consecutive GCD cycling. 3-D porous hierarchical CQDs were hybridized with M_xO_y (M= Co, Ni) by thermolysis of corresponding $CQDs/M(OH)_y$ [82]. CQDs tuned morphologies of the $M(OH)_y$ in the resulting composites. The porous hierarchical $CQDs/Co_3O_4$ architectures displayed high C_{sp} 1603 Fg^{-1} at 1 Ag^{-1} with excellent cycling durability (97.0% capacity retention over 2000 cycles). In addition, fabricated ASSC $(CQDs/Co_3O_4//AC)$ displayed a high C_{sp} of 210.4 Fg^{-1} with excellent energy density 74.8 Wh kg^{-1}, indicating their strong potential during electrochemical analysis.

In another work, CQDs composited with porous $NiCo_2O_4$ sphere was synthesized by reflux synthesis route and utilized as an electrode for ASSC [83]. Resulted composite, $CQDs/NiCo_2O_4$, owing to their unique properties, exhibits high C_{sp} (856 Fg^{-1} at 1 Ag^{-1}) and excellent rate capability. Their cycling durability is further evidenced by ~99% capacitive retention after 10,000 cycles at 5 Ag^{-1}. Their asymmetric assembly

AC//CQDs/NiCo$_2$O$_4$ demonstrates a high energy density (27.8 Wh kg^{-1}) at a power density of 128 W kg^{-1}, demonstrating potential of the CQDs/NiCo$_2$O$_4$porous composites in practical applications. In addition, tremella-like NiCo$_2$O$_4$ with GQDs as prepared where NiCo$_2$O$_4$ is efficiently encapsulated by GQDs [84]. In the composite, GQDs have π-conjugated core with abundant edge sites making it favorable for SC application. An excellent C$_{sp}$ of 1242 Fg^{-1} at very high current density ~30 Ag^{-1} was achieved, larger than that of bare pseudocapacitive NiCo$_2$O$_4$. Their 99% capacitive retention after 4000 cycles made them highly favorable towards electrochemical evaluation. Besides, their asymmetric assembly with AC (NiCo$_2$O$_4$@GQDs//AC), offer high energy density 38 Wh kg^{-1} at power density of 800 W kg^{-1}. Furthermore, Fe$_2$O$_3$QDs were also decorated on functionalized graphene sh0eets and utilized in fabricating ASSCs in 1M Na$_2$SO$_4$ aqueous electrolyte. Fe$_2$O$_3$/FGS//MnO$_2$/FGS ASSC showed high energy density of 50.7 Wh kg^{-1} at a power density of 100 W kg^{-1} in 2 V potential window as well as excellent cycling stability and power capability [19].

3.2.2 Carbon QDs and polymer nanocomposites

Asymmetric assembly of GQDs and PANI nanofibers were employed to fabricate micro-SC on interdigital finger gold electrodes [85]. The electrochemical measurements displayed excellent rate capability with faster power response capability in aqueous electrolyte. Moreover, their all-solid-state assembly also showed encouraging results i.e. C$_{sp}$ 210 μF cm^{-2} at 15 μA cm^{-2} along with long term cycling stability ~97% retention after 1500 cycles. Authors ascribed these excellent capabilities to synergistic effects of GQDs and PANI providing full accessibility for facilitated charge transfer through the active material.

Conclusion and outlook

Quantum dots (QDs) possess several unique properties which arise from their nanoscale size. Consequently, QDs could be used in future optoelectronic, imaging or energy storage applications. In order to utilize the QDs in various fields, it is necessary to find low-cost, and simple mass production method with high yield uniform products. Uniform properties are generated from uniform sized and functional group distributed materials. Recent advances in the growth of QDs have demonstrated potential for their exploration for high performance supercapacitors (SCs). QDs based SCs possess ultrafast charge/discharge rates and relatively large cycling life at an elevated power density, positioning itself between conventional capacitors and batteries in term of energy and power density. Subsequently, QDs based SCs can work as standalone units in satisfying the continually rising demands for safe energy. Despite the rapid growth in the field of

QDs based SCs, numerous challenges are essentially to be fixed before QDs can comprehend extensive adoption. Although few facile synthetic methods are advanced for the growth of CDs, TMO QDs, TMD QDs and their nanocomposites, research related to GQDs and their nanocomposites with other form of QDs is still at a more nascent stage. For example, research related to development of an atom-precise and uniform structures have not yet been achieved, consequently the relationships between structure and properties are yet to be precisely understood. Furthermore, more studies should emphasize to address the functionalization of QDs with various functional groups and their hybridization with other materials, which can essentially enhance their active role SCs.

In addition, although promising strategies are developed with elaborate designs of micro-device structures for practical applications, further study is needed to innovate and fabricate novel designs and configuration strategies. Despite these challenges, we are confident that advance research can effectively unravel the problems and the future will be magnificent for QDs based devices.

References

[1] M. F. El-Kady, Y. Shao, R. B. Kaner, Graphene for batteries, supercapacitors and beyond, Nat. Rev. Mater. 1 (2016) 16033. https://doi.org/10.1038/natrevmats.2016.33

[2] M. Salanne, B. Rotenberg, K. Naoi, K. Kaneko, P.-L. Taberna, C. P. Grey, B. Dunn, P. Simon, Efficient storage mechanisms for building better supercapacitors, Nat Energy 1 (2016) 16070. https://doi.org/10.1038/nenergy.2016.70

[3] H. Wang, Y. Yang, L. Guo, Nature-inspired electrochemical energy-storage materials and devices, Adv. Energy Mater. 7 (2017) 1601709. https://doi.org/10.1002/aenm.201601709

[4] C. Zhong, Y. Deng, W. Hu, J. Qiao, L. Zhang, J. Zhang, A review of electrolyte materials and compositions for electrochemical supercapacitors, Chem. Soc. Rev. 44 (2015) 7484–7539. https://doi.org/10.1039/C5CS00303B

[5] P. Ahuja, S.K. Ujjain, Graphene-Based Materials for Flexible Supercapacitors, in: Inamuddin, B. Satyanarayan, A. M. Asiri, (Eds.), Self-standing Substrates, Springer International Publishing, Switzerland AG, 2020, pp. 297-326

[6] M. Semeniuk, Z. Yi, V. Poursorkhabi, J. Tjong, S. Jaffer, Z.H. Lu, M. Sain, Future perspectives and review on organic carbon dots in electronic applications, ACS Nano 13 (2019) 6224−6255. https://doi.org/10.1021/acsnano.9b00688

[7] P. Ahuja, S.K. Ujjain, R. Kanojia, MnO$_x$/C nanocomposite: an insight on high-performance supercapacitor and non-enzymatic hydrogen peroxide detection, Appl. Surf. Sci. 404 (2017) 197-205. https://doi.org/10.1016/j.apsusc.2017.01.300

[8] V. Ganesh, S. Pitchumani, V. Lakshminarayanan, New symmetric and asymmetric supercapacitors based on high surface area porous nickel and activated carbon, J. Power Sources 158 (2006) 1523-1532. https://doi.org/10.1016/j.jpowsour.2005.10.090

[9] S. Bak, D. Kim, H. Lee, Graphene quantum dots and their possible energy applications: A review, Curr. Appl. Phys. 16 (2016) 1192-1201. https://doi.org/10.1016/j.cap.2016.03.026

[10] Y.-Y. Song, Z.-D. Gao, J.-H. Wang, X.-H. Xia, R. Lynch, Multistage coloring electrochromic device based on TiO$_2$ nanotube arrays modified with WO$_3$ nanoparticles, Adv. Funct. Mater. 21 (2011) 1941- 1946. https://doi.org/10.1002/adfm.201002258

[11] M. R. J. Scherer, L. Li, P. M. S. Cunha, O. A. Scherman, U. Steiner, Enhanced electrochromism in gyroid-structured vanadium pentoxide, Adv. Mater. 24 (2012) 1217- 1221. https://doi.org/10.1002/adma.201104272

[12] H. S. Choi, W. Liu, P. Misra, E. Tanaka, J. P. Zimmer, B. Itty Ipe, M. G. Bawendi, J. V. Frangioni, Renal clearance of quantum dots, Nat. Biotechnol. 25 (2007) 1165-1170. https://doi.org/10.1038/nbt1340

[13] S. Cong, Y. Tian, Q. Li, Z. Zhao, F. Geng, Single-crystalline tungsten oxide quantum dots for fast pseudocapacitor and electrochromic applications, Adv. Mater. 26 (2014) 4260-4267. https://doi.org/10.1002/adma.201400447

[14] W. Yin, D. He, X. Bai, W. W. Yu, Synthesis of tungsten disulfide quantum dots for high-performance supercapacitor electrodes, J. Alloys Compd. 786 (2019) 764-769. https://doi.org/10.1016/j.jallcom.2019.02.030

[15] M. Chhowalla, H. S. Shin, G. Eda, L.-J. Li, K. P. Loh, H. Zhang, The chemistry of two-dimensional layered transition metal dichalcogenide nanosheets, Nat. Chem. 5 (2013) 263-275. https://doi.org/10.1038/nchem.1589

[16] L. Lin, Y. Xu, S. Zhang, I. M. Ross, A. C. M. Ong, D. A. Allwood, Fabrication of luminescent monolayered tungsten dichalcogenides quantum dots with giant spin-valley coupling, ACS Nano 7 (2013) 8214-8223. https://doi.org/10.1021/nn403682r

[17] M. Jing, C. Wang, H. Hou, Z. Wu, Y. Zhu, Y. Yang, X. Jia, Y. Zhang, X. Ji, Ultrafine nickel oxide quantum dots embedded with few-layer exfoliative graphene for

an asymmetric supercapacitor: Enhanced capacitances by alternating voltage, J. Power Sources 298 (2015) 241-248. https://doi.org/10.1016/j.jpowsour.2015.08.039

[18] Y. Li, H. Zhang, S. Wang, Y. Lin, Y. Chen, Z. Shi, N. Li, W. Wang, Z. Guo, Facile low-temperature synthesis of hematite quantum dots anchored on three-dimensional ultra-porous graphene-like framework as advanced anode materials for asymmetric supercapacitors, J. Mater. Chem. A 4 (2016) 11247-11255. https://doi.org/10.1039/C6TA02927B

[19] H. Xia, C. Hong, B. Li, B. Zhao, Z. Lin, M. Zheng, S. V. Savilov, S. M. Aldoshin, Facile synthesis of hematite quantum-dot/functionalized graphene-sheet composites as advanced anode materials for asymmetric supercapacitors, Adv. Funct. Mater. 25 (2014) 627-635. https://doi.org/10.1002/adfm.201403554

[20] L. Liu, J. Lang, P. Zhang, B. Hu, X. Yan, Facile synthesis of Fe_2O_3 nano-dots@nitrogen-doped graphene for supercapacitor electrode with ultralong cycle life in KOH electrolyte, ACS Appl. Mater. Interfaces 8 (2016) 9335−9344. https://doi.org/10.1021/acsami.6b00225

[21] S. Liu, J. Zhou, Z. Cai, G. Fang, Y. Cai, A. Pan, S. Liang, Nb_2O_5 Quantum dots embedded in MOF derived nitrogen-doped porous carbon for advanced hybrid supercapacitors applications, J. Mater. Chem. A 4 (2016) 17838-17847. https://doi.org/10.1039/C6TA07856G

[22] V. Bonu, B. Gupta, S. Chandra, A. Das, S. Dhara, A.K. Tyagi, Electrochemical supercapacitor performance of SnO_2 quantum dots, Electrochim. Acta 203 (2016) 230−237. https://doi.org/10.1016/j.electacta.2016.03.153

[23] Y. Wu, F. Ran, Vanadium nitride quantum dot/nitrogen-doped microporous carbon nanofibers electrode for high-performance supercapacitors, J. Power Sources 344 (2017) 1-10. https://doi.org/10.1016/j.jpowsour.2017.01.095

[24] J. Wang, W. Dou, X. Zhang, W. Han, X. Mu, Y. Zhang, X. Zhao, Y. Chen, Z. Yang, Q. Su, E. Xie, W. Lan., X. Wang, Embedded Ag quantum dots into interconnected Co_3O_4 nanosheets grown on 3D graphene networks for high stable and flexible supercapacitors, Electrochim. Acta 224 (2017) 260−268. https://doi.org/10.1016/j.electacta.2016.12.073

[25] A. Ghorai, A. Midya, S. K. Ray, Superior charge storage performance of WS_2 quantum dots in a flexible solid state supercapacitor, New J. Chem. 42 (2018) 3609-3613. https://doi.org/10.1039/C7NJ03869K

[26] X. Feng, J. Wu, M. Ai, W. Pisula, L. Zhi, J. P. Rabe, K. Müllen, Triangle-shaped polycyclic aromatic hydrocarbons, Angew. Chem. 119 (2007) 3093−3096. https://doi.org/10.1002/ange.200605224

[27] X. Yan, X. Cui, L. Li, Synthesis of large, stable colloidal graphene quantum dots with tunable size, J. Am. Chem. Soc. 132 (2010) 5944−5945. https://doi.org/10.1021/ja1009376

[28] S. Qu, X. Wang, Q. Lu, X. Liu, L. Wang, A biocompatible fluorescent ink based on water-soluble luminescent carbon nanodots, Angew. Chem. 124 (2012) 12381−12384. https://doi.org/10.1002/ange.201206791

[29] D. Qu, M. Zheng, P. Du, Y. Zhou, L. Zhang, D. Li, H. Tan, Z. Zhao, Z. Xie, Z. Sun, Highly luminescent S, N Co-doped graphene quantum dots with broad visible absorption bands for visible light photocatalysts, Nanoscale 5 (2013) 12272−12277. https://doi.org/10.1039/C3NR04402E

[30] X. Zhang, J. Wang, J. Liu, J. Wu, H. Chen, H. Bi, Design and preparation of a ternary composite of graphene oxide/carbon dots/polypyrrole for supercapacitor application: Importance and unique role of carbon dots, Carbon 115 (2017) 134-146. https://doi.org/10.1016/j.carbon.2017.01.005

[31] X. Jian, H.-M. Yang, J.-G. Li, E.-H. Zhang, L.-L. Cao, Z.-H. Liang, Flexible all-solid-state high-performance supercapacitor based on electrochemically synthesized carbon quantum dots/polypyrrole composite electrode, Electrochim. Acta 228 (2017) 483–493. https://doi.org/10.1016/j.electacta.2017.01.082

[32] Y. Xie, H. Du, Electrochemical capacitance of a carbon quantum dots–polypyrrole/titania nanotube hybrid, RSC Adv. 5 (2015) 89689-89697. https://doi.org/10.1039/C5RA16538E

[33] X. Jian, J.-G. Li, H.-M. Yang, L. –L. Cao, E. –H. Zhang, Z.-H. Liang, Carbon quantum dots reinforced polypyrrole nanowire via electrostatic self-assembly strategy for high-performance supercapacitors, Carbon 114 (2017) 533-543. https://doi.org/10.1016/j.carbon.2016.12.033

[34] Z. Zhao, Y. Xie, Enhanced electrochemical performance of carbon quantum dots-polyaniline hybrid, J. Power Sources 337 (2017) 54-64. https://doi.org/10.1016/j.jpowsour.2016.10.110

[35] Y. Zhou, Y. Xie, Enhanced electrochemical stability of carbon quantum dots-incorporated and ferrous-coordinated polypyrrole for supercapacitor, J. Solid State Electr. 22 (2018) 2515–2529. https://doi.org/10.1007/s10008-018-3964-5

[36] S. N. Baker, G. A. Baker, Luminescent carbon nanodots: Emergent nanolights. Angew. Chem., Int. Ed. 49 (2010) 6726−6744. https://doi.org/10.1002/anie.200906623

[37] K. A. S. Fernando, S. Sahu, Y. Liu, W. K. Lewis, E. A. Guliants, A. Jafariyan, P. Wang, C. E. Bunker, Y. Sun, Carbon quantum dots and applications in photocatalytic energy conversion, ACS Appl. Mater. Interfaces 7 (2015) 8363−8376. https://doi.org/10.1021/acsami.5b00448

[38] L. Wang, X. Chen, Y. Lu, C. Liu, W. Yang, Carbon quantum dots displaying dual-wavelength photoluminescence and electrochemiluminescence prepared by high-energy ball milling, Carbon 94 (2015) 472−478. https://doi.org/10.1016/j.carbon.2015.06.084

[39] H. Hou, C. E. Banks, M. Jing, Y. Zhang, X. Ji, Carbon quantum dots and their derivative 3D porous carbon frameworks for sodium-ion batteries with ultralong cycle life, Adv. Mater. 27 (2015) 7861−7866. https://doi.org/10.1002/adma.201503816

[40] H. Li, Z. Kang, Y. Liu, S. Lee, Carbon nanodots: Synthesis, properties and applications, J. Mater. Chem. 22 (2012) 24230−24253. https://doi.org/10.1039/C2JM34690G

[41] Y. Xu, J. Liu, C. Gao, E. Wang, Applications of carbon quantum dots in electrochemiluminescence: A mini review, Electrochem. Commun. 48 (2014) 151−154. https://doi.org/10.1016/j.elecom.2014.08.032

[42] H. Sun, L. Wu, W. Wei, X. Qu, Recent advances in graphene quantum dots for sensing, Mater. Today 16 (2013) 433−442. https://doi.org/10.1016/j.mattod.2013.10.020

[43] L. Ponomarenko, F. Schedin, M. Katsnelson, R. Yang, E. Hill, K. Novoselov, A. Geim, Chaotic Dirac billiard in graphene quantum dots, Science 320 (2008) 356−358. https://doi.org/10.1126/science.1154663

[44] S. Zhu, Y. Song, J. Wang, H. Wan, Y. Zhang, Y. Ning, B. Yang, Photoluminescence mechanism in graphene quantum dots: quantum confinement effect and surface/edge state, Nano Today 13 (2017) 10−14. https://doi.org/10.1016/j.nantod.2016.12.006

[45] X. Li, M. Rui, J. Song, Z. Shen, H. Zeng, Carbon and graphene quantum dots for optoelectronic and energy devices: A Review, Adv. Funct. Mater. 25 (2015) 4929−4947. https://doi.org/10.1002/adfm.201501250

[46] N. G.- Bretesche, O. Crosnier, G. Buvat, F. Favier, T. Brousse, Electrochemical study of aqueous asymmetric $FeWO_4/MnO_2$ supercapacitor, J. Power Sources, 326 (2016) 695-701. https://doi.org/10.1016/j.jpowsour.2016.04.075

[47] P. Ahuja, V. Sahu, S. K. Ujjain, R. K. Sharma, G. Singh, Performance evaluation of asymmetric supercapacitor based on cobalt manganite modified graphene nanoribbons, Electrochim. Acta 146 (2014) 429-436. https://doi.org/10.1016/j.electacta.2014.09.039

[48] S. K. Ujjain, P. Ahuja, R. Bhatia, P. Attri, Printable multi-walled carbon nanotubes thin film for high performance all solid state flexible supercapacitors, Mater. Res. Bull. 83 (2016) 167-171. https://doi.org/10.1016/j.materresbull.2016.06.006

[49] S. K. Ujjain, R. Bhatia, P. Ahuja, P. Attri, Highly conductive aromatic functionalized multi-walled carbon nanotube for inkjet printable high performance supercapacitor electrodes, PloS one 10 (2015), e0131475. https://doi.org/10.1371/journal.pone.0131475

[50] P. Ahuja, S. K. Ujjain, R. K. Sharma, G. Singh, Enhanced supercapacitor performance by incorporating nickel in manganese oxide, RSC Adv. 4 (2014) 57192-57199. https://doi.org/10.1039/C4RA09027F

[51] S. K. Ujjain, P. Ahuja, R. K. Sharma, Graphene nanoribbon wrapped cobalt manganite nanocubes for high performance all-solid-state flexible supercapacitors, J. Mater. Chem. A 3 (2015) 9925-9931. https://doi.org/10.1039/C5TA00653H

[52] S. K. Ujjain, V. Sahu, R. K. Sharma, G. Singh, High performance, all solid state, flexible supercapacitor based on ionic liquid functionalized graphene, Electrochim. Acta 157 (2015) 245-251. https://doi.org/10.1016/j.electacta.2015.01.061

[53] K. Deori, S. K. Ujjain, R. K. Sharma, S. Deka, Morphology controlled synthesis of nanoporous Co_3O_4 nanostructures and their charge storage characteristics in supercapacitors, ACS Appl. Mater. Interfaces 5 (2013) 10665-10672. https://doi.org/10.1021/am4027482

[54] S. K. Ujjain, G. Singh, R. K. Sharma, Co_3O_4@ reduced graphene oxide nanoribbon for high performance asymmetric supercapacitor, Electrochim. Acta 169 (2015) 276-282. https://doi.org/10.1016/j.electacta.2015.03.141

[55] P. Ahuja, S. K. Ujjain, R. Kanojia, Electrochemical behaviour of manganese & ruthenium mixed oxide@ reduced graphene oxide nanoribbon composite in symmetric and asymmetric supercapacitor, Appl. Surf. Sci. 427 (2018) 102-111. https://doi.org/10.1016/j.apsusc.2017.08.028

[56] A. Manikandan, Y. -Z. Chen, C. -C. Shen, C. -W. Shen, H. -C. Kuo, Y. -L. Chueh, A critical review on two-dimensional quantum dots (2D QDs): From synthesis toward applications in energy and optoelectronics, Prog. Quant. Electron. 68 (2019) 100226. https://doi.org/10.1016/j.pquantelec.2019.100226

[57] A. Borenstein, O. Hanna, R. Attias, S. Luski, T. Broussi, D. Aurbach, Carbon-based composite materials for supercapacitor electrodes: a review, J. Mater. Chem A. 5 (2017) 12653-12672. https://doi.org/10.1039/C7TA00863E

[58] L. Lv, Y. Fan, Q. Chen, Y. Zhao, Y. Hu, Z. Zhang, N. Chen, L. Qu, Three-dimensional multichannel aerogel of carbon quantum dots for high-performance supercapacitors, Nanotechnology 25 (2014) 235401. https://doi.org/10.1088/0957-4484/25/23/235401

[59] M. Xu, Q. Huang, R. Sun, X. Wang, Simultaneously obtaining fluorescent carbon dots and porous active carbon for supercapacitors from biomass, RSC Adv. 6 (2016) 88674–88682. https://doi.org/10.1039/C6RA18725K

[60] Y. Dong, H. Pang, H. Bin Yang, C. Guo, J. Shao, Y. Chi, C. M. Li, T. Yu, Carbon-based dots co-doped with Nitrogen and Sulfur for high quantum yield and excitation-independent emission, Angew. Chem. Int. Ed. 52 (2013) 7800-7804. https://doi.org/10.1002/anie.201301114

[61] V. B. Kumar, A. Borenstein, B. Markovsky, D. Aurbach, A. Gedanken, M. Talianker, Z. Porat, Activated carbon modified with carbon nanodots as novel electrode material for supercapacitors, J. Phys. Chem. C 120 (2016) 13406-13413. https://doi.org/10.1021/acs.jpcc.6b04045

[62] V. C. Hoang, L. H. Nguyen, V. G. Gomes, High efficiency supercapacitor derived from biomass based carbon dots and reduced graphene oxide composite, J. Electroanal. Chem. 832 (2019) 87–96. https://doi.org/10.1016/j.jelechem.2018.10.050

[63] V. C. Hoang, V. G. Gomes, High performance hybrid supercapacitor based on doped zucchini-derived carbon dots and graphene, Mater. Today Energy 12 (2019) 198–207. https://doi.org/10.1016/j.mtener.2019.01.013

[64] Y. Hu, Y. Zhao, G. Lu, N. Chen, Z. Zhang, H. Li, H. Shao, L. Qu, Graphene quantum dots–carbon nanotube hybrid arrays for supercapacitors, Nanotechnology 24 (2013) 195401. https://doi.org/10.1088/0957-4484/24/19/195401

[65] J. Huang, B. G. Sumpter, V. Meunier, Theoretical model for nanoporous carbon supercapacitors, Angew. Chem. Int. Ed. 47 (2008) 520-534. https://doi.org/10.1002/ange.200703864

[66] Q. Chen, Y. Hu, C. Hu, H. Cheng, Z. Zhang, H. Shao, L. Qu, Graphene quantum dots–three-dimensional graphene composites for high-performance supercapacitors, Phys. Chem. Chem. Phys. 16 (2014) 19307-19313. https://doi.org/10.1039/C4CP02761B

[67] W. -W. Liu, Y. -Q. Feng, X. -B. Yan, J. -T. Chen, Q. -J. Xue, Superior micro-supercapacitors based on graphene quantum dots, Adv. Funct. Mater. 23 (2013) 4111-4122. https://doi.org/10.1002/adfm.201203771

[68] K. Bhattacharya, P. Deb, Hybrid nanostructured C-Dot decorated Fe_3O_4 electrode materials for superior electrochemical energy storage performance, Dalton Trans. 44 (2015) 9221−9229. https://doi.org/10.1039/C5DT00296F

[69] Y. Zhu, X. Ji, C. Pan, Q. Sun, W. Song, L. Fang, Q. Chen, C. E. Banks, A carbon quantum dot decorated RuO_2 network: outstanding supercapacitances under ultrafast charge and discharge, Energy Environ. Sci. 6 (2013) 3665−3675. https://doi.org/10.1039/C3EE41776J

[70] G. Wei, X. Xu, J. Liu, K. Du, J. Du, S. Zhang, C. An, J. Zhang, Z. Wang, Carbon quantum dots decorated hierarchical $Ni(OH)_2$ with lamellar structure for outstanding supercapacitor, Mater. Lett. 186 (2017) 131−134. https://doi.org/10.1016/j.matlet.2016.09.126

[71] J. Xu, Y. Xue, J. Cao, G. Wang, Y. Li, W. Wang, Z. Chen, Carbon quantum dots/nickel oxide (CQDs/NiO) nanorods with high capacitance for supercapacitors, RSC Adv. 6 (2016) 5541-5546. https://doi.org/10.1039/C5RA24192H

[72] R. Narayanan, Single step hydrothermal synthesis of carbon nanodot decorated V_2O_5 nanobelts as hybrid conducting material for supercapacitor application, J. Solid State Chem. 253 (2017) 103-112. https://doi.org/10.1016/j.jssc.2017.05.035

[73] J. Wei, H. Ding, P. Zhang, Y. Song, J. Chen, Y. Wang, H.-M. Xiong, Carbon Dots/$NiCo_2O_4$ nanocomposites with various morphologies for high performance supercapacitors, Small 12 (2016) 5927−5934. https://doi.org/10.1002/smll.201602164

[74] Y. Wei, X. Zhang, X. Wu, D. Tang, K. Cai, Q. Zhang, Carbon quantum dots/Ni–Al layered double hydroxide composite for high-performance supercapacitors, RSC Adv. 6 (2016) 39317-39322. https://doi.org/10.1039/C6RA02730J

[75] J. Liu, M. Zheng, X. Shi, H. Zeng, H. Xia, Amorphous FeOOH quantum dots assembled mesoporous film anchored on graphene nanosheets with superior electrochemical performance for supercapacitors, Adv. Funct. Mater. 26 (2016) 919-930. https://doi.org/10.1002/adfm.201504019

[76] Y. Huang, T. Shi, Y. Zhong, S. Cheng, S. Jiang, C. Chen, G. Liao, Z. Tang, Graphene-quantum-dots induced $NiCo_2S_4$ with hierarchical-like hollow nanostructure for supercapacitors with enhanced electrochemical performance, Electrochim. Acta 269 (2018) 45-54. https://doi.org/10.1016/j.electacta.2018.02.145

[77] H. Lv, Y. Yuan, Q. Xu, H. Liu, Y.-G. Wang, Y. Xi, Carbon quantum dots anchoring MnO_2/graphene aerogel exhibits excellent performance as electrode materials for supercapacitor, J. Power Sources 398 (2018) 167-174.
https://doi.org/10.1016/j.jpowsour.2018.07.059

[78] B. De, J. Balamurugan, N. H. Kim, J. H. Lee, Enhanced electrochemical and photocatalytic performance of core–shell CuS@carbon quantum dots@carbon hollow nanospheres, ACS Appl. Mater. Interfaces 9 (2017) 2459-2468.
https://doi.org/10.1021/acsami.6b13496

[79] B. Unnikrishnan, C.-W. Wu, I.-W. P. Chen, H.-T. Chang, C.-H. Lin, C.-C. Huang, Carbon dot-mediated synthesis of manganese oxide decorated graphene nanosheets for supercapacitor application, ACS Sustainable Chem. Eng. 4 (2016) 3008-3016.
https://doi.org/10.1021/acssuschemeng.5b01700

[80] S. N. J. S. Z. Abidin, Md. S. Mamat, S. A. Rasyid, Z. Zainal, Y. Sulaiman, Electropolymerization of poly(3,4-ethylenedioxythiophene) onto polyvinyl alcohol-graphene quantum dot-cobalt oxide nanofiber composite for high-performance supercapacitor, Electrochim. Acta, 261 (2018) 548-556.
https://doi.org/10.1016/j.electacta.2017.12.168

[81] H. Lv, X. Gao, Q. Xu, H. Liu, Y.-G. Wang, Y. Xia, Carbon quantum dot-induced MnO_2 nanowire formation and construction of a binder-free flexible membrane with excellent superhydrophilicity and enhanced supercapacitor performance, ACS Appl. Mater. Interfaces 9 (2017) 40394-40403. https://doi.org/10.1021/acsami.7b14761

[82] G. Wei, X. Zhao, K. Du, Z. Wang, M. Liu, S. Zhang, S. Wang, J. Zhang, C. An, A general approach to 3D porous $CQDs/M_xO_y$ (M = Co, Ni) for remarkable performance hybrid supercapacitors, Chem. Eng. J. 326 (2017) 8-67.
https://doi.org/10.1016/j.cej.2017.05.127

[83] Y. Zhu, Z. Wu, M. Jing, H. Hou, Y. Yang, Y. Zhang, X. Yang, W. Song, X. Jia, X. Ji, Porous $NiCo_2O_4$ spheres tuned through carbon quantum dots utilised as advanced materials for an asymmetric supercapacitor, J. Mater. Chem. A 3 (2015) 866-877.
https://doi.org/10.1039/C4TA05507A

[84] J. Luo, J. Wang, S. Liu, W. Wu, T. Jia, Z. Yang, S. Mu, Y. Huang, Graphene quantum dots encapsulated tremella-like $NiCo_2O_4$ for advanced asymmetric supercapacitors, Carbon 146 (2019) 1-8. https://doi.org/10.1016/j.carbon.2019.01.078

[85] W. Liu, X. Yan, J. Chen, Y. Feng, Q. Xue, Novel and high-performance asymmetric micro-supercapacitors based on graphene quantum dots and polyaniline nanofibers, Nanoscale 5 (2013) 6053-6062. https://doi.org/10.1039/C3NR01139A

Quantum Dots – Properties and Applications
Materials Research Foundations 96 (2021) 191-215

Materials Research Forum LLC
https://doi.org/10.21741/9781644901250-8

Chapter 8

Quantum Dots Based Material for Drug Delivery Applications

Himani Tiwari, Neha Karki, Monika Matiyani, Gaurav Tatrari, Anand Ballabh Melkani,
Nanda Gopal Sahoo*

Prof. Rajendra Singh Nanoscience and Nanotechnology Centre, Department of Chemistry, DSB
Campus, Kumaun University, Nainital, Uttarakhand, India

* ngsahoo@yahoo.co.in

Abstract

Quantum dots (QDs) have shown promising potential to many biomedical and biological
applications, mainly in drug delivery or activation and cellular imaging. These
semiconductor nanoparticles, QDs, whose particle size is in the range of 2-10 nanometer
with unique photo-chemical and -physical properties that are not possessed by any other
isolated molecules, have become one of the distinct class of imaging probes and
worldwide platforms for manufacturing of multifunctional nanodevices. In this chapter,
properties, applications of QDs, and importance in the biomedical field especially in drug
delivery is presented.

Keywords

Quantum Dots, Nanoparticles, Toxicity, Drug Delivery, Imaging

Contents

1.   Introduction

In recent years, nanoscience and nanotechnology came into the limelight for the development of novel and fascinating nanostructures. These nanostructures have been attained widespread attention due to their exceptional physical, chemical and mechanical properties compared to macro- and micro-structured materials. These nanostructures such as graphene, fullerenes, carbon nanotubes (CNT), nanoshells, dendrimers, quantum dots (Qdots or QDs), superparamagnetic nanoparticles, Au, and Ag nanoparticles exhibit interesting and potential applications in the field of biomedical, electronics, sensors, nanocomposites etc. [1].

The gap between the macro and micro levels can be bridged by nanostructure material and helms to exclusively vernal paths for applications, particularly in the field of optoelectronics and biology. In past few years, nanotechnology has alluded towards the field of medicine and has shown great potential with uses in diagnostic and therapeutic agents for cancer cell focusing and imaging [2-4]. Particle size less than 10^2 nm may be a nanostructure material and can be classified as thin sheets or quantum wells (two dimension), quantum chains (dimension.) and dots (zero dimension) [5]. Fig. 1 [6] shows the optical properties of cadmium-selenium (CdSe) QDs at four different sizes (2.2, 2.9, 4.1, and 7.3 in nm) and are semiconductor nanocrystals with size of about 2 to 100 nm with unique electrical and optical properties presently applied in biomedical imaging and many other electronic industries.

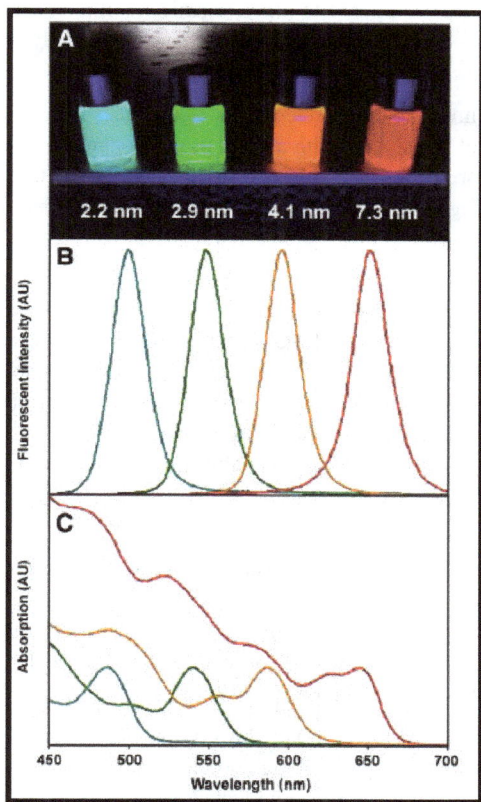

Fig.1. Size-dependent optical properties of cadmium selenide QDs (A) Fluorescence image of four vials of monodisperse QDs with sizes (diameter) ranging about 2.2-7.3 nm, (B) Fluorescence spectra of the same four QD samples, and (C) Absorption spectra of the same four QD samples. Adapted with permission from Smith et al. [6]

The properties of quantum dots are depended on several factors, such as shape, size, defect and impurities. On changing the size, quantum dots display various color of emission due to the change in the surface-to-volume ratio and quantum confinement effect.

They are more preferable compared to other organic fluorophores because QDs show a broad series of size wise tuned colors and at the same time only with a single laser, a bundle of colored dots can be activated accordingly [7, 8].

Quantum dots offer invaluable societal profits like cancer cell targeting and imaging. They may also harmful to human health as well as environment in definite circumstances [9]. For instance, fluorescent QDs may also be bind with bioactive molecules to target specific biologic strategies and cellular structures, such as labeling neoplastic cells and cell membrane receptors.

In this new era of nanotechnology, carbon nanodots (C-dots) have become a promising photo-luminescent (PL) nanomaterial in a variety of applications [10]. In organism, industry and society, carbon is one of the essential elements and always occupies a vital position in the development of modern science and technology. Carbon-based nanomaterials such as graphene QDS (GQDs) are novel type of nanomaterials. Due to their unique optical properties, these particles are considered as the future of new green nanomaterials with vast capacity as bio-compatible imaging probes [11, 12].

It is very well known that in the treatment of many diseases alternative side effects also occur because of indiscriminate allocation of relative therapeutics to the diseased as well as healthy cells. To overcome this, researchers are trying to the formulation of nanocarrier, a carrier, which carries drug and helps to deliver drug to the specific target cells and reduces side effects or death of healthy cells. Fluorescent properties can be enhanced by wider band-gap semiconductor shell material when QDs are used as core–shell structures in biological applications [13-17].

Recently, some research has been going towards the developing of delivery systems, particularly for anticancer therapies. Wang and co-workers developed a ligand-modified graphene quantum dots (GQDs) and conjugated with folic acid (FA) to demonstrate highly selective and specific tumor cell imaging [18]. Nigam et al. reported nano-delivery vehicle for the pancreatic cancer cells from graphene quantum dots conjugated them with HA functionalized HSA-NPs for preparing a novel and efficient specific drug delivery and bio-imaging [19]. Jichuan et al. [20] showed that the feasibility of using GQDs as noticeable drug delivery systems with the ability for the pH-triggered delivery of drugs into target cells. Quantum dots are depended upon crystal lattice and composition of compound or solid [5]. A variety of binary combinations of semiconductor materials (PbTe, CdS, ZnS, CdTe, CdSe, InP, PbS,) are typically used in the synthesis of Quantum dot nanocrystals and provide a full range of emission wavelengths (λ_{max}) from the UV to the near IR region [21].

Quantum dots can become the excellent 'prototype', for the development and optimization of nano carriers in those biocompatible nano vehicles of analogous sizes and suitable surface properties can be used in medical fields. QD surface may be covalently linked with the targeting and therapeutic compounds via feeble bonds, therefore firstly bio-conjugate moieties are avoided to renal filtration due to large enough size, but when the moieties are braked they are little sufficient to clear out of the bio system [22]. QDs as prototype materials for nanocarrier engineering and a stupendous understanding tool for the screening of drug and corroboration, as they have been applied as drug carriers to small animals as well as cells.

In recent times, different types of methods have been discovered for the synthesis of carbon dots, such as, hydrothermal treatment, microwave radiation, laser ablation, arc discharge, oxidation and many other [23]. In which, laser ablation and arc discharge methods are limited by costly, complex and energy-intensive equipment and strong acids needed for chemical oxidation. But microwave radiation and hydrothermal treatment are easy and common methods for synthesis of CDs. In microwave radiation method, the reaction procedure takes very short time period and due to the high energy it could be completed only in few minutes. Hydrothermal method is straightforward and possesses the benefit of very low-cost and easy to handle reaction conditions.

To tackle the present challenges in cancer therapies, the unique incorporation of drug targeting and visualization shows high potential. The most significant and widely used emerging advantages of QDs appear to be *visible* or *detectable* drug delivery system, because of the excellent tendency to expound the pharma-co-kinetics as well as pharma-co-dynamics of drug candidates and to endow with the design values for nanovehicle preparing [24].

2.   Synthesis and properties of quantum dots

2.1   Synthesis of QDs

A number of methods have been developed to synthesize quantum dots. These various methods are broadly divided into two categories: Top-down and Bottom-up approaches. Top-down method involves the breakdown of macroscopic material into desired nano-sized quantum dots while in bottom-up approach, small molecules undergo carbonization and polymerization in order to form quantum dots Fig. 2.

Materials Research Forum LLC
https://doi.org/10.21741/9781644901250-8

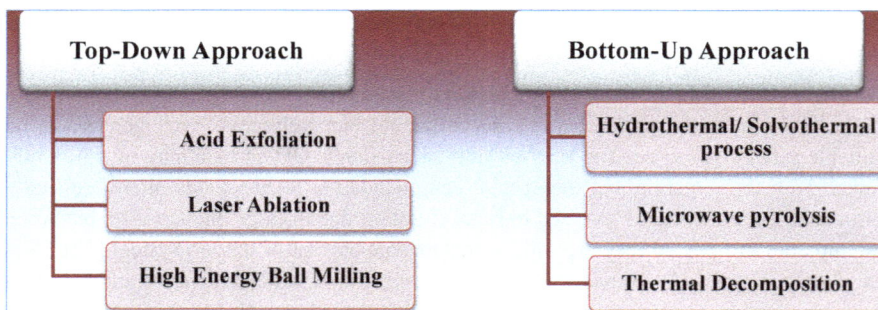

Fig. 2. Different methods of synthesis of quantum dots

In acid exfoliation method, highly concentrated acids are used to exfoliate the bulk material into small i.e. nanosize fragments and surface modification of these fragments are done by oxidation process which are introduced hydrophilic groups i.e. hydroxyl and carboxylic groups and therefore to obtain quantum dots [25-27]. Ruquan and co-authors [28] developed a flexible method to fabricate different graphene QDs with the help of various coals as the precursors. In a distinctive procedure, coal was dissolved in a solution of concentrated sulphuric acid (H_2SO_4) and nitric acid (HNO_3), followed by continuous sonication for 2 hours at room temperature. Then the reaction mixture was stirred with temperature range at 100-120 °C for 24 hours. The resulting mixture was allowed to cooled at 25-30 °C after that this mixture poured into a container of 100 mL of ice. After neutralizing the solution by sodium hydroxide, the resulting solution was filtered and then dialyzed against water for 5 days to obtain graphene QDs. These graphene QDs exhibit pH-sensitive photo-luminescence characteristic in aqueous medium. Laser ablation method has been widely used to develop pure quantum dots. Neither the requirement of any external chemicals nor the formations of any by-product are the major advantages of this method [29-33]. In this method, the surface of solid target material was irradiated by a high energy laser pulse source in a reactor at a high temperature. As the vaporized material condenses on the cooler surface of the reactor, the desired quantum dots are obtained. Ajimsha and coworkers [34] reported a facile laser ablation approach to synthesize chemically pure, biocompatible and highly luminescent quantum dots in various liquid media such as water, methanol, and ethanol. Furthermore, Hu et al. [35], performed one step synthesis of QDs using laser irradiation method in polymer solution. The photoluminescence behavior of as-obtained quantum dot shows dependence on size distribution and wavelength of excitation. Under the method of high energy ball milling, bulk precursor broken down into nano range materials with the help of stainless-steel balls placed in a stainless-steel vessel. Wang and coworkers [36] found that the QDs prepared via high energy ball milling method display dual-wavelength photoluminescence and electro chemiluminescence property. Microwaves are the form of electromagnetic radiation having wavelength in the range of one millimeter to one meter. Microwave provides intensive and efficient energy to break the chemical bonds in the molecule within a short period of time. In this process, the homogeneous transparent solution of precursor heated in a microwave oven. During the heating, the homogeneous solutions undergo thermal carbonization followed by nucleation. Finally, the nuclei diffused with other surrounding molecules to develop the desired quantum dots. On this note, microwave technique is well thought-out as the simple, fast as well as eco-friendly method to synthesize quantum dots with uniform size distribution [37-43]. Zhu et al. [44], firstly suggested a convenient and cost-effective microwave assisted method to formulate highly luminescent carbon quantum dots with the help of various composition of saccharide and polymeric form of ethylene glycol i.e. PEG-200. The appropriate amounts are dissolved in double distilled H_2O to prepare a crystal-clear solution. The resulting transparent solution is treated over 2-10 minutes in the microwave oven. As the reaction proceeds, the apparent variation in color of the solution occurs which indicates the development of carbon QDs. Later, Hou et al. [45] experimented for a novel and single-step microwave technique for the mass production of hydrophilic carbon QDs with high photoluminescence behavior. In this typical process, a transparent solution is prepared by dissolving Triammonium citrate (ACA) and soluble phosphate in dd H_2O. Further, heating the transparent solution in a microwave oven for 2-3 minutes to get solid powder which is brownish-yellow in colour, confirms the formation of carbon dots.

Recently, bottom-up approaches draw immense consideration owing the precise control and low cost for the modification of QDs. Thermal decomposition is one of the bottom-up strategies in which precursor is directly undergone carbonization under controlled reaction conditions. This technique offers various advantages of low cost, easy operation, short reaction time, large scale production and solvent-free approach. Ma et al. [46] formulate different chemically functionalized graphene QDs by using various precursors. The synthesized graphene quantum dots show better pH stability and bright fluorescence and also the TEM analysis confirms the uniform particle size distribution which helps them to be a promising material for optoelectronic and bio-labeling applications.

Materials Research Forum LLC
https://doi.org/10.21741/9781644901250-8

2.2 Properties of QDs

2.2.1 Quantum confinement effect and band-gap

The band gap is generally widened with smaller size QDs due to the effect of quantum confinement. The band gap of an element represents the energy needed to produce a hole and an electron at rest and at the distance where the Coulombic attraction between a hole and an electron is insignificant. If one carrier comes near to the other, they generate the electron-hole pair (i.e. excitation), having energy few meV less than the band-gap.

The distance between the hole and electron is known as the exciton Bohr radius. If the radius of a QD approaches the Bohr radius, then the motion of the electrons and holes is limited locally to the QD level, as a result, excitonic transition energy increases in the band gap of QDs as well as observed blue shift in the QDs luminescence. When the radius of QD is small then the confinement effect becomes significant and Bohr radius of exciton represents the threshold value. For small size QD, exciton and biexciton binding energy are extremely larger than mass materials [47].

2.2.2 Luminescence property

After being energized by external energy, electrons and holes are held in high energy due to the transfer of electrons from the lower energy state to the higher energy state. The electronic structure of the particular moiety plays a crucial role to measure the energies involved in such kind of optical absorption. The electron can re-adjust to the hole followed by the relaxation in a low energy state, eventually attaining the ground state. The overload energy ensuing from recombining and relaxing can be either non-radiative or radiative. Radiative relaxation results in impulsive luminescence from QDs. Such luminescence can occur from band-edge or near-band-edge conversion or from error and / or quantum states of the activator.

The band-edge and nearest band-edge emission are the most familiar radiative relaxation course of action in insulators and intrinsic semiconductors. One of these is the re-adjustment of excited electrons in a conduction band and hole in the valence band is called band-edge emission.

The presence of impurities and/or activator quantum states is also responsible for radiative emissions from QDs within the band-gap. The defect states can serve as a donor (electron rich) or an acceptor (electron deficient) depending upon the kind of impurities and defects [48]. Because of the Coulombic attraction, electrons or holes are drawn to these sites with lack or additional local charge. These defects states can be classified into either shallow or deep levels. In these, the shallow defects states having the energy close to the valence band-edge or conduction band. Generally, shallow defect shows the radiative relaxations at adequately low temperatures whereas deep level shows nonradiative relaxations.

3. Quantum dots as a nanovehicle for drug loading and drug delivery

A fluorescent semiconductor nanoparticle, QDs, represents a versatile platform for design and engineering of nano drug delivery (NDD) carriers. Combination of unique physical, chemical and optical properties, associated with QDs, facilitate in detail study of interactions of biological systems with nanocarrier through real-time monitoring of NP bio-distribution, intracellular uptake, drug release, and long-term nanocarrier fate. Because of their compact size and compatibility with a number of surface modification strategies, QDs enable to substitute a variety of bio active molecules within single drug delivery vehicles. Therefore, QDs based nanovahicles offer a powerful tool for designing and studying the actions of drug carriers and optimization of physicochemical characteristics as well as functionalities with the corresponding moieties to specific and targeted drug delivery applications [21].

3.1 QDs in nonoparticle-mediated drug delivery

Monitoring nanoparticles cellular uptake and determining intracellular fate are the most important parts of the delivery system. The unique information provided by the QDs in each configuration is discussed along with how they contribute to the overall understanding of nanoparticles mediated drug delivery (NMDD) [21]. Mechanisms of nanoparticle cellular delivery; namely, specific targeting, uptake, intracellular fate i.e. overall mechanism are the most vital factors for the successful accomplishment of NMDD [49]. Being smaller than other organic or inorganic (e.g. condensed DNA or gold respectively) nanoparticles cores of interest, QDs can be easily reciprocated for the purpose of studying nanovehicle actions and characteristics. Simultaneously, for examining biological distribution and intracellular trafficking as tracers, QDs can be non-invasively incorporated within drug delivery carriers. At the last, individually released QDs can imitate redistribution and potential clearance of therapeutic agent [50]. QDs have already made a contribution to this area, the first report was that CdSe/Zns core–shell quantum dots labeled with the Fe-transport protein transferrin could bind to its receptor and afterward undergo cellular uptake [51] and many other researchers have tested a variety of similar QD cellular targeting strategies. Rather than focusing on understanding nanoparticles mediated drug delivery, this was instead forced by an interest in testing and demonstrating the utility of new QD preparation or bio-conjugation methods [52], or, conversely, using QD fluorophores to expound aspects of the cellular delivery equipment [53].

In some cases, at single cell resolution QDs is allowed to uptake to be monitored real time [53]. Various different QDs surface modification strategies, from small dihydrolipoic acid capping molecules [54] to multilayer polymeric encapsulation [53], to better loading and release of bio active molecules, are also reported. Even though diversity of the QD related materials and different cellular interactions, almost all the QD conjugates become visible to endocytic uptake and intracellular sequestration in endosomes or vesicles. QD endocytosis is also held true across many different cellular phenotypes, including different type of cancer cell lines as well as liver and neural cells. Endocytosis is the primary method by which almost all membrane-bound moieties and nutrients are internalized and sorted by cells [49, 55].

Since many additional nanoparticles of different sizes are made hydrophilic using surface modifications like functionalization strategies and are amenable to bio-conjugation with the same ligands, it is reasonable to expect some similar results when using them.

Farokhzad et al. [56] reported QDs based ternary system for targeted drug delivery of doxorubicin (Dox) drug for in vitro imaging and biological sensing. The Fig. 3 [56] illustrates the QDs conjunction along with aptamers to process on related targeting site. The incorporation of nucleic acid with doxorubicin altered the several changes in the system. Fig. 4 [6] describes the advanced QD based probes mostly used for bio imaging applications. The biological QDs are prepared by various methods, which are depicted. As shown (Fig 4.) [6], the QDs is usually coated with single layers of thiols viz. mercaptoacetic acid, commonly stabilized in solution by ionization. The silane biochemistry can be implemented for the synthesis of cross-linked silica shell based QDs. Similarly, the encapsulated polymeric QDs are highly stable and having micelle-like structures, these QDs can be certainly modified likewise by polyethylene glycol (PEG) which somehow decrease their surface charge with increase in colloidal stability.

The Fig. 4 [6] (middle), represents the surface functionalization of QDs through streptavidin which may be easily bound to several biotinylated moeities with high affinity and used in immunofluorescence labeling of cancer marker Her2. Whereas QDs bind to antibodies shows selectivity towards the numerous antigens and usually prepared by the interaction of reduced antibody fragments with maleimide-activated QDs-PEG [6].

3.2   Monitoring drug release

The QDs monitored drug delivery is quite unique system of monitored drug delivery, as it provides an extension to the site for the safer, much easier, tracking of molecules and cells as they have biologically compatible nature along with excellent optical and fluorescent properties. QDs have proven strongly useful in the development of new advanced nano vehicles which can be achieved targeted and traceable drug delivery, and also for cellular studies [57-60]. Now a day's, numbers of techniques have been used to monitor and label cells internally with QDs, by using chemical transfection, passive uptake, mechanical delivery and receptor-mediated internalization. Through endocytosis, QDs passively grafted with active cells by exploiting the innate capacity to uptake their extracellular space [61-63]. Behavior of the cell is profoundly depending on the chemical and physical cues adapted from the limited environment. Pretension of these cues with *in vitro* biologically active cell imaging is restricted by the inability to accurate modernize the complex whereas examination of cells with *in vivo* is interrupted by lack of resolution and poor sensitivity of predictable bio imaging methods [64].

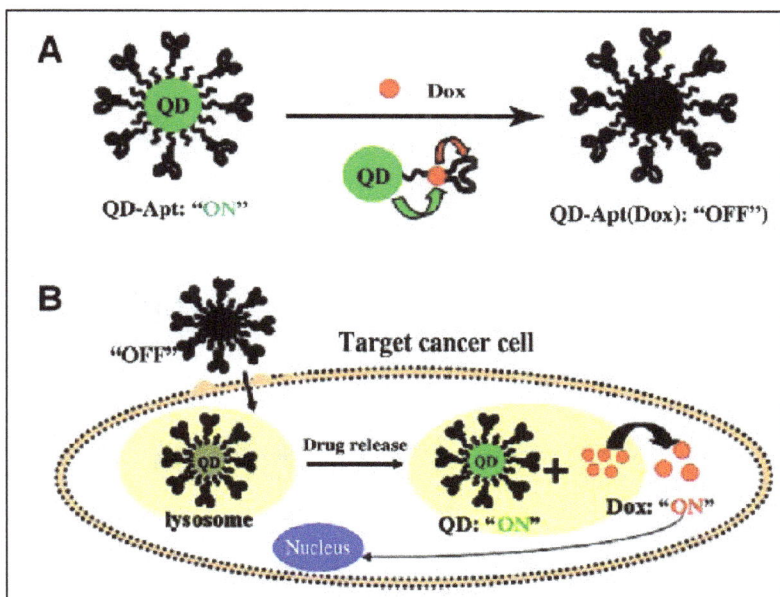

Fig. 3. Description of quantum dots, Aptamer–Dox system for targeted drug delivery via endocytosis mediated receptors. (A) study of quantum dot based conjugates fluorescence property mediated by doxorubicin and its further quenching by FRET. (B) biological internalization of conjugate, release profile of doxorubicin. Adapted with permission from Bagalkot et al. [56].

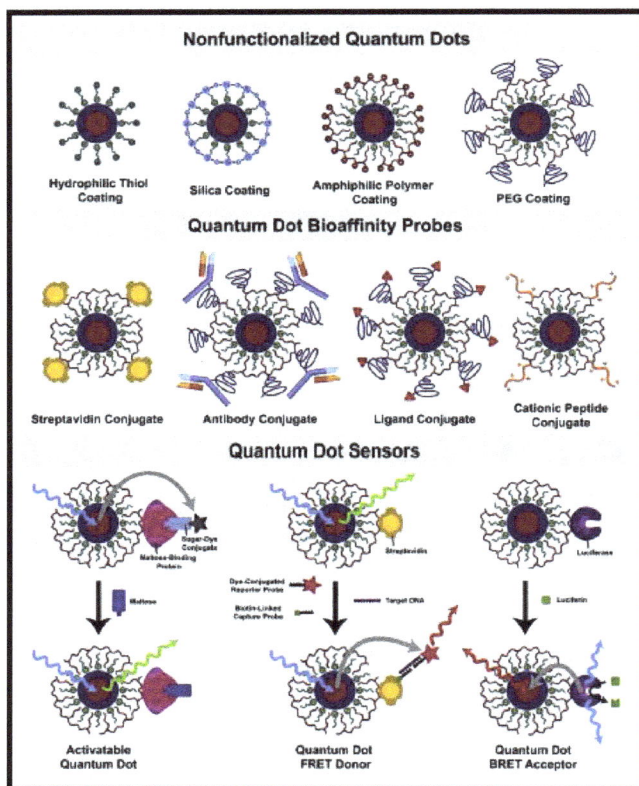

Fig. 4. Describes the advanced QD based probes mostly used for bio imaging applications. Adapted with permission from Smith et al. [6]

3.3   Photo-physical properties for traceable or visible drug delivery

QDs use as the special fluorescent agents for tracking to drug release monitoring because of their optical properties, fluorescence properties and optimized band gaps along with tenability in size. High-quality monodisperse QDs release instance luminosity in a definite as well as fine spectral range, and also the λ_{max} of this fluoresce light straightly proportional to the size of NPs core [65-68]. The absorbance wavelength of light, for QDs starts from ultra violet to specific emission frequency of the molecule [69]. These unique features made QDs superior candidates for biological imaging and other analogous applications, which include single source excitations in UV light for cellular tracking of various multi-phase nano carriers and detection of their comparative behaviors for the same. For example, Kobayashi and co authors revealed that QDs between 565 nm to 800 nm worked as multicolor tracers, to verify lymphatic transportation of nanoparticles and their further accommodation to the sentinel lymph of five different lymph drainage basis of the mouse [70]. Similarly, Popović et al. [71] had mentioned intravital microscopy by taking multicolored QDs for the dual-phase monitoring through extravasation of 125, 60 and 12 nm sized nano-carriers in the region of similar vascular system, which further defined unbiased extravasation along with proper diffusion via cytoplasmic matrix for nanoparticle of smaller sizes than that of larger ones. The benefits of multicolor probing through QDs have been reported by Delehanty et al. [72] where they mentioned various routes for intracellular delivery such as amino acid sequence-based endocytosis, polymer mediated transfection and microinjection delivery of cytosol in live cells. This report outlines the huge potential of inquiries starting for single cell uptake to whole system uptake, which can be easily done by polycolor imaging. Study and in vivo imaging of nanocarrier bio-distribution significantly profits by the large QD Stokes shift of 300 to 400 nm [52, 73]. As biologically active cargo tends to emit number of illumination in the range of violet-green spectra, changing of QD emission in the direction of red or near-IR region reflects clear distinction between tissue auto-fluorescence and QD signals, while continuous permitting proficient excitation by the violet-green luminosity.

4. Toxicity

There are various types of QDs and each type has its own properties along with potential toxicity/non toxicity. Currently, the availability of literature information on QDs based toxicity is not much reliable due to some factors such as: QDs concentration used during the process to their diversity in physiochemical properties [1,9]. QDs based toxicity can be outlined by description of core material study, material coating and core shell studies etc.

4.1 Core toxicity

Kirchner et al. [74] reported that cadmium based core of QDs nanoparticles have been mentioned as toxic, whereas the QDs with non-coated and free cadmium having colloidal suspension have affected intra cellular activity of the biological cell. Further they found that cadmium based QDs were cytotoxic for pheochromocytoma cells of rat with the concentration level of 1 µg/mL which induced apoptosis cells death by condensing the chromatin and cell membrane of the same. This is due to the uncoated nanoparticles incubation in rat hepatocytes tissues that released the Cd by surface oxidation along with biological degradation of the cell [75]. The electronically sensitive QDs formulated from cadmium-tellurium (CdTe) and cadmium-selenium (CdSe) which are suitable to light as well as air oxidation, that promotes formation of free-radical preferentially, and which can initiate cytotoxicity [76]. The free Cd ion promotes oxidative stress, while free ions are not capable to form free radicals instead that the core of QD does participate in the generation of free radicals. Cho et al. [77] showed that dyes that are Cd based in cell culture evaluated that the cytotoxic effect was not due to removal of Cd instead it was a counterpart of free radical generation in the process. The DNA damage was also reported as a consequence of radical generation in presence and absence of photonic activation [78].

4.2   Encapsulated QDs and their toxicity

The research on encapsulated QDs and their cytotoxic effects has revealed that the toxicity of QDs is reduced after encapsulating by zinc sulfate or by some other metals.

Bakalova et al. [79] reported that human breast cancer cells i.e. MCF-7 cell showed the destruction of CdTe Cd cadmium toxicity. The CdTe/ZnS shell and their capping with suitable chemicals such as *n*-acetylcysteine, mercaptopropionic acid and cysteamine showed capping materials and ZnS are effective and minimize the several effects because ZnS reduces the process of apoptosis with the elimination of released cadmium ion in aqueous medium. Previous research data revealed that the decrease in fluorescence property of CdSe/ZnS alongside the blue spectra in live-cell overtime which indicates the biodegradability of the ZnS core-shell with the intracellular system [1]. Some studies also reported that in the presence of air or photooxidation ZnS shell is not eliminated cytotoxicity and for that CdSe/ZnS-QD promoted free radical production. Kirchner et al. [74] showed that ZnS shell protects the core of CdSe and prevents from oxidation and suggested that cells could be impaired if precipitation of nanoparticles occur on the surface of cell with the release of cadmium ion from the layer of CdSe/ZnS. These collected data suggests that stability of capping materials and toxicity behavior requires intensive investigation for different QD preparations [6].

5.   Quantum dots in bioimaging applications

The recent studied showed the extensive utilization of optical photographing, magnetic resonance imaging (MRI) along with nuclear based imaging for various bio-imaging applications [80]. These basic technologies vary in their rate of sensitivity, resolution time, structural complexions along with cost of operation to acquisition time. In spite of having such variations, usually these technologies behave complementary inter alia. Various studies have been made by research groups for the explanations of their technical aspects [80, 81] which includes instruments [82,83] to their performance measurements [84, 85].

The significance of QDs in this field is due to its excellence in optical and florescent properties [85-89]. The dyes were the integral part of the photo light imaging [90, 91], whereas some exclusive flaws are still there with this technique most of them are along with their use. Similarly, the florescence in normal light has a well-established use in bio-imaging [92], commercially which can be explained as (1) the automatic florescence leads signaling from labeled synthetic dye molecules; (2) the less stability of synthetic dyes under the photon radiation is excellently established, that also results shorter examination period; (3) commonly, usual dyes have lower excitation possibility that limits the excitations of multi dye simultaneously; (4) the different pH conditions are the major sensitive environmental tool that affects dyes; (5) usually the synthetic dyes have reddish emission spectrum that affects the detection channels and quantity of variations in various probes.

Biological importance of QDs is special, mostly due to excel of extinction coefficient, excellent quantum yields, lower photo bleaching, tunability of photo excitation by controlling the sizes, very less interference while using simultaneously two or more QDs, lower toxicity in most cases and due to the excess into ease of functionalities via various biological and organic molecules. The NIR near Infra-Red (NIR) emission quantum dots have been mostly used to avoid the interactions of interface for various cases especially automatic fluorescence because the presence of water and Hb (hemoglobin) inside cellular systems have lower absorption and scattering coefficient (near infra-red of 650 – 900 nm). While visible light is the part of routine process for intra vital microscopy, but imagine of tissue in the range of 500 μm to 1 cm requires near infra-red light [93].

Metal or non-metal based quantum dots were reported to have excellent photo stable under different excitation especially ultraviolet excitation in comparison to other synthetic molecules i.e. organic QDs. The comparative studies of various nanomaterials have shown that the QDs and their tunability of size offers excellent tunability of emission colors along with its variations which further is a potential candidate that can develop some multi-color optical coding techniques, basically by the surface modifications of these QDs.

Conclusions

The QDs were recognized as the fundamental probes for the diagnosis of distinct disease in biological cells and related in-vitro studies mostly owing to their outstanding optical, structural and morphological characteristics. The QDs were enlightened as the flexible linkage and doping tools as they have large surface area to volume ratio, along with broad spectral possibilities of attached functional moieties. This opens up a huge possibility for their utilization in traceable targeted drug delivery. The heavy metal based QDs for near and far infrared region with contracted emission and good quantum yield are mainly reported with low cyclic toxicity till now. Instead of having such possible limitations, QDs have been utilized widely for biological cells to animals as nano carriers. The QDs are also utilized as the screening and validation tools for various drug industries and biotechnological engineering. The silicon and carbon based QDs were reported as lower toxicity QDs or as QDs with no toxicity at all along with excellent quality and quantum yield, thus they can be used for further successive utilization to the field. The prime most challenge of drug delivery is going to be the acute concentration of drug used into the targeted cell for the prevention of cell toxicity. But this window is still to be opened up as far as QDs are concerned, that requires their managed evaluation and systematic study. The concentration of drug can be maintained to scale up the unwanted effects and the circulation time for plasma, with possible stability of compounds by engineering the QDs. The therapeutic biomolecules and QDs are bind through the covalent bonding and pi-pi interactions that can avoid the renal filtration and can also easily liberate smaller moieties from the host body after detachment of ligands. Their non-toxic nature, multi-functionality, development of new QDs based targeting ligands and nano materials, are some of the few possibilities for the futuristic utilization of QDs to expand their reach and potential in the field of drug delivery.

Acknowledgments

National Mission of Himalayan Studies, GBPIHED, Kosi Kataramal, Almora, India (Ref. No.:NMHS/MG- 2016/002/8503-7).

References

[1]   S. Ghaderi, B. Ramesh, A.M. Seifalian, Fluorescence nanoparticles "quantum dots" as drug delivery system and their toxicity: A review, J. Drug Target. 19 (2011) 475-486. https://doi.org/10.3109/1061186X.2010.526227

[2]   D. Bera, L. Qian, P.H. Holloway, Phosphor Quantum Dots, John WIley & Sons, Ltd: West Sussex, UK, 2008.

[3]   P. Walter, E. Welcomme, P. Hallegot, N.J. Zaluzec, C. Deeb, J. Castaing, P. Veyssiere, R. Breniaux, J.L. Leveque, G. Tsoucaris, Early use of PbS nanotechnology for an ancient hair dyeing formula, Nano Lett. 6 (2006) 2215-2219. https://doi.org/10.1021/nl061493u

[4]   D.Y. Lee, K. C.P. Li, Molecular theranostics: a primer for the imaging professional, AJR Am. J. Roentgenol. 197 (2011) 318-324. https://doi.org/10.2214/AJR.11.6797

[5]   D. Bera, L. Qian, T.K. Tseng, P.H. Holloway, Quantum dots and their multimodal applications: a review, Materials. 3 (2010) 2260-2345. https://doi.org/10.3390/ma3042260

[6]   A.M. Smith, H. Duan, A.M. Mohs, S. Nie, Bioconjugated quantum dots for in vivo molecular and cellular imaging, Adv. Drug Del. Rev. 60 (2008) 1226–1240. https://doi.org/10.1016/j.addr.2008.03.015

[7]   Q. Yuan, S. Hein, R.D.K. Misra, New generation of chitosan-encapsulated ZnO quantum dots loaded with drug: synthesis, characterization and in vitro drug delivery response, Acta Biomater. 6 (2010) 2732-2739. https://doi.org/10.1016/j.actbio.2010.01.025

[8]   T. Jamieson, R. Bakshi, D. Petrova, R. Pocock, M. Imani, A.M. Seifalian, Biological applications of quantum dots, Biomaterials. 28 (2007) 4717–4732. https://doi.org/10.1016/j.biomaterials.2007.07.014

[9]   R. Hardman, A toxicologic review of quantum dots: toxicity depends on physicochemical and environmental factors, Environ. Health Perspect. 114 (2005) 165-172. https://doi.org/10.1289/ehp.8284

[10]   C.L. Li, C.M. Ou, C.C. Huang, W.C. Wu, Y.P. Chen, T.E. Lin, Lin-Chen Ho, C.W. Wang, C.C. Shih, H.C. Zhou, Y.C. Lee, W.F. Tzeng, T.J. Chiou, S.T. Chu, J. Cangm H.T. Chang, Carbon dots prepared from ginger exhibiting efficient inhibition of human hepatocellular carcinoma cells, J. Mat. Chem. B. 2 (2014) 4565-4571. https://doi.org/10.1039/C4TB00216D

[11]   J. Shen, Y. Zhu, X. Yang, C. Li, Graphene quantum dots: emergent nano lights for bioimaging, sensors, catalysis and photovoltaic devices, ChemComm. 48 (2012) 3686–3699. https://doi.org/10.1039/C2CC00110A

[12]   S.J. Zhu, J.H. Zhang, C.Y. Qiao, Strongly green-photoluminescent graphene quantum dots for bioimaging applications, ChemComm. 47 (2011) 6858–6860. https://doi.org/10.1039/C1CC11122A

[13] C.B. Murray, C.R. Kagan, M.G. Bawendi, Synthesis and characterization of monodisperse nanocrystals and close-packed nanocrystal assemblies, Ann. Rev. Mater. Sci. 30 (2000) 545-610. https://doi.org/10.1146/annurev.matsci.30.1.545

[14] A.P. Alivisatos, W. Gu, C.A. Larabell, Quantum dots as cellular probes, Ann Rev Biomed Eng. 7 (2005) 55-76. https://doi.org/10.1146/annurev.bioeng.7.060804.100432

[15] I. Medintz, H. Uyeda, E. Goldman, H. Mattoussi, Quantum dot bioconjugates for imaging, labeling and sensing, Nat. Mater. 4 (2005) 435-446. https://doi.org/10.1038/nmat1390

[16] J.M. Klostranec, W.C.W. Chan, Quantum dots in biological and biomedical research: recent progress and present challenges, Adv. Mater. 18 (2006) 1953-1964. https://doi.org/10.1002/adma.200500786

[17] B.O. Dabbousi, V. Rodriguez-Viejo, F.V. Mikulec, J. R. Heine, H. Mattoussi, R. Ober, K.F. Jensen, M.G. Bawendi, (CdSe)ZnS core-shell quantum dots: synthesis and characterization of a size series of highly luminescent nanocrystallites, J. Phys. Chem. B. 101 (1997) 9463-9475. https://doi.org/10.1021/jp971091y

[18] X. Wang, X. Sun, J. Lao, H. He, Tiantian Cheng, Mingqing Wang, S. Wang, F. Huang, Multifunctional graphene quantum dots for simultaneous targeted cellular imaging and drug delivery, Colloids Surf. B. 122 (2014) 638–644. https://doi.org/10.1016/j.colsurfb.2014.07.043

[19] P. Nigam, S. Waghmode, M. Louis, S, Wangnoo, P. Chavan, D. Sarkar, Graphene quantum dots conjugated albumin nanoparticles for targeted drug delivery and imaging of pancreatic cancer, J. Mater. Chem. B. 2 (2014) 3190-3195. https://doi.org/10.1039/C4TB00015C

[20] J. Qiu, R. Zhang, J. Li, Y. Sang, W. Tang, P. R. Gil, Hong Liu, Fluorescent graphene quantum dots as traceable, pH-sensitive drug delivery systems, Int. J. Nanomedicine. 10 (2015) 6709–6724. https://doi.org/10.2147%2FIJN.S91864

[21] J.B. Delehanty, K. Boeneman, C.E. Bradburne, K. Robertson, I.L. Medintz, Quantum dots: a powerful tool for understanding the intricacies of nanoparticle-mediated drug delivery. 6 (2009) 1091-1112. https://doi.org/10.1517/17425240903167934

[22] H.S. Choi , W. Liu, P. Misra, E. Tanaka, J.P. Zimmer, B.L. Lpe, M.G. Bawendi, J.V. Franjioni, Renal clearance of quantum dots, Nat. Biotechnol. 25 (2007) 1165 - 1170. https://doi.org/10.1038/nbt1340

[23]   L. Lia, L. Lia, C.P. Chena, F. Cui, Green synthesis of nitrogen-doped carbon dots from ginkgo fruits and the application in cell imaging, Inorg. Chem. Commun. 86 (2017) 227-231. https://doi.org/10.1016/j.inoche.2017.10.006

[24]   L. Qi, X. Gao, Emerging application of quantum dots for drug delivery and therapy, Expert Opin. Drug Deliv. 5 (2008) 263-267. https://doi.org/10.1517/17425247.5.3.263

[25]   H. Tao, K. Yang, Z. Ma, J. Wan, Y. Zhang, Z. Kang, Z. Liu, In vivo NIR fluorescence imaging, biodistribution, and toxicology of photoluminescent carbon dots produced from carbon nanotubes and graphite, Small. 8 (2012) 281-290. https://doi.org/10.1002/smll.201101706

[26]   P. Shen, Y. Xia, Synthesis-modification integration: one-step fabrication of boronic acid functionalized carbon dots for fluorescent blood sugar sensing, Anal. Chem. 86 (2014) 5323-5329. https://doi.org/10.1021/ac5001338

[27]   Q. Zhang, X. Sun, H. Ruan, K. Yin, H. Li, Production of yellow-emitting carbon quantum dots from fullerene carbon soot, Sci. China Mater. 60 (2017) 141-150. https://doi.org/10.1007/s40843-016-5160-9

[28]   R. Ye, C. Xiang, J. Lin, Z. Peng, K. Huang, Z. Yan, N. P. Cook, E.L. Samuel, C.C. Hwang, G. Ruan, G. Ceriotti, Coal as an abundant source of graphene quantum dots, Nat. Commun. 4 (2013) 2943. https://doi.org/10.1038/ncomms3943

[29]   D. Reyes, M. Camacho, M. Camacho, M. Mayorga, D. Weathers, G. Salamo, Z. Wang, Neogi, Laser ablated carbon nanodots for light emission, Nanoscale Res. Lett. 11 (2016) 424. https://doi.org/10.1186/s11671-016-1638-8

[30]   N. Tarasenka, A. Stupak, N. Tarasenko, S. Chakrabarti, D. Mariotti, Structure and optical properties of carbon nanoparticles generated by laser treatment of graphite in liquids, ChemPhysChem. 18 (2017) 1074-1083. https://doi.org/10.1002/cphc.201601182

[31]   X. Li, H. Wang, Y. Shimizu, A. Pyatenko, K. Kawaguchi, N. Koshizaki, Preparation of carbon quantum dots with tunable photoluminescence by rapid laser passivation in ordinary organic solvents, ChemComm. 47 (2010) 932-934. https://doi.org/10.1039/C0CC03552A

[32]   V. Nguyen, L. Yan, J. Si, X. Hou, Femtosecond laser-induced size reduction of carbon nanodots in solution: effect of laser fluence, spot size, and irradiation time, J. Appl. Phys. 117 (2015) 084304. https://doi.org/10.1063/1.4909506

Materials Research Forum LLC
https://doi.org/10.21741/9781644901250-8

[33] S.L. Hu, K.Y. Niu, J. Sun, J. Yang, N.Q. Zhao, X.W. Du, One-step synthesis of fluorescent carbon nanoparticles by laser irradiation. J. Mater. Chem. 19 (2009) 484-488. https://doi.org/10.1039/B812943F

[34] R.S. Ajimsha, G. Anoop, A. Aravind, M.K. Jayaraj, Luminescence from surfactant-free ZnO quantum dots prepared by laser ablation in liquid, Electrochem Solid St. 11 (2008) 14-17. https://doi.org/10.1149/1.2820767

[35] S. Hu, J. Liu, J. Yang, Y. Wang, S. Cao, Laser synthesis and size tailor of carbon quantum dots, J. Nanopart. Res. 13 (2011) 7247-7252. https://doi.org/10.1007/s11051-011-0638-y

[36] L. Wang, X. Chen, Y. Lu, C. Liu, W. Yang, Carbon quantum dots displaying dual-wavelength photoluminescence and electrochemiluminescence prepared by high-energy ball milling, Carbon. 94 (2015) 472-478. https://doi.org/10.1016/j.carbon.2015.06.084

[37] Q. Lu, C. Wu, D. Liu, H. Wang, W. Su, H. Li, Y. Zhang, S. Yao, A facile and simple method for synthesis of graphene oxide quantum dots from black carbon, Green Chemistry. 19 (2017) 900-904. https://doi.org/10.1039/C6GC03092K

[38] G. Wang, Q. Guo, D. Chen, Z. Liu, X. Zheng, A. Xu, S. Yang, G. Ding, Facile and highly effective synthesis of controllable lattice sulfur-doped graphene quantum dots via hydrothermal treatment of durian, ACS Appl. Mater. Inter. 10 (2018) 5750-5759. https://doi.org/10.1021/acsami.7b16002

[39] Q. Wang, X. Liu, L. Zhang, Y. Lv, Microwave-assisted synthesis of carbon nanodots through an eggshell membrane and their fluorescent application, Analyst. 137 (2012) 5392-5397. https://doi.org/10.1039/C2AN36059D

[40] L. Tang, R. Ji, X. Cao, J. Lin, H. Jiang, X. Li, K. S. Teng, C. M. Luk, S. Zeng, J. Hao, S. P. Lau, Deep ultraviolet photoluminescence of water-soluble self-passivated graphene quantum dots, ACS nano. 6 (2012) 5102-5110. https://doi.org/10.1021/nn300760g

[41] X. Zhai, P. Zhang, C. Liu, T. Bai, W. Li, L. Dai, W. Liu, Highly luminescent carbon nanodots by microwave-assisted pyrolysis. ChemComm. 48 (2012) 7955-7957. https://doi.org/10.1039/C2CC33869F

[42] C. Liu, P. Zhang, F. Tian, W. Li, F. Li, W. Liu, One-step synthesis of surface passivated carbon nanodots by microwave assisted pyrolysis for enhanced multicolor photoluminescence and bioimaging, J. Mater. Chem. 21 (2011) 13163-13167. https://doi.org/10.1039/C1JM12744F

[43]   P.C. Hsu, H. T. Chang, Synthesis of high-quality carbon nanodots from hydrophilic compounds: role of functional groups, ChemComm. 48 (2012) 3984-3986. https://doi.org/10.1039/C2CC30188A

[44]   H. Zhu, X. Wang, Y. Li, Z. Wang, F. Yang, X. Yang, Microwave synthesis of fluorescent carbon nanoparticles with electrochemiluminescence properties, ChemComm. 34 (2009) 5118-5120. https://doi.org/10.1039/B907612C

[45]   J. Hou, J. Yan, Q. Zhao, Y. Li, H. Ding, L. Ding, A novel one-pot route for large-scale preparation of highly photoluminescent carbon quantum dots powders, Nanoscale. 5 (2013) 9558-9561. https://doi.org/10.1039/C3NR03444E

[46]   C.B. Ma, Z.T. Zhu, H.X. Wang, X. Huang, X. Zhang, X. Qi, H.L. Zhang, Y. Zhu, X. Deng, Y. Peng, Y. Han, A general solid-state synthesis of chemically-doped fluorescent graphene quantum dots for bioimaging and optoelectronic applications, Nanoscale. 7 (2015) 10162-10169. https://doi.org/10.1039/C5NR01757B

[47]   V.I. Klimov, Mechanisms for photogeneration and recombination of multiexcitons in   semiconductor nanocrystals: implications for lasing and solar energy conversion, J. Phys. Chem. B. 110 (2006) 16827-16845. https://doi.org/10.1021/jp0615959

[48]   A. Issac, C. V. Borczyskowski, F. Cichos, Correlation between photoluminescence intermittency of CdSe quantum dots and self-trapped states in dielectric media, Phys. Rev. B. 71 (2005) 161302. https://doi.org/10.1103/PhysRevB.71.161302

[49]   O.H. Frenkel, Y. Altschuler, S. Benita, Nanoparticle-cell interactions: drug delivery implications. Crit. Rev. Ther. Drug Carrier Syst. 25 (2008) 485-544. https://doi.org/10.1615/CritRevTherDrugCarrierSyst.v25.i6.10

[50]   C.E. Probst, P. Zrazhevskiy, V. Bagalkot, X. Gao, Quantum dots as a platform for nanoparticle drug delivery vehicle design, Adv. Drug Deliv. 65 (2013) 703-718. https://doi.org/10.1016/j.addr.2012.09.036

[51]   W.C.W. Chan, S. Nie, Quantum dot bioconjugates for ultrasensitive nonisotopic detection, Science. 281(1998) 2016-2018. https://doi.org/10.1126/science.281.5385.2016

[52]   M.C. Mancini, B.A. Kairdolf, A.M. Smith, S.M. Nie, Oxidative quenching and degradation of polymer-encapsulated quantum dots: new insights into the long-term fate and toxicity of nanocrystals in vivo, J. Am. Chem. Soc. 130 (2008) 10836-10837. https://doi.org/10.1021/ja8040477

[53]   D. S. Lidke, P. Nagy, R. Heintzmann, Quantum dot ligands provide new insights into erbB/HER receptor-mediated signal transduction, Nat Biotechnol. 22 (2004) 198-203. https://doi.org/10.1038/nbt929

[54]   J.B. Delehanty, I.L. Medintz, T. Pons, Self-assembled quantum dot-peptide bioconjugates for selective intracellular delivery, Bioconjug. Chem. 17 (2006) 920-927. https://doi.org/10.1021/bc060044i

[55]   N.E. Bishop, Dynamics of endosomal sorting, in: K.W. Jeon (Ed.), International review of cytology, Elsevier Academic Press, London, 2003, pp. 1-57.

[56]   V. Bagalkot, L. Zhang, E. Levy-Nissenbaum, S. Jon, P.W. Kantoff, R. Langer, O.C. Farokhzad, Quantum dot–aptamer conjugates for synchronous cancer imaging, therapy, and sensing of drug delivery based on bi-fluorescence resonance energy transfer, Nano Lett. 7 (2007) 3065–3070. https://doi.org/10.1021/nl071546n

[57]   Y. Wu, Uptake and intracellular fate of multifunctional nanoparticles: a comparison between lipoplexes and polyplexes via quantum dot mediated Forster resonance energy transfer, Mol. Pharm. 8 (2011) 1662–1668. https://doi.org/10.1021/mp100466m

[58]   K.C. Weng, Targeted tumor cell internalization and imaging of multifunctional quantum dot-conjugated immunoliposomes in vitro and in vivo, Nano Lett. 8 (2008) 2851–2857.

[59]   S. Liu, Bortezomib induces DNA hypomethylation and silenced gene transcription by interfering with Sp1/NF-kappaB-dependent DNA methyltransferase activity in acute myeloid leukemia, Blood. 111 (2008) 2364–2373. https://doi.org/10.1182/blood-2007-08-110171

[60]   Y. Liu, Y. Mi, J. Zhao, S.S. Feng, Multifunctional silica nanoparticles for targeted delivery of hydrophobic imaging and therapeutic agents, Int. J. Pharm. 421 (2011) 370–378. https://doi.org/10.1016/j.ijpharm.2011.10.004

[61]   K. Hanaki, A. Momo, T. Oku, A. Komoto, S. Maenosono, Y. Yamaguchi, K. Yamamoto, Semiconductor quantum dot/albumin complex is a long-life and highly photostable endosome marker, Biochem. Biophys. Res. Commun. 302 (2003) 496–501. https://doi.org/10.1016/S0006-291X(03)00211-0

[62]   J.K. Jaiswal, H. Mattoussi, J.M. Mauro, S.M. Simon, Long-term multiple color imaging of live cells using quantum dot bioconjugates, Nat. Biotechnol. 21 (2003) 47–51. https://doi.org/10.1038/nbt767

[63] W.J. Parak, R. Boudreau, M. Le Gros, D. Gerion, D. Zanchet, C.M. Micheel, S.C. Williams, A.P. Alivisatos, C. Larabell, Cell motility and metastatic potential studies based on quantum dot imaging of phagokinetic tracks, Adv. Mater. 14 (2002) 882–885. https://doi.org/10.1002/1521-4095

[64] P. Zrazhevskiy, M. Sena, X. Gao, Designing multifunctional quantum dots for bioimaging, detection, and drug delivery Chem. Soc. Rev. 39 (2010) 4326–4354. https://doi.org/10.1039/B915139G

[65] M. Bruchez, Semiconductor nanocrystals as fluorescent biological labels, Science. 281 (1998) 2013–2016. https://doi.org/10.1126/science.281.5385.2013

[66] A.P. Alivisatos, Semiconductor clusters, nanocrystals, and quantum dots, Science, 271 (1996) 933–937. https://doi.org/10.1126/science.271.5251.933

[67] I.L. Medintz, Quantum dot bioconjugates for imaging, labelling and sensing, Nat. Mater. 4 (2005) 435–446. https://doi.org/10.1038/nmat1390

[68] A.P. Alivisatos, Perspectives on the physical chemistry of semiconductor nanocrystals, J. Phys. Chem. 100 (1996) 13226–13239. https://doi.org/10.1021/jp9535506

[69] W.C. Chan, D.J. Maxwell, X. Gao, R.E. Bailey, M. Han, S. Nie, Luminescent quantum dots for multiplexed biological detection and imaging, Curr. Opin. Biotechnol. 13 (2002) 40–46. https://doi.org/10.1016/S0958-1669(02)00282-3

[70] H. Kobayashi, Simultaneous multicolor imaging of five different lymphatic basins using quantum dots, Nano Lett., 7 (2007) 1711–1716. https://doi.org/10.1021/nl0707003

[71] Z. Popovic, A nanoparticle size series for in vivo fluorescence imaging, Angew. Chem. Int. Ed. 122 (2010) 8831–8834. https://doi.org/10.1002/ange.201003142

[72] J.B. Delehanty, Spatiotemporal multicolor labeling of individual cells using peptide functionalized quantum dots and mixed delivery techniques, J. Am. Chem. Soc. 133 (2011) 10482–10489. https://doi.org/10.1021/ja200555z

[73] X. Gao, In vivo cancer targeting and imaging with semiconductor quantum dots, Nat. Biotechnol. 22 (2004) 969–976. https://doi.org/10.1038/nbt994

[74] C. Kirchner, T. Liedl, S. Kudera, T. Pellegrino, A. M. Javier, H. E. Gaub, S. Sto1lzle, N. Fertig, W. J. Parak, Cytotoxicity of colloidal CdSe and CdSe/ZnS nanoparticles. Nano Lett. 5 (2005) 331–338. https://doi:10.1021/nl047996

[75]   A.M. Derfus, W.C.W. Chan, S. N. Bhatiya, Intracellular delivery of quantum dots for live cell labeling and organelle tracking, Adv. Matar. 16 (2004) 961-966. https://doi.org/10.1002/adma.200306111

[76]   B.I. Ipe, M. Lehnig, C. M. Niemeyer, On the generation of free radical species, from quantum dots, Small. 7 (2005) 706-709. https://doi: 10.1002/smll.200500105

[77]   S.J. Cho, D. Maysinger, M. Jain, B. Roder, S. Hackbarth, F.M. Winnik, Long – term exposure to CdTe quantum dot causes functional impairments in live cell, Langmuir. 23 (2007) 1974-1980. https://doi.org/10.1021/la060093j

[78]   M. Green, E. Howman, Semiconductor quantum dots and free radical induced DNA nicking. Chem Commun. 1 (2005)121–123. https://doi: 10.1039/b413175d.

[79]   R. Bakalova, Z. Zhelev, R. Jose, T. Nagase, H. Ohba, M. Ishikawa, Y. Baba, Role of free cadmium and selenium ions in the potential mechanism for the enhancement of photoluminescence of CdSe quantum dots under ultraviolet irradiation, J. Nanosci. Nanotechnol. 5 (2005) 887–894. https://doi:10.1166/jnn.2005.117

[80]   P.J. Cassidy, G.K. Radda, Molecular imaging perspectives, J. R. Soc. Interface. 2 (2005) 133–144. https://doi.org/10.1098/rsif.2005.0040

[81]   O. Schillaci, R. Danieli, F. Padovano, A. Testa, G. Simonetti, Molecular imaging of atheroslerotic plaque with nuclear medicine techniques (Review), Int. J. Mol. Med. 22 (2008) 3–7. https://doi.org/10.3892/ijmm.22.1.3

[82]   M. Lecchi, L. Ottobrini, C. Martelli, A.D. Sole, G. Lucignani, Instrumentation and probes for molecular and cellular imaging, Q. J. Nucl. Med. Mol. Imaging. 51 (2007) 111–126.

[83]   B.J. Pichler, H.F. Wehrl, M.S. Judenhofer, Latest advances in molecular imaging instrumentation, J. Nucl. Med. 49 (2008). https://doi:10.2967/jnumed.108.045880

[84]   M. Levenson, D.T. Lynch, H. Kobayashi, J.M. Backer, M.V. Backer, Multiplexing with multispectral imaging: from mice to microscopy, Ilar J. 49 (2008) 78–88. https://doi.org/10.1093/ilar.49.1.78

[85]   P. Sharma, S. Brown, G. Walter, S. Santra, B. Moudgil, Nanoparticles for bioimaging, Adv. Colloid Interface Sci. 123 (2006) 471–485. https://doi.org/10.1016/j.cis.2006.05.026

[86]   A.M. Smith, X.H. Gao, S.M. Nie, Quantum dot nanocrystals for in vivo molecular and cellular imaging, Photochem. Photobiol. 80 (2004) 377–385. https://doi.org/10.1111/j.1751-1097.2004.tb00102.x

Materials Research Forum LLC

https://doi.org/10.21741/9781644901250-8

[87] R.E. Bailey, A.M. Smith, S.M. Nie, Quantum dots in biology and medicine. Physica E. 25 (2004) 1–12. https://doi.org/10.1016/j.physe.2004.07.013

[88] S. Santra, J.S. Xu, K.M. Wang, W.H. Tan, Luminescent nanoparticle probes for bioimaging, J. Nanosci. Nanotech. 4 (2004) 590–599. https://doi.org/10.1166/jnn.2004.017

[89] P. Zrazhevskiy, X. Gao, Quantum dots for cancer molecular imaging. Minerva Biotecnologica. 21 (2009) 37–52.

[90] E.M.C. Hillman, Optical brain imaging in vivo: techniques and applications from animal to man, J. Biomed. Optics. 21 (2007) 051402. https://doi.org/10.1117/1.2789693

[91] G.D. Luker, K. E. Luker, Optical imaging: current applications and future directions, J. Nucl. Med. 49 (2008) 1–4. https://doi.org/10.2967/jnumed.107.045799

[92] W.W. Wu, A.D. Li, Optically switchable nanoparticles for biological imaging, Nanomedicine. 2 (2007) 523–531. https://doi.org/10.2217/17435889.2.4.523

[93] R. Weissleder, A clearer vision for in vivo imaging, Nat. Biotechnol. 19 (2001) 316–317. https://doi.org/10.1038/86684

[94] Z. Medarova, W. Pham, C. Farrar, In vivo imaging of siRNA delivery and silencing in tumors, Nat Med. 13 (2007) 372 -377. https://doi.org/10.1038/nm1486

Quantum Dots – Properties and Applications
Materials Research Foundations **96** (2021) 216-250

Materials Research Forum LLC
https://doi.org/10.21741/9781644901250-9

Chapter 9

Quantum Dots based Materials for New Generation Supercapacitors Application: A Recent Overview

Gaurav Tatrari[1], Manoj Karakoti[1], Mayank Pathak[1], Anirban Dandapat[1], Tanmoy Rath[2], Nanda Gopal Sahoo[1]*

[1] Prof. Rajendra Singh Nanoscience and Nanotechnology Lab, Department of Chemistry, Department of Chemistry, DSB Campus, Kumaun University, Nainital, Uttrakhand, India

[2] Motihari College of Engineering (Aryabhatta Knowledge University) Motihari, Bihar, India

* ngsahoo@yahoo.co.in

Abstract

The need of energy storage and related devices are increasing day by day, due to the expansion of global population. To deal with such universal crisis, current energy storage devices like supercapacitors need to be improved in their performances and qualities. In this regard, quantum dots (QDs) are extensively being studied, especially due to their excellent properties. The utilization of QDs in supercapacitors is huge as electrode material as well as for fluorescent electrolytes. Various QDs based composites have been made for the same, which includes doping with various metals, non-metals and carbon nanomaterials (CNMs) like graphene, carbon nanotubes (CNTs) etc. In the present chapter the current advancement and futuristic possibilities of supercapacitors have been mentioned extensively.

Keywords

Carbon Nanomaterials, Graphene Quantum Dots, Heteroatom Doping, Energy Storage Devices, Supercapacitors

Contents

1.   Introduction

As the global population is increasing, the demand of energy storage devices is also increasing. The increasing demands of energy can only be fulfilled by low-cost production of energy storage devices. To deal with such possible crisis we need to developed portable, light weighted and flexible electronic energy storage systems with excellent energy density along with power density [1-4]. In this regard ultracapacitors have been recognized as the excellent candidates as they have excellent properties like fastest known charge-discharge cycles, excellent power density along with amazing high energy density in comparison to the traditional batteries and other energy storage devices. The difference in energy storage mechanics of ultracapacitors have made them a favorite for the energy industries they offer to store the essence of power density in short time duration which enhances their utility as portable electronic goods [5,6]. The non-faradic charge storage mechanism with the presence of van der Wall interactions in between electrode's active material and electrolyte, is the basic difference amid to the electric double layer supercapacitors (EDLCs) and traditional batteries [7-11].

The excellence of supercapacitors are vital due to their utilization in every required form of human life because of their extraordinary properties like having high power density along with long lifecycles, light weighted structure, cost effective and eco-friendly production along with flexibility in shape, higher thermal range [9,10]. Thus supercapacitors are the basic toolkits as energy regaining materials and also as the power booster for most of the automobile sector machines (i.e. heavy electric buses, high load holder cranes, electric trucks etc.) along with other portable electronic appliances [11,12]. The basic unit of a supercapacitor is comprised of two electrodes along with electrolyte (which may be of aqueous, jelly substance or organic solutions etc.) and an insulator that works as the insulating materials, which leads the separation of the charges or ions mutually in between electrodes [13,14]. On overall performance and structural alignment basis the effective part for the enhancement of performance in supercapacitor is their electrode material. On the basis of variation in the structural proportion of electrode materials the ultracapacitors have been sub divided into three important categories i.e. hybrid capacitors (HCs), pseudocapacitors and electro-chemical double layer supercapacitors (EDLCs). EDLCs are the electrically double layered capacitors having the most profound utilization in the field of energy specially they work on the utilization of porous carbonic nano-materials as the electrode materials [15], where the charge can only be stored via physical interactions or by non-faradic pathway in between electrodes and electrolyte materials ion, that not only allows to build up the excellent capacitance and huge power density but also responsible for extraordinary life cycles of thus formed supercapacitors [16,17].

While the operating voltage and the obtained specific capacitance are the major factors that are responsible for the electrochemical storage behavior of the EDLCs supercapacitors. The conductive properties of EDLCs are mainly due to the outcome of high surface area materials along with conduction of electron between the same and different layers of the electrode materials. Similarly, the porosity of used electrode materials is another alternative source for the material selection to hold up the variations in their electrochemical properties that utilize as one of the prime requirements for EDLCs supercapacitor electrode materials. The separation of differently charged ions of material that accumulates around the surface of separator aligned towards the opposite electrodes is the result of variation in the applications of voltage and potential difference for the same [18,19]. The proper selection of these qualities with specifications of different electrolytes i.e. aqueous, aero-gel electrolytes, solid and organic electrolytes etc. define the capacitive behavior to the extreme. The EDLCs mainly uses different carbon based materials like activated carbonaceous materials, carbon nanotubes (CNTs), graphene and its counterparts based composites for generation of high power and energy density of formed device [20-24].

Another category of SCs i.e. pseudocapacitors (PCs) which can be differentiated on the basis of methodologic differences of charge storage mechanism of PCs i.e. redox reaction which is basically the faradic charge transfer thus responsible for high capacitance and energy density while the PCs mainly finds themselves as low power density and cycle life

Materials Research Forum LLC
https://doi.org/10.21741/9781644901250-9

in comparison to EDLCs. Usually when the rate of charging is faster the power density is also found to be higher for ultra-capacitors, whereas the charge storage capability of any electrode material is responsible for their energy density. The diffusion of ions between electrode and electrolyte materials along with the excellence of charge storage defines the electrochemical performance of pseudocapacitors. The faradic reaction is somehow limit the pseudocapacitors to find the faster rate of charge transfer, which is ultimately the reason behind their lower power density and excellently higher energy density in comparison to EDLCs. The materials used in pseudocapacitors are mostly the metal oxides of higher and lower transition metals, conducting polymers and their nanocomposites etc., that act as the basic unit materials for ultracapacitors [25-31].

Hybrid ultracapacitors are another groups of ultracapacitors also known as asymmetric supercapacitors, which is bit like a mixture of both EDLCs and pseudocapacitors as they have two different electrodes having faradic reaction and capacitance storage process [32-33]. These differences in the mechanism of electrochemical energy storage and charging behaviors make them achieve the status of excellently higher power density like EDLCs, exceptional energy density like traditional batteries and pseudocapacitors. Thus obtained nature is the main reason behind their name viz. hybrid supercapacitors that are the main alternative route for the energy saving due to their excellent cell potential, energy density, power density and excellent overall performance in comparison to EDLCs and traditional batteries [34-36]. The balanced ratio of various materials and their comparative particle size along with morphological innovation that creates higher surface area and excellent alignment with the electrolyte system can effectively enhance their electrochemical utility [36]. The excellent power density, long-term cycle life and extraordinary charging–discharging tendency of ultracapacitors have allowed them to establish their market thoroughly in research to various industries [37,38].

From the last couple of decades, carbon based nanomaterials (CNMs) i.e. graphene and its counterparts have been consistently used as the base material for various kinds of scientific applications viz. supercapacitors mainly due to their excellent electric conductivity, higher possibilities of morphological variations, exceptional mechanical stiffness, highly elastic nature, good thermal stability, amazing molecular transparency and light-weighted structure along with the tremendous potential of doping capabilities [39,40]. The CNMs family can be described as 0D (fullerenes), 2D (carbon nanotubes),3D (graphite) on the basis of different dimensional arrangements of layers in accordance to their relative layered structure and sizes of particles [23,24]. Graphene and its counterparts have been highlighting the research potential mostly due to their excellent surface area and flexibility along with thermochemical stability and accessible materialistic mobility [41,42].

Whereas the graphene shows agglomeration in different solvents along with scattering nature [43], thus it is more compatible to alter the graphene into sheets of graphene having size lesser than 10 nm i.e. graphene quantum dots(GQDs) [44]. GQDs sheets are made up of carbon, hydrogen and oxygen based functional groups i.e. hydroxyl, epoxy, carboxylic acids and carbonyl [45]. Usually, GQDs are the sheets of graphene having size in between 3–20 nm i.e. few or 2-5 layers of graphene, while the measured height is about 0.5 to 5nm along with the dimensions of 1.5 to 100 nm range [46,47]. While the geometry of such sheets of graphene mainly categorized as hexagonal quantum dots, quadratic quantum dots, round quantum dots and elliptical quantum dots [47-49]. Similarly, the quantum dots based on carbonic skeleton as the basic structural unit is basically known as carbon quantum dots (CDs), show more or less similar properties as GQDs [50].

Due to the presence of such exceptional properties quantum dots especially GQDs sought huge research attention and have become the next level of interest to scientific community [51]. This graphene-based sheets i.e. GQDs have some excellent properties like stable photo-luminescence, better dispersion in various nano hybrids and conducting polymers along with comprehensive biomedical utility, thus GQDs have been extensively used into the field of electrochemical researches especially supercapacitors [52-53], polymer nano-composites for various structural and mechanical application [54], solar cells [55-56] and bio-imaging applications [57-58]. While as graphene itself lack the bandgaps but by controlling the morphological nature of graphene i.e. various structural modifications or converting into GQDs with altering their sizes, one can easily adjust the magnitude of bandgaps [59,60], the GDQs having sp^2 carbonic bonds inside their layers display variations in most of the properties in comparison to graphene might be due to the presence of edge and quantum defects [61,62]. Recently, GQDs have become the most prior scientific attention basically due to some of their excellent properties like eco-friendly nature, stable fluorescent and photo-luminescence activity, huge surface area, highly transparent nature including an amazing hole transporting tendency etc. whereas, the morphological grafting can be utilized to stimulate their potential uses in the field of polymer nano-composite to biomedical applications i.e. bio-imaging and drug delivery and their biologically compatible nature ensures them to be one of the best vehicle for drug delivery application [63-66].

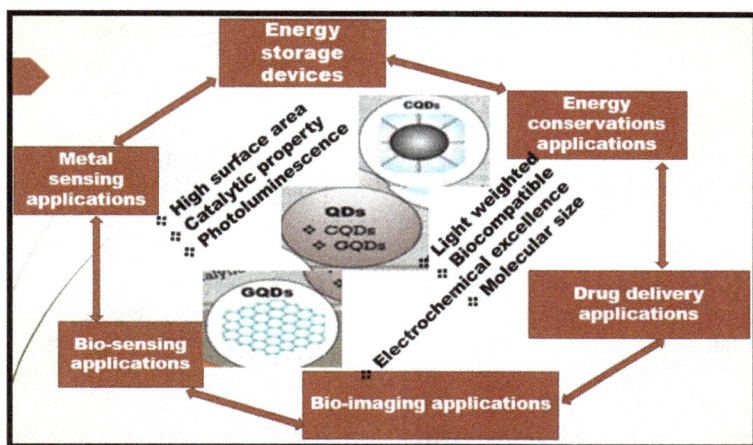

Fig. 1 *Illustration of the main applications and properties of QDs.*

According to report [42] the GQDs will enhance the optoelectrical and physio-chemical properties of nanohybrids by simple incorporations and doping with different members of CNMs family or conductive polymers. Whereas the GQDs have shown their exceptional transparency, huge surface area and excellently high conductivity in comparison to that of oxidic form of graphene i.e. GO. Along with this, the presence of graphene as the matrix structure of few layers make them appropriate option in comparison to carbon quantum dots (CQDs) for energy storage applications specially for the preparation of electrode and electrolytic materials [67-69]. These are some of the basic characteristics those are responsible for their excellent candidatures as electrode materials for various energy storage to energy conservation applications like for ultra-capacitors, traditional batteries and Li-ion batteries [70-75].

This book chapter contains the recent overview of various methods (up to down and down to up approaches) to synthesize quantum dots (QDs) i.e. specially carbon quantum dots (CQDs) and graphene quantum dots (GQDs) and quantum dots based materials and their hybrid composites for supercapacitors applications.

2.   Synthesis of quantum dots via different methods

Usually, QDs are synthesized by two approaches viz. top-down approach and bottom up approach [76]. The bottom-up methodology is mostly preferred over the top down and its precursors are CNMs like GO, CNTs etc. while up to down approach is basically cutting bigger sheets of carbonic materials into smaller ones i.e. QDs. These both methodologies have their own pros and cons [77].

2.1   Top to down approach

In the top to down approaches usually the bulk phase degradation of carbonic clustered material into their smaller counterparts takes place, where the initial raw material is carbon. This includes various experimental methodologies of QDs synthesis viz oxidation, hydro-solvothermal treatments [78,79], microwave irradiations [80], sonication [81], electrochemical exfoliations [82-84], stronger acidic treatments [85,86], lithographic methodologies and mechanical grinding etc. [87,88]. The oxidative treatments of carbonic culture lead CQDs and GQDs by surface oxidation of different carbonic materials [89]. The top to down approach is mainly used for the bulk phase production of QDs. The oxidative form of graphene i.e. GO is an excellent raw candidate for this but due to complications in the bulk production of GO itself pulls it out from such opportunity [90,91].

However, the natural precursors like agriculturally based carbonic materials are mostly converted by their oxidative cleavage for such bulk production of QDs. Similarly, the production of QDs from starting from graphite is also not that high as it lacks functionality [92,93]. This approach leads the maximum yield by approaching cost effective treatments viz. cheaper carbonic raw materials [60]. The oxidative treatments of acids lead hydrophilic sites of QDs, which can be due to the structural defects present around the successive layers of graphite structure. Whereas the basic medium like NaOH has been used for the reduction of QDs [60].

While the cost-effective and greener pathways, bulk production scale are some of the biggest plus for top to down approaches but simultaneously there are some associated drawbacks likewise non uniform size of synthesized QDs, difficulty in controlling the methodology, less yield relative to the starting materials etc. [94].

2.1.1   Hydrothermal /solvothermal treatment

Hydrothermal treatment is the cheaper, environmentally friendly, nontoxic technique and simple method for the synthesis of carbon dots and graphene quantum dots from the various carbon sources as the starting materials. In this process, the solution of organic

Materials Research Forum LLC
https://doi.org/10.21741/9781644901250-9

precursor sealed in a Teflon lined autoclave reactor followed by the heating at the low to high temperature. Mostly rGO is the common precursor that is being used during the hydrothermal treatments. rGO sheets are commonly hot treated with oxidizing acids viz nitric acids to provide a better site, while the basic medium provides the final selective reduction of the surface [95-98].

In this regard, Yin et al. [99] reported the synthesis of carbon quantum dots from the fresh pepper via two-step process: (i) first, low temperature carbonization of fresh pepper with water as the solvent in the Teflon linked autoclave at 180 °C for 5 h; (ii) second, collection of carbon dots by removing the large particles via centrifugation process. Further, the obtained products were dispersed in the water and dialyzed for 20 h for getting high quality carbon dots. Whereas Zhu et al. [96] outlined the sonication based solvothermal method for the production of GQDs from graphene oxide. Pan et al. [95] have mentioned the solution processed procedure for the synthesis of QDs, where the mixture was placed in Teflon based hydrothermal autoclave at 200 °C for several hours. Here the production of QDs has been enhanced by the attachments of more oxidative functional moieties into the oxidizing form of starting materials. In this procedure, different types of solvents have been taken to change the tuning of their structure and morphology, as a result photoluminescence peaks of QDs has also been tuned. The solvothermal approach leads the alteration into the edges size of the QDs [100]. GQDs with excellent thermal solubility in polar solvents have been synthesized by solvothermal approach, which showed the green emission peaks of photoluminescence (PL) due to the excellence of QDs in hetero atom capture from the polar solvents [101].

Mehta et al. [102] reported the synthesis of carbon dots from *Saccharum officinarum* juice via the hydrothermal method. In this method juices of *Saccharum officinarum* mixed with the ethanol then it was poured into the Teflon reactor, that was kept at 120°C for 180 minutes. They have obtained the dark brown color solution, which was further washed with the dichloromethane for the removal of uncreative organic species. After this, they have obtained aqueous solution through centrifugation process and separated the less fluorescent deposit. Further, the excess of acetone was added into the obtained solution then centrifuged for 15 minutes, as a result they have obtained the high fluorescent carbon dots of ~2.5–3.0 nm size range. Fig. 2 [103] depicts the better visualization of the hydrothermal processing on the sheets of carbonic compound.

Sahu et al. [104] reported the synthesis of highly luminescent carbon dots from orange juice via hydrothermal process at 120°C for 2.5 hours. Further, these were separated into two types of quantum dots via optimization of the centrifugation process. First, separated were highly fluorescent carbon dots almost monodisperse (CD) at 10,000 rpm and second, less fluorescent carbon dots coarse particles (CP) at 3000 rpm.

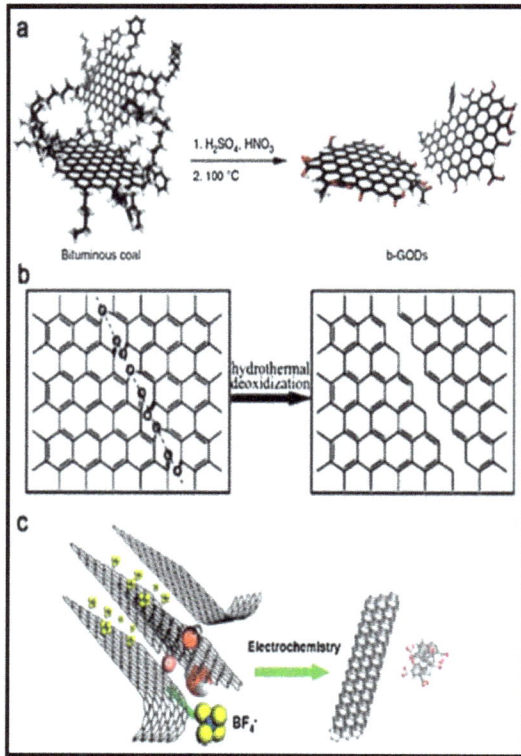

Fig. 2 (a) Chemical oxidation. (b) Hydrothermal synthesis approach. (c) Electrochemical exfoliation [103].

The synthesis of carbon dots was reported with the cabbage, aloe, papaya, tulsi leaves, pseudo-stem of banana plant, sweet potato, rose-heart radish, and rice flour via hydrothermal methods [105-109]. The solvothermal method is a bit like a combination of aerogel and hydrothermal methodology for the synthesis of various QDs. The nitrogen incorporated nanomaterials based on iron trioxide and graphene nanoparticles were reported with controlled size, shape and crystal like nature [110], where the even distribution of nitrogen on the layers of graphene was outlined by experimental data [59]. Similarly, several other researches have been made for the synthesis and utilization of

QDs by solvothermal method just to interlink the possible benefit of for futuristic research [111,112].

2.1.2 Microwave irradiation

The microwave irradiation method utilizes the microwave heat that is emitted by the molecular rotation of polar solvents having dipole moment [12]. This is a well-developed method for the synthesis of carbon quantum dots due to its fast processing, which is easy and quick preparation technique for QDs (Fig. 3) [69]. In this regard, Milosavljevic et al. [113] reported the synthesis of carbon quantum dots from the combining poly(ethylene glycol) (PEG200) and a saccharide (glucose, fructose, etc.) in water and then heated in the microwave oven at 120 °C to 140 °C.

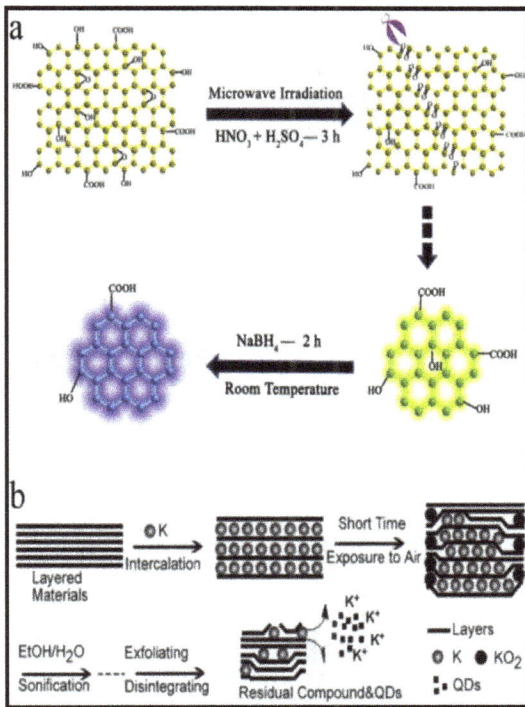

Fig. 3 *(a) Microwave synthesis of GQDs (b) GQDs synthesis by exfoliation of CNTs and graphite [69].*

Umrao et al. [114] reported the synthesis of green and blue graphene quantum dots via microwave carbonization of aromatization acetylacetone as a starting organic solvent with switchable and size tunable by alteration in the size of quantum dots. They reported the two step microwave radiation process. First step for the synthesis of green quantum dots at 200 °C and power of 800 W were performed for 5 minutes and subsequently, in second step power of 900 W was applied to get blue-GQDs. Further, Gu et al. [115] reported the synthesis of nitrogen doped fluorescent carbon dots from the lotus roots via microwave method.

2.2 Bottom to up approach

In bottom to up approach the smaller organic molecules i.e. micro sized molecules are basically used as the precursor for the synthesis of various QDs. Extensive research has been done mostly by using benzene and its derivatives for the same. Earlier, Zheng et al. [116] have reported bottom to up approach while using chemical synthesis route for the cage opening of the carbonic allotrope i.e. fullerene (C_{60}) for the synthesis of CQDs (Fig.4) [80]. However, they have utilized stepwise organic synthesis approach to reduce the presence of various defects starting from edge dislocation to several other 3D defects of the geometry of QDs.

Thus, synthesized QDs were found to have excellent morphology even at lower yield and difficulty in aggregation of molecules due to their intermolecular ionic interactions [113]. Similarly, the pyrolytic heating of smaller carbonic materials can be done for their production with the utilization of bottom to up approach. Mostly, in this approach the organic molecules are usually being melted above their melting point and further nucleated for the agitation process to form QDs. The fullerene based GQDs synthesis by the following bottom to up approach, however, provides well defined structure of GQDs but still requires some control over the difficulty of its processing along with the simplifications of each complicated synthesis step [112].

According to Yan et al. [91] the oxidative agitation of aromatic moieties in polyphenylene based precursor leads the formation of GQDs with the carbonic cluster as $C_{168,}$ $C_{132,}$ C_{170} atoms. For the thermal stabilization of QDs the covalent linkage has been used as the basis of agitation in between graphene and trialkyl phenyl groups [110].

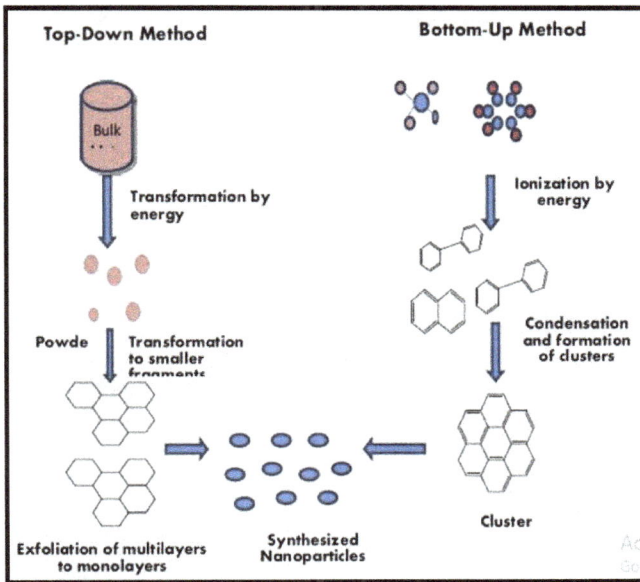

Fig. 4 *The bottom to up and top to down approach for the synthesis of quantum dots [80].*

2.2.1 Chemical method

The chemical synthesis method is basically used for the production of QDs with excellent quality and specific size like 2 to 3 nm range. The opening of the fullerene cage by taking it as a precursor has been reported by Chua et al. [117] where they have reported the oxidative cutting of the source carbon material followed by the modified Hummers method. They have introduced oxidative groups such as hydroxyl group or carboxyl group in quantum dots by using highly oxidative reagent such as concentrate HNO_3 and H_2SO_4. The extensive oxidation increases the solubility of QDs in water and enhances the fluorescence properties of quantum dots. Likewise, trifluoroacetic anhydride and 4-nitrobenzoyl chloride were used for the chemical tailoring of GQDs, which showed the fine changes in photoluminescence of QDs. Here the dialysis of the solution was done by the help of a membrane having cutoff below 1kDa. Umrao et al. [114] have reported molecular aggregation of oxygen, carbon and hydrogen for the synthesis of GQDs. Thus, obtained GQDs showed an increase in luminescence intensity i.e. 460 nm intensity at 340 nm wavelength.

Zhang et al. [118] reported the synthesis of uniform graphene quantum dots from the extensive oxidation of graphene oxide via HNO_3 followed by the hydrothermal process at 100 °C and carbonized at 800 °C for 4 hours. The hetero atom i.e. sulphur, nitrogen and fluorine doped GQDs were synthesized by Kundu et al. [119] having the average diameter of 2 nm. They have taken CNTs as the precursors with microwave heating of ionic medium, which further results in excellent yield of GQDs i.e. 85%. Thus, doped GQDs were found to have excellent luminescence properties and they have also reported doping of three hetero atoms enhances the electrochemical and catalytical properties of QDs. While, nitrogen doped GQDs were reported by Ju et al. [120] with incorporation of the Au nanoparticles, where citric acid along with dicyandiamide (DCD) was fabricated via the hydrothermal route in bottom to up approach. Thus, obtained materials have been shown to have excellent electro-chemical properties for various energy to live cell imaging applications. While the incorporation of Au nanoparticle with nitrogen-doped GQDs, is found to have significant detection rate for cervical cancer in human.

2.2.2 Carbonization procedure

The synthesis of QDs by the pyrolytic carbonization process basically uses the organic source as the precursor for their synthesis such as peels and juices of fruits, food items, plants leave, grass, etc. This process is very facile, greener and cost-effective for the synthesis of QDs. In this process, the source materials are carbonized at a certain temperature in the presence of air. Further, the obtained carbonized products are dispersed in the water via sonication followed by filtration and centrifugation, which forms the luminescent C- dots (CD). After this process, to obtain the high quality and strong luminescent C- dots, this solution is dialyzed with the water with more spherical shape and water-soluble C- dots (CD) [120]. In this regard, Zhou et. al [121] reported the synthesis of carbon quantum dots from the carbonization of fresh watermelon peel at 220 °C for 2 hours under the air atmosphere. Most organic salts like octadecylammonium citrate, diethylglycoammonium citrate, citric acids, ethylenediaminetetraacetate (EDTA) are some of the reported precursors for the production of QDs via this methodology [121-125].

These both top down and bottom-up approaches for the synthesis of QDs i.e. CQDs and GQDs basically follows either aggregation or degradation methodology. Thus, these two approaches can be utilized for their potential use for the synthesis and structural tailoring of obtained QDs. While both of them have their associated advantages and disadvantages, which mostly includes variations of experimental procedure and conditions, chemicals used, equipment along with various difficulties in synthesizing steps. Thus, there is still a need to look after a better methodology in terms of ease of procedure and production yield.

3.   Quantum dots based materials for ultracapacitors application

Due to the excellence in structural, electrochemical to mechanical properties, the QDs and their counterparts have been used as the premium candidates for various industrial and scientific applications likewise bio-imaging [57], fuel cells and supercapacitors [126] etc. The QDs are basically utilized as electrode or electrolytic material into the field of ultracapacitors (UCs). Whereas the doping of metals, non-metals and various carbonic clusters i.e. graphene, graphite, CNTs, fullerenes, activated carbon etc. have different effects on their structures to their direct application in the field of supercapacitors (SCs) [78]. In this chapter the electrode materials based on different QDs have been discussed thoroughly.

3.1   QDs based hybrid composites

An electrode material with high surface area, high electric conductivity, excellent thermal stability, high power density and porosity are the basic requirement to get the high performance of supercapacitors. By taking this as a basis of the division, supercapacitors are further subdivided into three sub classes viz pseudocapacitors (PCs), hybrid capacitors and electric double layer capacitors (EDLCs). The pseudocapacitors follow faradic i.e. redox mechanism for the charge storage, while the non-faradic mechanics for charge storage has been defined as the process of the electric double layer capacitors. In order to achieve excellent outputs for energy storage devices, various researchers have outlined the QDs as brighter candidates for such electrochemical studies. Their utilization into electrochemical capacitors is extensive known due to their excellent electrochemical properties [86,126].

The incorporation of conductive polymers, metal oxides and heteroatoms with the QDs moiety reflects the extension of operating potential window along with the improved capacitance performance rate of QDs based electrode materials by enhancing their surface area and compatibility with electrolytes. While the anchoring at the surface structure of QDs based electrode material also affects the performance of the fabricated device extensively. The polyaniline and GQDs based hybrid composites have been prepared by chemical oxidation of aniline for pseudocapacitors as reported by Mondal et al. [86]. The prepared GQDs were sized around ~6.0 nm and obtained from oxidation of GO flakes. Moreover, the morphological diameter of GQDs could be controlled by the weight ratio of both aniline and GQDs. Thus obtained composite showed the huge

capacitance value of ~1044 Fg^{-1} at the current density of 1 Ag^{-1}, where the cyclic
capacitance showed the 80.1% retention after almost 3000 cycles (Fig. 5) [127].

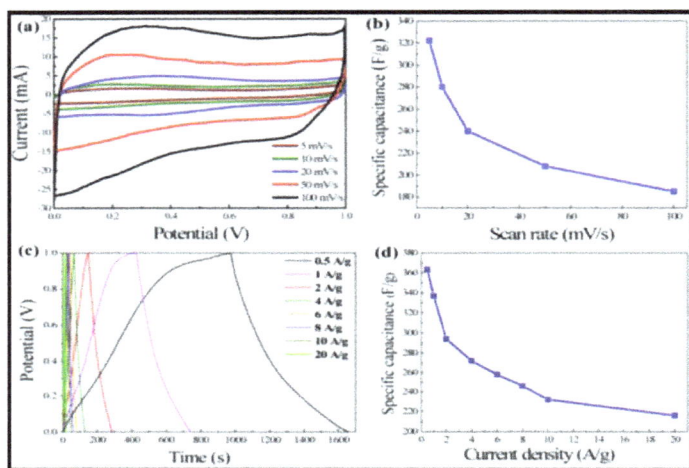

Fig. 5 *(a) CV plots (b) specific capacitances of GQD based composites (c)
charge–discharge curves, (d) Capacitances of GQD based nanocomposites at different
current densities [127].*

The gold metals electrode based symmetrical ultracapacitors have been fabricated by a
complete electrode deposition method by Liu et al., [128]. The device showed excellent
rate potency of 1000 V s^{-1} with variations of electrolytes. They have reported good
retention of specific capacitance for 1500 life cycles. In another experiment,
ultracapacitor electrode based on conductive polymer aligned GQDs have been fabricated
via the electrochemical deposition method. The potential candidature of the above
electrode was also tested and showed good specific capacity and thermal stability. RuO_2
incorporated QDs having porous and multichannel structures reported with excellent
specific capacitance, where the thermal stability of capacitance was found to be 100%
after 5000 cycles. The incorporation of QDs with such metal shows the relative increase
in electrical properties by enhancing the crystalline nature of the material with good rate
[80]. The GQDs and PANI hybrid composite was fabricated for asymmetric device,
where PANI i.e. polyaniline nanomaterials were prepared via electrochemical route

afterwards the deposition of GQDs were made possible through electrical deposition method. Here PANI/GQDs hybrid showed excellent rate with shorter relaxation time i.e. 115.9ms [128]. Qing et al. [129] have improved the capacitative performance of the supercapacitor by creating inclusive conductive networks using GQDs.

While the GQD/GQD based symmetrical capacitor was developed by electrical deposition method and ionic liquid was taken as the electrolyte for further investigation. Along with this, a GQDs/MnO_2 based asymmetric ultracapacitor was also fabricated by the same researchers, where MnO_2 nano tubes were utilized as positive and GQDs as their counter electrode materials in ionic liquid electrolyte. Here the GQDs elaborate the rate of ionic movability along with surface area, thus fabricated GQDs/MnO_2 asymmetric device showed two times better performance in comparison to GQDs based symmetric capacitor [130].

The electrical performance of the asymmetric ultracapacitor was found to have excellent rate capability on the scanning rate of 1000 Vs^{-1} along with the good power density having the relaxation time of 103.6 ms, whereas the specific capacitance of 468.1 m Fcm^{-2} was reported for the same device in aqueous solution of 0.5M Na_2SO_4 [130]. The CQDs based ultracapacitors were synthesized by the simultaneous deposition of pyrrole and CQDs in aqueous solution. The role of deposition time was mentioned as the crucial link for the same. The symmetric capacitor with CQDs/pyrrole hybrid electrode was fabricated by taking gel electrolyte of polyvinyl alcohol/LiCl as the separator of electrodes. The fabricated device showed excellent capacitance of 315 mF cm^{-2} at current density of 0.2 mA cm^{-2}, whereas the excellent cycle life with 85.7% retention was reported after 2000 cycles [131].

3.2   QDs based carbonic materials

QDs i.e. CDs and GQDs have been recognized as noticeably special materials for the advancement of the various industrial and scientific applications for the current worlds need, which includes supercapacitors, solar cell, electrochemical batteries, drug delivery to polymer-composites due to their extensively lucid properties [126]. To improve the performance of various applications, the fabricated device based on QDs can be aligned with various other forms of carbon skeleton likewise CNT, fullerene, graphene etc. [132-133].

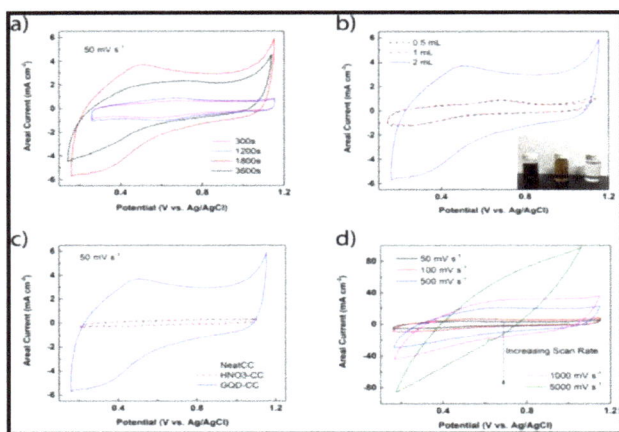

Fig. 6 *CV of GQD/Carbon cloth (a) Composite of GQDs with different deposition times and scan rates, (b) Composite electrodes with GQDs, (c) Comparison of HNO₃-treated and plain GQD at different scan rates, (d) CV curves with different scan [134].*

Hu et al. [100] have reported the electrical method of CNTs deposition for their alignment on GQDs. They obtained specific capacitance of 44 mF cm^{-2}, and reported the better performance of CNT based electrodes over previously studied CNT devices. The presence of the porous nature of surface morphology of QDs has major advantages over their other counterparts of same family [107]. As the meso or micro porous nature of the QDs facilitates easy passage for ions transportation in between the layers, which ultimately excel the thermal stability and electrical properties of electrode materials thus accordingly the process of charge storage mechanism varies in each kind of supercapacitors [108-109]. The GQDs/halloysite nanotubes (HNT) composite was reported to show excellent capacitive properties for the ultracapacitors application [135]. The amide bonding based homogeneous sliding of GQDs on the surface of HNTs was reported. The hybrid was allowed to go through the cyclic voltammetry (CV) study. The GQDs/halloysite nanotubes (HNT) composite had showed excellent results in 1MNa₂SO₄ electrolyte at voltage rates of 5 to 100 mV/s, where the rectangular shape indicated the EDLC capacitor with better cyclic stability and rate capacity. They have reported the specific electrochemical capacitance of 323F/g at the rate of 5 mV/s which on increasing the rate to 100 mV/s was found to be at 186 F/g.

The overall capacitance showed the retention of 58%. In another experiment, the GQDs was filmed over the glassy carbon materials electrode CV experiment showed the specific conductance of 108 F/g, but the composite of GQD/HNT was found to have upper hand over the plain GQDs based electrode. The galvanometric charge-discharge experiment was performed in 1m Na_2SO_4 solution with the current densities ranging from 0.5 A/g to 20 A/g, where the resulting curve showed symmetric nature of capacitor [136]. Fig. 7 [69] clearly shows the fabrication and photographic imaging including SEM pictures of QDs.

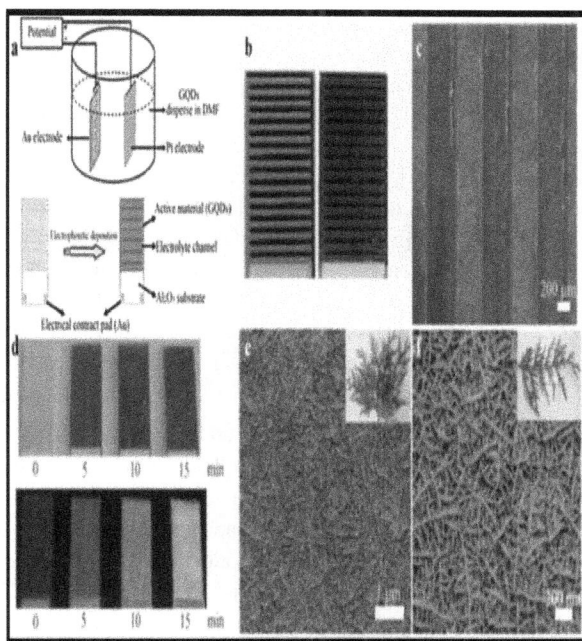

Fig. 7 *(a) electrophoretic deposition of GQDs on GQDs (b) Digital image of GQDs//MnO₂ asymmetric supercapacitor, (c) SEM image of GQDs and MnO₂ materials, (d) Digital images of GQDs with variations in deposition time, (e, f) SEM images of MnO₂ grafted material [69].*

Similarly, Ricky et al. [136] reported a flexible supercapacitor based on graphene quantum dots (GQDs) and carbon cloth. The hydrothermal methodology followed by electrochemical deposition was used for the synthesis and fabrication of ultracapacitors. The carbon cloth based electrodes were reported as flexible electrodes with excellent specific capacitance of 70 mF cm^{-2} at the rate of 50 mVs^{-1}. The cyclic voltammetry curves were reported in rectangular shape, where the excellent electrochemical behavior was supported by specific capacitance of 24 mF cm^{-2} even at the high scan rates i.e. varying from 1 Vs^{-1} to 50 mVs^{-1}.

Su Zhang et al. [137] have reported solid and solution type GQDs as electrolytic having the excessive number of oxygen containing functional groups which mainly includes carboxylic acids (-COOH) and alcoholic (-OH) groups for the supercapacitors applications. They showed the conductive properties of obtained GQDs can be improved by the simple solvation of potassium hydroxide. Thus, neutralized form of GQDs were reported to have excellent electrolytic properties of various supercapacitors and the reason for such enhancement in properties ascribed by complete ionization of the oxygen containing functional groups due to the neutralization process. The Fig. 8 [69] depicts the CV based curves of symmetric ultracapacitors, GQDs/MnO$_2$ supercapacitor at different voltage ranges. Zhang et al. [132] reported the CQDs/GO/polypyrrole based hybrid as the electrode material where the carbon dots were intertwined in between GO and polypyrrole layers. The huge aspect ratio and extensive surface area of CQDs were reported to enhance the ionic transport and reduction of resistance for the same. The cyclic voltammetry results showed huge capacitance of 576 F gm^{-1} at the current density of 0.5 A gm^{-1} while the excellent cyclic life stability of 5000 cycles were obtained.

The enhancement of surface area for NiMoO$_4$ QDs@nanohole rGO composite was reported by Ganganboina et al. [138]. The obtained device displayed the excellent specific capacitance of 2327.3 F gm^{-1} with high cyclic stability when compared to NiMoO$_4$@rGO composite which showed the capacitance of 1801.2 F gm^{-1}. Xia et al. [139] have described the development of iron trioxide QDs based composite of graphene as ultracapacitor electrode material that showed much higher density of 50.7 Wh Kg^{-1} and power density of 100 Wkg^{-1} with excellent cycle stability. Xuemei Zhou et al. [134] have reported the MWCNT/GQDs composite which exhibits the enhancement of ORR activity by increasing the catalytic activity of Pt/C catalyst. They have mentioned the reported composite enhances the electrochemical properties along with life cycles in comparison to other commercial available catalysts viz. Pt/C catalysts. Yan Qing et al. [129] reported GQDs based activated carbon materials for the ultracapacitors device, where the GQDs were found to be embedded into matrix of activated carbon to enhance the capacitative performance of fabricated device. The microporous nature of the obtained materials showed excellent cyclic stability with huge surface area of 2829 m^2 g^{-1}. Due to the high surface area the capacitance also enhanced to 388 Fg^{-1} with good power density and high energy density of 13.47 Wh Kg^{-1}.

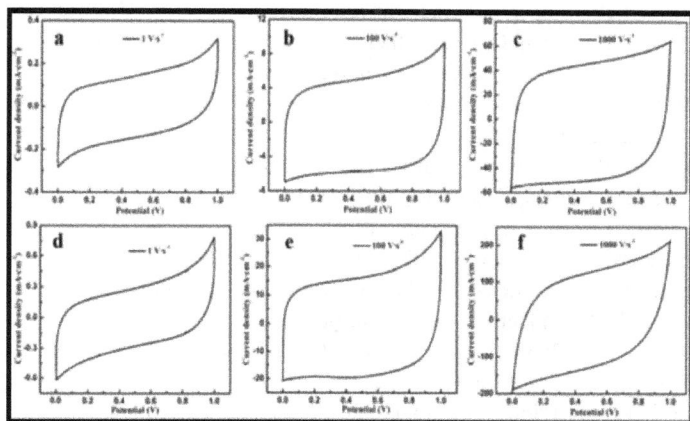

Fig. 8 *(a, b, c) CV curves of GQDs based symmetric supercapacitor (a) 1 Vs_1, (b) 100 Vs_1 (c) 1000 Vs_1. (d,e, f) GQDs/MnO₂supercapacitor in 0.5 M Na₂SO₄ at (a) 1 Vs_1, (b) 100 Vs_1, (c) 1000 Vs_1 [69].*

Conclusive overview and future scope

The universal energy crisis with increasing global population has scripted the attention of researcher towards the new wonder of nanoscience i.e. QDs and their counterparts based various hybrid nano-composites for most of the industrial applications of energy storage devices. The huge attention of researchers on the QDs is due to their excellent properties and has created new chapter into the direction of research to find a unique solution for such huge global energy crisis related needs. This chapter briefly mentioned the various approaches to synthesize various QDs specially carbon quantum dots (CQDs), graphene quantum dots(GQDs) and based various composites, along with this the chapter discusses the most recent developments on the QDs based hybrid composites including their potential candidature for energy storage devices and various tailoring approaches for their method of synthesis have also been discussed broadly, the chapter also covers up the various new polishing methodologies that can enhance the performance of various electrochemical devices.

We expect this chapter acts as an excellent modular to generate some of the advanced ideas for enhancing the electrochemical performance of various energy storage devices (especially for supercapacitors applications) along with this it opens up a door for newer materials to the field of energy storage for futuristic ultracapacitor devices. But still there are many issues regarding QDs and their counterparts as the electrode materials for supercapacitors that need to be dealt likewise energy density, charge-discharge cycle, power density, cyclic and thermal stability etc. In order to make their realistic commercial mass scale candidature, QDs need a dedicated approach for their promotion to the next level of nano-science and technology also for their commercial use into the field of supercapacitors applications. This chapter elaborates the current knowledge, efforts, advancements and new finding into the tailoring and doping of various QDs for the good understanding along with futuristic utilization of their structural and morphological changes in the electrochemical behavior of the thus formed composites of QDs. The excellent conductive, optical, bio-compatibility, low toxicity, huge surface area, mechanical and fluorescent properties of QDs have attracted the scientific world towards its application potential to the great extent. Due to the easy synthesis approaches and simple fabrication including photoluminescent advantages of QDs make them extraordinary materials for the ultracapacitors.

In spite of having such a huge advantage over other existing materials of the same family, QDs have some drawbacks also associated with them, which leads several direct effects into their capacitive properties and developed fabricated devices for the supercapacitors. Some of the major associated disadvantages with QDs that need to be corrected for the betterment of nanoscience in the applied field are explained as follow:

(1) The presence of defective sites on to the surface of QDs works as the poison for their catalytic, optical and electrochemical activities. But the techniques with excellent accuracy for depleting of defects in their shape and size are still unavailable. Thus, there is urgent need to develop a technique which can bring up more accuracy in the synthesis of QDs along with the depletion of associated defects.

(2) Due to availability of excessive oxygen containing functional groups, the functionalization potential of QDs are huge but the simplest and specifically targeted controlled functionalization along with doping of non-metals, metals have its own potential that can elaborate the direction of research in the field, so this could bring up new opening with good efficiency for various industrial applications and this still need to be dealt. Nevertheless, of such associated issues QDs i.e. GQDs & CQDs are still in excellent run due to their immense candidature and potential, which need to be upgraded to their divergent potential by implementing some good scientific advancements and technological innovations to the field [132]. Thus discussed disadvantages can be ruled

out with the controlled doping of various materials viz CNMs, conductive polymers, heteroatoms or metal oxide that will ultimately result the exile of defects from their morphology, moreover it may work as the modulator for their electrochemical properties. It is expected that this chapter could enlighten the hidden aspects of the topic and enriches the current knowledge on the versatility of QDs and based electrochemical materials for the development of more advanced supercapacitors to coming future generations.

Acknowledgments

The work is financially supported by NMHS, Kosi Kataramal, Almora, Uttrakhand, India (Ref No.: GBPNI/NMHS-2019-20/MG) and Department of Science and Technology INSPIRE division (Ref No.: IF180347), New Delhi, India.

References

[1]   D.J. Lipomi, Z. Bao, Stretchable, elastic materials and devices for solar energy conversion, Energy Environ. Sci. 4 (2011) 3314–3328. https://doi.org/10.1039/c1ee01881g

[2]   H. Nishide, K. Oyaizu, Toward flexible batteries, Science 80 (2008) 737–738. https://doi.org/10.1126/science.1151831

[3]   J.M. Tarascon, M. Armand, Issues and challenges facing rechargeable lithium batteries, Mater. Sustain. Energy (2010) 171–179.https://doi.org/10.1142/9789814317665_0024

[4]   J.A. Rogers, T. Someya, Y. Huang, Materials and mechanics for stretchable electronics, Science 327 (2010) 1603–1607. https://doi.org/10.1126/science.1182383

[5]   D. Chao, C. Zhu, X. Xia, J. Liu, X. Zhang, J. Wang, P. Liang, J. Lin, H. Zhang, Z.X. Shen, Graphene quantum dots coated VO_2 arrays for highly durable electrodes for Li and Na ion batteries, Nano Lett., 15 (2014) 565-573. https://doi.org/10.1021/nl504038s

[6]   J.R. Miller, P. Simon, Electrochemical capacitors for energy management, Sci. Magazine, 321 (2008) 5889 651–652. http://dx.doi.org/10.1126/science.1158736

[7]   M.D. Stoller, S. Park, Y. Zhu, J. An, R.S. Ruoff, Graphene-based ultracapacitors, Nano Lett. 8 (2008) 3498–3502. https://doi.org/10.1021/nl802558y

[8]   H.P. Wu, D.W. He, Y.S. Wang, M. Fu, Z.L. Liu, J.G. Wang, H.T. Wang, Graphene as the electrode material in supercapacitors, Proc. 2010 8th Int. Vac. Electron Sources Conf. Nanocarbon, IVESC (2010) 465–466. https://doi.org/10.1109/IVESC.2010.5644267

[9]   J. Zhao, G. Chen, L. Zhu, G. Li, Graphene quantum dots-based platform for the fabrication of electrochemical biosensors, Electrochem. Commun. 13 (2011) 31–33. https://doi.org/10.1016/j.elecom.2010.11.005

[10]   M. Hassan, E. Haque, K.R. Reddy, A.I. Minett, J. Chen, V.G. Gomes, Edge enriched graphene quantum dots for enhanced photo-luminescence and supercapacitance, Nanoscale, 6 (2014) 11988-11994. https://doi.org/10.1039/C4NR02365J

[11]   Y. Wang, Z. Shi, Y. Huang, Y. Ma, C. Wang, M. Chen, Y. Chen, Supercapacitor devices based on graphene materials. J. Phys. Chem. C.113 (2009) 13103-13107. https://doi.org/10.1021/jp902214f

[12]   N.L. Wu, Nanocrystalline oxide supercapacitors, Mater. Chem. Phys. 75 (2002) 6–11. https://doi.org/10.1016/S0254-0584(02)00022-6

[13]   Z. S. Iro, C. Subramani, S.S. Dash, A brief review on electrode materials for supercapacitor, Int. J. Electro chem. Sci.11 (2016) 10628 – 10643. https://doi.org/10.20964/2016.12.50

[14]   C.A. Downing, D.A. Stone, M.E. Portnoi, Zero-energy states in graphene quantum dots and rings. Phys. Rev. B. 84 (2011) 155437. https://doi.org/10.1103/PhysRevB.84.155437

[15]   M. Salanne, B. Rotenberg, K. Naoi, K. Kaneko, P. L. Taberna, C. P. Grey, B. Dunn, P. Simon, Efficient storage mechanisms for building better supercapacitors, Nat. Energy. 1 (2016) 16070. https://doi.org/10.1038/nenergy.2016.70

[16]   M.V. Kiamahalleh, S.H.S. Zein, G. Najafpour, S.A. Sata, S. Buniran, Multiwalled carbon nanotubes based nanocomposites for supercapacitors : a review of electrode materials, Nano. 7 (2012) 1230002. https://doi.org/10.1142/S1793292012300022

[17]   M. Jayalakshmi, K Balasubramanian, Simple capacitors to supercapacitors-an overview, Int. J. Electrochem. Sci. 3 (2008) 1196-1217.

[18]   M.S. Halper, J. C. Ellenbogen, Supercapacitors: a brief overview, The MITRE Corporation, McLean, VA. 1–34.

[19]   H. Choi, H. Yoon, Nanostructured electrode materials for electrochemical capacitor applications. Nanomaterials 5 (2015) 906-936. https://doi.org/10.3390/nano5020906

[20]   J. Gamby, P.L. Taberna, P. Simon, J.F. Fauvarque, M. Chesneau, Studies and characterisations of various activated carbons used for carbon/carbon supercapacitors, J. Power Sources 101 (2001) 109–116. https://doi.org/10.1016/S0378-7753(01)00707-8

[21]   M. D. Stoller, S. Park, Y. Zhu, J. An, R. S. Ruoff, Graphene-based ultracapacitors, Nano Lett. 8 (2008) 3498–3502. https://doi.org/10.1021/nl802558y

[22]   C. Liu, Z. Yu, D. Neff, A. Zhamu, B. Z. Jang, Graphene-based supercapacitor with an ultrahigh energy density, Nano Lett. 10 (2010) 4863–4868. https://doi.org/10.1021/nl102661q

[23]   A. I. Najafabadi, S. Yasuda, K. Kobashi, T. Yamada, D. N. Futaba, H. Hatori, M. Yumura, S. Iijima, K. Hata, Extracting the full potential of single-walled carbon nanotubes as durable supercapacitor electrodes operable at 4 V with high power and energy density, Adv. Mater. 22 (2010) 235-241. https://doi.org/10.1002/adma.200904349

[24]   H. Zhou, G. Han, One-step fabrication of heterogeneous conducting polymers-coated graphene oxide/carbon nanotubes composite films for high-performance supercapacitors, Electro. Acta. 192 (2016) 448-455. https://doi.org/10.1016/j.electacta.2016.02.015

[25]   S. A. Hashmi, H. M. Upadhyaya, Polypyrrole and poly (3-methyl thiophene)-based solid state redox supercapacitors using ion conducting polymer electrolyte, Solid State Ions. 152 (2002) 883–889. https://doi.org/10.1016/S0167-2738(02)00390-9

[26]   J.H. Park, J. M. Ko, O.O.Park, D. W. Kim, Capacitance properties of graphite/polypyrrole composite electrode prepared by chemical polymerization of pyrrole on graphite fiber, J. Power Sources 105 (2002) 20–25. https://doi.org/10.1016/S0378-7753(01)00915-6

[27]   T.P. Gujar, W.Y. Kim, I. Puspitasari, K.D. Jung, O.S. Joo, Electrochemically deposited nanograin ruthenium oxide as a pseudocapacitive electrode, J. Phys. Chem. C120 (2016) 2036–2046. https://doi.org/10.1021/acs.jpcc.5b09078

[28] Q. Cheng, J. Tang, J. Ma, H. Zhang, N. Shinya, L.C. Qin, Polyaniline-coated electro-etched carbon fiber cloth electrodes for supercapacitors, J. Phys. Chem. C 115 (2011) 23584–23590. https://doi.org/10.1021/jp203852p

[29] H.Y. Wu, H.W. Wang, Electrochemical synthesis of nickel oxide nanoparticulate films on nickel foils for high-performance electrode materials of supercapacitors, Int. J. Electrochem. Sci. 7 (2012) 4405–4417.

[30] Z. Cai, L. Li, J. Ren, L. Qiu, H. Lin, H. Peng, Flexible, weavable and efficient microsupercapacitor wires based on polyaniline composite fibers incorporated with aligned carbon nanotubes, J. Mater. Chem. A 1 (2013) 258–261. https://doi.org/10.1039/C2TA00274D

[31] R. Jalili, J.M. Razal, G.G. Wallace, Wet-spinning of PEDOT: PSS/functionalized-SWNTs composite: a facile route toward production of strong and highly conducting multifunctional fibers, Sci. Rep. 3 (2013) 3438. https://doi.org/10.1038/srep03438

[32] P. Simon, Y. Gogotsi, Materials for electrochemical capacitors, J. Nanosci. Nanotechno. 320–329. https://doi.org/10.1142/9789814287005_0033

[33] G. Yu, X. Xie, L. Pan, Z. Bao, Y. Cui, Hybrid nanostructured materials for high-performance electrochemical capacitors, Nano Energy 2 (2012) 213–234. https://doi.org/10.1016/j.nanoen.2012.10.006

[34] J. Zhang, X.S. Zhao, Conducting polymers directly coated on reduced graphene oxide sheets as high-performance supercapacitor electrodes, J. Phys. Chem. C 116 (2012) 5420-5426. https://doi.org/10.1021/jp211474e

[35] J. Zhang, P. Yedlapalli, J. W. Lee, Thermodynamic analysis of hydrate-based pre-combustion capture of CO_2, Chem. Eng. Sci. 64 (2009) 4732–4736. https://doi.org/10.1016/j.ces.2009.04.041

[36] G. Yu, X. Xie, L. Pan, Z. Bao, Y. Cui, Hybrid nanostructured materials for high-performance electrochemical capacitors, Nano Energy 2 (2013) 213–234. https://doi.org/10.1016/j.nanoen.2012.10.006

[37] P. Simon, Y. Gogotsi, B. Dunn, Where do batteries end and supercapacitors begin?, Science 343 (2014) 1210-1211. https://doi.org/10.1126/science.1249625.

[38] P. Simon, Y. Gogotsi, Materials for electrochemical capacitors, Nat. mater. 7 (2008) 845-854. https://doi.org/10.1142/9789814287005_0033

[39]   A.K. Geim, K.S. Novoselov, The rise of graphene, J. Nanosci. Nanotechno. (2010) 11-19. https://doi.org/10.1142/9789814287005_0002

[40]   A.K. Geim, Graphene: status and prospects, Science 324 (2009) 1530-1534. https://doi.org/10.1126/science.1158877

[41]   Y. Shao, J. Wang, H. Wu, J. Liu, I.A. Aksay, Y. Lin, Graphene based electrochemical sensors and biosensors: a review, Electroanalysis 22 (2010) 1027–36. https://doi.org/10.1002/elan.200900571

[42]   A.T. Yousefi, S. Bagheri, S. Shahnazar, M.H. Rahman, N.A. Kadri, Computational local stiffness analysis of biological cell: high aspect ratio single wall carbon nanotube tip, Mater. Sci. Eng. C 59 (2016) 636-642. https://doi.org/10.1016/j.msec.2015.10.041

[43]   R. Atif, F. Inam, Reasons and remedies for the agglomeration of multilayered graphene and carbon nanotubes in polymers. Beilstein J. Nanotechnol. 7 (2016) 1174-1196. https://doi.org/10.3762/bjnano.7.109

[44]   P. Tian, L. Tang, K.S. Teng, S.P. Lau, Graphene quantum dots from chemistry to applications. Mater. Today Chem. 10 (2018) pp.221-258. https://doi.org/10.1016/j.mtchem.2018.09.007

[45]   H. Sun, L. Wu, W Wei, X. Qu, Recent advances in graphene quantum dots for sensing, Mater. Today Chem. 16 (2013) 433–442. https://doi.org/10.1016/j.mattod.2013.10.020

[46]   L. Li, G. Wu, G. Yang, J. Peng, J. Zhao, J.J. Zhu, Focusing on luminescent graphene quantum dots: current status and future perspectives, Nanoscale 5 (2013) 4015–39. https://doi.org/10.1039/C3NR33849E

[47]   S. Kim, S.W. Hwang, M.K. Kim, D.Y. Shin, D.H. Shin, C.O. Kim, S. B. Yang, J. H. Park, E. Hwang, S. H. Choi, G. Ko, S. Sim, C. Sone, H.J. Choi, S. Bae, B.H. Hong, Anomalous behaviors of visible luminescence from graphene quantum dots: interplay between size and shape, ACS Nano. 6 (2012) 8203–8208. https://doi.org/10.1021/nn302878r

[48]   J. Peng, W. Gao, B.K. Gupta, Z. Liu, R. R. Aburto, L. Ge, L. Song, L. B. Alemany, X. Zhan, G. Gao, S. A. Vithayathil, B. A. Kaipparettu, A.A. Marti, T. Hayashi, J. J. Zhu, P. M. Ajayan, Graphene quantum dots derived from carbon fibers. Nano Lett. 12(2012) 844–849. https://doi.org/10.1021/nl2038979

[49]   J.S. Wei, H. Ding, P. Zhang, Y.F. Song, J. Chen, Y.G. Wang, H.M. Xiong, Carbon dots/$NiCo_2O_4$ nanocomposites with various morphologies for high performance supercapacitors, Small 12 (2016) 5927-5934. https://doi.org/10.1002/smll.201602164

[50]   Q. Liu, J. Sun, K. Gao, N. Chen, X. Sun, D. Ti, C. Bai, R. Cui, L. Qu, Graphene quantum dots for energy storage and conversion: from fabrication to applications, Mater. Chem. Front 4 (2020) 421-436. https://doi.org/10.1039/C9QM00553F

[51]   Y. Huang, L. Lin, T. Shi, S. Cheng, Y. Zhong, C. Chen, Z. Tang, Graphene quantum dots-induced morphological changes in $CuCo_2S_4$ nanocomposites for supercapacitor electrodes with enhanced performance. Appl. Surf. Sci.463 (2019) 498-503. https://doi.org/10.1016/j.apsusc.2018.08.247

[52]   L. Xu, C. Cheng, C. Yao, X. Jin, Flexible supercapacitor electrode based on lignosulfonate-derived graphene quantum dots/graphene hydrogel, Org. Electron.78 (2020) 105407. https://doi.org/10.1016/j.orgel.2019.105407

[53]   S. Zheng, Z. Jin, C. Han, J. Li, H. Xu, S. Park, J.O. Park, E. Choi, K. Xu, Graphene quantum dots-decorated hollow copper sulfide nanoparticles for controlled intracellular drug release and enhanced photothermal-chemotherapy, J. Mater. Sci. 55 (2020) 1184-1197. https://doi.org/10.1007/s10853-019-04062-x

[54]   V. Prévot, E.B. Lami, Recent advances in layered double hydroxide/polymer latex nanocomposites: from assembly to in situ formation, LDH Polymer Nanocomposite. (2020) 461-495. https://doi.org/10.1016/B978-0-08-101903-0.00011-8

[55]   C. Luk, L. Tang, W. Zhang, S. Yu, K. Teng, S. Lau, An efficient and stable fluorescent graphene quantum dot–agar composite as a converting material in white light emitting diodes, J. Mater. Chem. 22 (2012) 22378–22381. https://doi.org/10.1039/C2JM35305A.

[56]   V. Gupta, N. Chaudhary, R. Srivastava, G.D. Sharma, R. Bhardwaj, S. Chand Luminscent graphene quantum dots for organic photovoltaic devices, J. Am. Chem. Soc 133 (2011) 9960–9963. https://doi.org/10.1021/ja2036749

[57]   C. Tewari, G. Tatrari, M. Karakoti, S. Pandey, M. Pal, S. Rana, B.S. Bhushan, A.B. Melkani, A. Srivastava, N.G. Sahoo, A simple, eco-friendly and green approach to synthesis of blue photoluminescent potassium-doped graphene oxide from agriculture waste for bio-imaging applications, Mater. Sci. Eng. C 104 (2019) 109970. https://doi.org/10.1016/j.msec.2019.109970

[58]   D.I. Son, B.W. Kwon, D.H. Park, W.S. Seo, Y. Yi, B. Angadi, C.L. Lee, W.K. Choi, Emissive ZnO graphene quantum dots for white-light-emitting diodes, Nat. nanotechno. 7 (2012) 465-471. https://doi.org/10.1038/nnano.2012.71

[59]   C.X. Guo, H.B. Yang, Z.M. Sheng, Z.S. Lu, Q.L. Song, C.M. Li, Layered graphene/quantum dots for photovoltaic devices, Angew. Chem. Int. Ed., 49 (2010) 3014-3017. https://doi.org/10.1002/anie.200906291

[60]   O. Koshy, Y. B. Pottathara, S. Thomas, B. Petovar, M. Finsgar, A flexible, disposable hydrogen peroxide sensor on graphene nanoplatelet-coated cellulose. Curr. Anal. Chem. 13 (2017) 480–487. https://doi.org/10.2174/1573411013666170427121958

[61]   H. Fei, R. Ye, G. Ye, Y. Gong, Z. Peng, X. Fan, E.L. Samuel, P.M. Ajayan, J.M. Tour, Boron-and nitrogen-doped graphene quantum dots/graphene hybrid nanoplatelets as efficient electrocatalysts for oxygen reduction. ACS Nano. 8 (2014) 10837-10843. https://doi.org/10.1021/nn504637y

[62]   A. Ananthanarayanan, X. Wang, P. Routh, B. Sana, S. Lim, D. H. Kim, J. Li, P. Chen, Facile synthesis of graphene quantum dots from 3D graphene and their application for Fe^{3+} sensing, Adv Funct. Mater. 24 (2014) 3021–3026. https://doi.org/10.1002/adfm.201303441

[63]   S.L. Ting, S.J. Ee, A. Ananthanarayanan, K.C. Leong, P. Chen, Graphene quantum dots functionalized gold nanoparticles for sensitive electrochemical detection of heavy metal ions, Electrochem. Acta. 172 (2015) 7–11. https://doi.org/10.1016/j.electacta.2015.01.026

[64]   D. Qu, M. Zheng, J. Li, Z. Xie, Z. Sun, Tailoring color emissions from N-doped graphene quantum dots for bioimaging applications, Light Sci. Appl. 4 (2015) 364-364.https://doi.org/10.1038/lsa.2015.137

[65]   A. Brotchie, Graphene quantum dots: It's all in the twist, Nat. Rev. Mater. 1(2016) 1-1. https://doi.org/10.1038/natrevmats.2016.6

[66]   M. Bacon, S. J. Bradley, T. Nann, Graphene quantum dots, Part. Syst. Char. 31 (2014) 415-428. https://doi.org/10.1002/ppsc.201300252

[67]   M. Arvand, S. Hemmati, Magnetic nanoparticles embedded with graphene quantum dots and multiwalled carbon nanotubes as a sensing platform for electrochemical detection of progesterone, Sens. Actuators B: Chem. 238 (2017) 346–356. https://doi.org/10.1016/j.snb.2016.07.066

[68]   S. Bak, D. Kim, H. Lee, Graphene quantum dots and their possible energy applications: A review. Curr. Appl. Phys. 16 (2016) 1192-1201. https://doi.org/10.1016/j.cap.2016.03.026

[69]   Q. Liu, J. Sun, K. Gao, N. Chen, X. Sun, D. Ti, C. Bai, R. Cui, L. Qu, Graphene quantum dots for energy storage and conversion: from fabrication to applications, Mater. Chem. Front. 4 (2020) 421-436. https://doi.org/10.1039/C9QM00553F

[70]   R.A. Fisher, M.R. Watt, W.J. Ready, Functionalized carbon nanotube supercapacitor electrodes: a review on pseudocapacitive materials, Ecs J. Solid State Sc. 2(2013) 170. https://doi.org/10.1149/2.017310jss

[71]   G.A. Snook, P. Kao, A.S. Best, Conducting-polymer-based supercapacitor devices and electrodes, J. Power Sources 196 (2011) 1-12. https://doi.org/10.1016/j.jpowsour.2010.06.084

[72]   C.M. Luk, L.B. Tang, W. F. Zhang, S. F. Yu, K. S. Teng, S.P. Lau, An efficient and stable fluorescent graphene quantum dot–agar composite as a converting material in white light emitting diodes, J. Mater. Chem. 22 (2012) 22378–22381. https://doi.org/10.1039/C2JM35305A

[73]   L.L. Zhang, X. S. Zhao, Carbon-based materials as supercapacitor electrodes. Chem. Soc. Rev 38 (2009) 2520-2531. https://doi.org/10.1039/B813846J.

[74]   Y. Dong, J. Shao, C. Chen, H. Li, R. Wang, Y. Chi, X. Lin, G. Chen, Blue luminescent graphene quantum dots and graphene oxide prepared by tuning the carbonization degree of citric acid, Carbon 50 (2012) 4738–4743. https://doi.org/10.1016/j.carbon.2012.06.002

[75]   P. Zhang, X. Zhao, Y. Ji, Z. Ouyang, X. Wen, J. Li, Z. Su, G. Wei, Electrospinning graphene quantum dots into a nanofibrous membrane for dual-purpose fluorescent and electrochemical biosensors, J. Mater. Chem. B 3(2015) 2487–2496. https://doi.org/10.1039/C4TB02092H

[76]   J. Peng, W. Gao, B.K. Gupta, Z. Liu, R. R. Aburto, L. Ge, L. Song, L.B. Alemany, X. Zhan, G. Gao, S. A. Vithayathil, Graphene quantum dots derived from carbon fibers, Nano Lett. 12 (2012) 844-849. https://doi.org/10.1021/nl2038979

[77]   S. H. Choi, Unique properties of graphene quantum dots and their applications in photonic/electronic devices, J. Phys. D. Appl. Phys.50 (2017)103002.

[78]   S. Zhu, J. Zhang, X. Liu, B. Li, X. Wang, S. Tang, Q. Meng, Y. Li, C. Shi, R. Hu, B. Yang, Graphene quantum dots with controllable surface oxidation, tunable fluorescence and up-conversion emission. RSC Adv. 2 (2012) 2717-2720. https://doi.org/10.1039/C2RA20182H

[79]   C. Hu, Y. Liu, Y. Yang, J. Cui, Z. Huang, Y. Wang, L. Yang, H. Wang, Y. Xiao, J. Rong, One-step preparation of nitrogen-doped graphene quantum dots from oxidized debris of graphene oxide. J. Mater. Chem. B 1 (2013) 39–42. https://doi.org/10.1039/C2TB00189F

[80]   M. Kaur, M. Kaur, V. K. Sharma, Nitrogen-doped graphene and graphene quantum dots: A review on synthesis and applications in energy, Adv. Colloid. Interface. 259 (2018) 44-64. https://doi.org/10.1016/j.cis.2018.07.001

[81]   M. Hassan, E. Haque, K. R. Reddy. A.I. Minett, J. Chen, V.G. Gomes, Edge-enriched graphene quantum dots for enhanced photo-luminescence and supercapacitance. Nanoscale 6 (2014) 11988–11994. https://doi.org/10.1039/C4NR02365J

[82]   D. B. Shinde, V.K. Pillai, Electrochemical preparation of luminescent graphene quantum dots from multiwalled carbon nanotubes. Chem. Eur. J. 18 (2012) 12522–12528. https://doi.org/10.1002/chem.201201043

[83]   X. Tan, Y. Li, X. Li, S. Zhou, L. Fan, S. Yang, Electrochemical synthesis of small-sized red fluorescent graphene quantum dots as a bioimaging platform. Chem. Commun. 51 (2015) 2544–2546. https://doi.org/10.1039/C4CC09332A

[84]   S. Wei, R. Zhang, Y. Liu, H. Ding, Y.L. Zhang, Graphene quantum dots prepared from chemical exfoliation of multiwall carbon nanotubes: an efficient photocatalyst promoter, Catal. Commun. 74 (2016) 104–109. https://doi.org/10.1016/j.catcom.2015.11.010

[85]   J. M. Bai, L. Zhang, R.P. Liang, J.D. Qiu, Graphene quantum dots combined with europium ions as photoluminescent probes for phosphate sensing, Chem. Eur. J. 19 (2013) 3822–3826. https://doi.org/10.1002/chem.201204295

[86]   S. Mondal, U. Rana, S. Malik, Graphene quantum dot-doped polyaniline nanofiber as high performance supercapacitor electrode materials, Chem. Commun. 51 (2015) 12365-12368. https://doi.org/10.1039/C5CC03981A

[87]   Y. Zhang, C. Wu, X. Zhou, X. Wu X, Y. Yang, H. Wu, S. Guo, J. Zhang,
Graphene quantum dots/gold electrode and its application in living cell H_2O_2
detection, Nanoscale 5 (2013) 1816–1819. https://doi.org/10.1039/C3NR33954H

[88]   J. Kim, J. S. Suh, Size-controllable and low-cost fabrication of graphene quantum
dots using thermal plasma jet, ACS Nano. 8 (2014) 4190–4196.
https://doi.org/10.1021/nn404180w

[89]   X. Wu, F. Tian, W. Wang, J. Chen, M. Wu, J.X. Zhao, Fabrication of highly
fluorescent graphene quantum dots using L-glutamic acid for in vitro/in vivo imaging
and sensing, J. Mater. Chem. C 1 (2013) 4676–84.
https://doi.org/10.1039/C3TC30820K

[90]   P. Bondavalli, Graphene and Related Nanomaterials: Properties and Applications,
Elsevier. 2017.

[91]   C. Yan, X. Hu, P. Guan, T. Hou, P. Chen, D. Wan, X. Zhang, J. Wang, C. Wang,
Highly biocompatible graphene quantum dots: green synthesis, toxicity comparison
and fluorescence imaging. J. Mater. Sci. 55 (2020) 1198-1215.
https://doi.org/10.1007/s10853-019-04079-2

[92]   Y. Sun, S. Wang, C. Li, P. Luo, L. Tao, Y. Wei, G. Shi, Large scale preparation of
graphene quantum dots from graphite with tunable fluorescence properties, Phys.
Chem. Chem. Phys. 15 (2013) 9907e9913. https://doi.org/10.1039/C3CP50691F

[93]   Y. Shin, J. Lee, J. Yang, J. Park, K. Lee, S. Kim, Y. Park, H. Lee, Mass production
of graphene quantum dots by one-pot synthesis directly from graphite in high yield,
Small 10 (2014) 866-870. https://doi.org/10.1002/smll.201302286

[94]   M.J. Molaei, The optical properties and solar energy conversion applications of
carbon quantum dots: A review, J. Sol. Energy 196 (2020) 549-566.
https://doi.org/10.1016/j.solener.2019.12.036

[95]   D. Pan, J. Zhang, Z. Li, M. Wu, Hydrothermal route for cutting graphene sheets
into blueluminescent graphene quantum dots. Adv. Mater. 22 (2010) 734-738.
https://doi.org/10.1002/adma.200902825

[96]   S. Zhu, J. Zhang, C. Qiao, S. Tang, Y. Li, W. Yuan, B. Li, L. Tian, F. Liu, R. Hu,
H. Gao, H. Wei, H. Zhang, H. Sun, B. Yang, Strongly green-photoluminescent
graphene quantum dots for bioimaging applications, Chem. Commun. 477 (2011)
6858-6856. https://doi.org/10.1039/C1CC11122A

[97]   D. Pan, L. Guo, J. Zhang, C. Xi, Q. Xue, H. Huang, J. Li, Z. Zhang, W. Yu, Z. Chen, Z. Li, Cutting sp^2 clusters in graphene sheets into colloidal graphene quantum dots with strong green fluorescence. J. Mater. Chem. 22 (2012) 3314-3318. https://doi.org/10.1039/C2JM16005F

[98]   L. Lin, S. Zhang, Creating high yield water soluble luminescent graphene quantum dots via exfoliating and disintegrating carbon nanotubes and graphite flakes. Chem. Commun. 48 (2012)10177-10179. https://doi.org/10.1039/C2CC35559K

[99]   B. Yin, J. Deng, X. Peng, Q. Long, J. Zhao, Q. Lu, Q. Chen, H. Li, H. Tang, Y. Zhang, S. Yao, Green synthesis of carbon dots with down-and up-conversion fluorescent properties for sensitive detection of hypochlorite with a dual-readout assay, Analyst. 138 (2013) 6551-6557. https://doi.org/10.1039/C3AN01003A

[100] Y. Hu, Y. Zhao, G. Lu, N. Chen, Z. Zhang, H. Li, H. Shao, L. Qu, Graphene quantum dots–carbon nanotube hybrid arrays for supercapacitors, Nanotechnology 24 (2013) 195401. https://doi.org/10.1088/0957-4484/24/19/195401

[101] S. B. Martínez, M. Valcárcel, Graphene quantum dots as sensor for phenols in olive oil, Sensor Actuat. B-Chem. 197 (2014) 350-357. https://doi.org/10.1016/j.snb.2014.03.008

[102] V.N. Mehta, S. Jha, S.K. Kailasa, One-pot green synthesis of carbon dots by using Saccharum officinarum juice for fluorescent imaging of bacteria (Escherichia coli) and yeast (Saccharomyces cerevisiae) cells, Mater. Sci. Eng. C 38 (2014) 20–27. https://doi.org/10.1016/j.msec.2014.01.038

[103] P. Jegannathan, A.T. Yousefi, M.S.A. Karim, N.A. Kadri, Enhancement of graphene quantum dots based applications via optimum physical chemistry: A review, Biocybern. Biomed. Eng.38 (2018) 481-497. https://doi.org/10.1016/j.bbe.2018.03.006

[104] S. Sahu, B. Behera, T.K. Maiti, S. Mohapatra, Simple one-step synthesis of highly luminescent carbon dots from orange juice: application as excellent bio-imaging agents, Chem Commun. 48 (2012) 8835-8837. https://doi.org/10.1039/C2CC33796G

[105] H. Xu, X. Yang, G. Li, C. Zhao, X. Liao, Green synthesis of fluorescent carbon dots for selective detection of tartrazine in food samples, J. Agric. Food Chem. 63 (2015) 6707-6714. https://doi.org/10.1016/j.mtchem.2018.03.003

[106] N. Wang, Y. Wang, T. Guo, T. Yang, M. Chen, J. Wang, Green preparation of carbon dots with papaya as carbon source for effective fluorescent sensing of Iron (III)

and Escherichia coli, Biosens. Bioelectron. 85 (2016) 68-75.
https://doi.org/10.1016/j.bios.2016.04.089

[107] A. Kumar, A. Ray Chowdhuri, D. Laha, T.K. Mahto, P. Karmakar, S.K. Sahu, Green synthesis of carbon dots from Ocimum sanctum for effective fluorescent sensing of Pb^{2+} ions and live cell imaging,Sensor Actuat. B-Chem. 242 (2017) 679-686. https://doi.org/10.1016/j.snb.2016.11.109

[108] S.A.A. Vandarkuzhali, V. Jeyalakshmi, G. Sivaraman, S. Singaravadivel, K.R. Krishnamurthy, B. Viswanathan, Highly fluorescent carbon dots from Pseudo-stem of banana plant: applications as nanosensor and bio-imaging agents, Sensor Actuat. B-Chem. 252 (2017) 894e900. https://doi.org/10.1016/j.snb.2017.06.088

[109] J. Shen, S. Shang, X. Chen, Y. Cai, Facile synthesis of fluorescence carbon dots from sweet potato for Fe^{3+} sensing and cell imaging, Mater. Sci. Eng. C 76 (2017) 856e864. https://doi.org/10.1016/j.msec.2017.03.178

[110] Z. Li, X. Li, Y. Zong, G. Tan, Y. Sun, Y. Lan, M. He, Z. Ren, X. Zheng, Solvothermal synthesis of nitrogen-doped graphene decorated by super paramagnetic Fe_3O_4 nanoparticles and their applications as enhanced synergistic microwave absorbers. Carbon 115 (2017) 493–502. https://doi.org/10.1016/j.carbon.2017.01.036

[111] H. An, Y. Li, Y. Gao, C. Cao, J. Han, Y. Feng, W. Feng, Free-standing fluorine and nitrogen co-doped graphene paper as a high-performance electrode for flexible sodium-ion batteries, Carbon 116 (2017) 338–346. https://doi.org/10.1016/j.carbon.2017.01.101

[112] R. Yadav, C.K. Dixit, Synthesis, characterization and prospective applications of nitrogen- doped graphene: A short review, J. Sci. Adv. Mater. Devices 2 (2017) 141–149. https://doi.org/10.1016/j.jsamd.2017.05.007

[113] V. Milosavljevic, A. Moulick, P. Kopel, V. Adam, R. Kizek, Microwave preparation of carbon quantum dots with different surface modification, J. Metallomics Nanotechnol. 3 (2014)16-22.

[114] S. Umrao, M.H. Jang, J.H. Oh, G. Kim, S. Sahoo, Y.H. Cho, A. Srivastva, I.K. Oh, Microwave bottom-up route for size-tunable and switchable photoluminescent graphene quantum dots using acetylacetone: New platform for enzyme-free detection of hydrogen peroxide, Carbon81(2015) 514-524. https://doi.org/10.1016/j.carbon.2014.09.084

[115] D. Gu, S. Shang, Q. Yu, J. Shen, Green synthesis of nitrogen-doped carbon dots from lotus root for Hg (II) ions detection and cell imaging, Appl. Surf. Sci. 3 90 (2016) 38-42. https://doi.org/10.1016/j.apsusc.2016.08.012

[116] X. T. Zheng, A. Ananthanarayanan, K.Q. Luo, P. Chen, Glowing graphene quantum dots and carbon dots: properties, syntheses, and biological applications. Small 11(2015) 1620-1636. https://doi.org/10.1002/smll.201402648

[117] C.K. Chua, Z. Sofer, P. Simek, O. Jankovsky, K. Klimova, S. Bakardjieva, S.H. Kučková, M. Pumera, Synthesis of strongly fluorescent graphene quantum dots by cage-opening buckminsterfullerene, Acs Nano. 9 (2015) 2548-2555. https://doi.org/10.1021/nn505639q

[118] S. Zhang, L. Sui, H. Dong, W. He, L. Dong, L. Yu, High-Performance Supercapacitor of graphene quantum dots with uniform sizes, ACS Appl. Mater. Inter.10 (2018) 12983–12991. https://doi:10.1021/acsami.8b00323

[119] S. Kundu, R.M. Yadav, T. Narayanan, M.V. Shelke, R. Vajtai, P.M. Ajayan, V. K. Pillai, Synthesis of N, F and S co-doped graphene quantum dots,Nanoscale 7 (2015)11515–11519. https://doi.org/10.1039/C5NR02427G

[120] J. Ju, W. Chen, In situ growth of surfactant-free gold nanoparticles on nitrogen-doped graphene quantum dots for electrochemical detection of hydrogen peroxide in biological environments, Anal. Chem. 87(2015) 1903–1910. https://doi.org/10.1021/ac5041555

[121] J. Zhou, Z. Sheng, H. Han, M. Zou, C. Li, Facile synthesis of fluorescent carbon dots using watermelon peel as a carbon source, Mater. Lett. 66 (2012) 222-224. https://doi.org/10.1016/j.matlet.2011.08.081

[122] R. Das, R. Bandyopadhyay, P. Pramanik, Carbon quantum dots from natural resource: A review. Mater. Today Chem. 8(2018) 96-109. https://doi.org/10.1016/j.mtchem.2018.03.003

[123] A. B. Bourlinos, A. Stassinopoulos, D. Anglos, R. Zboril, M. Karakassides, E.P. Giannelis, Surface functionalized carbogenic quantum dots. Small 4 (2008) 455-458. https://doi.org/10.1002/smll.200700578

[124] P.C. Hsu, Z. Y. Shih, C.H. Lee, H.T. Chang, Synthesis and analytical applications of photoluminescent carbon nanodots. Curr. Green Chem. 14 (2012) 917-920. https://doi.org/10.1039/C2GC16451E

[125] C. W. Lai, Y.H. Hsiao, Y.K. Peng, P.T. Chou, Facile synthesis of highly emissive carbon dots from pyrolysis of glycerol; gram scale production of carbon dots/m SiO$_2$ for cell imaging and drug release, J. Mater. Chem. 22 (2012) 14403-14409. https://doi.org/10.1039/C2JM32206D

[126] Y. Yan, J. Gong, J. Chen, Z. Zeng, W. Huang, K. Pu, J. Liu, P. Chen, Recent advances on graphene quantum dots: from chemistry and physics to applications, Adv. Mater. 31(2019) 1808283. https://doi.org/10.1002/adma.201808283

[127] E. Ciotta, P. Prosposito, P. Tagliatesta, C. Lorecchio, I. Venditti, I. Fratoddi, I. R. Pizzoferrato, Sensitivity to heavy-metal ions of cage-opening fullerene quantum dots. M.D.P.I. proceedings 1 (2017) 475. https://doi.org/10.3390/proceedings 1040475

[128] W. Liu, X. Yan, J. Chen, Y. Feng, Q. Xue, Novel and high-performance asymmetric micro-supercapacitors based on graphene quantum dots and polyaniline nanofibers. Nanoscale 5 (2013) 6053-6062. https://doi.org/10.1039/C3NR01139A

[129] Y. Qing, Y. Jiang, H. Lin, L. Wang, A. Liu, Y. Cao, R. Sheng, Y. Guo, C. Fan, S. Zhang, D. Jia, Boosting the supercapacitor performance of activated carbon by constructing overall conductive networks using graphene quantum dot, J. Mater. Chem. A7 (2019) 6021-6027. https://doi.org/10.1039/C8TA11620B

[130] Y. Zhu, X. Ji, C. Pan, Q. Sun, W. Song, L. Fang, Q. Chen, C. E. Banks, A carbon quantum dot decorated RuO$_2$ network: outstanding supercapacitances under ultrafast charge and discharge. Energy Environ. Sci.6 (2013) 3665-3675. https://doi.org/10.1039/C3EE41776J

[131] X. Jian, H.M. Yang, J.G. Li, E.H. Zhang, Z.H. Liang, Flexible all-solid-state high-performance supercapacitor based on electrochemically synthesized carbon quantum dots/polypyrrole composite electrode, Electrochim. Acta. 228 (2017) 483-493. https://doi.org/10.1016/j.electacta.2017.01.082

[132] Z. Zhang, J. Zhang, N. Chen, L. Qu, Graphene quantum dots: an emerging material for energy-related applications and beyond, Energy Environ. Sci.5 (2012) 8869-8890. https://doi.org/10.1039/C2EE22982J

[133] W.W. Liu, Y.Q. Feng, X.B. Yan, J.T. Chen, Q.J. Xue, Superior micro-supercapacitors based on graphene quantum dots, Adv. Funct. Mater 23 (2013) 4111–22. https://doi.org/10.1002/adfm.201203771

[134] X. Zhou, Z. Tian, J. Li, H. Ruan, Y. Ma, Z. Yang, Y. Qu, Synergistically enhanced activity of graphene quantum dot/multi-walled carbon nanotube composites as metal-

free catalysts for oxygen reduction reaction. Nanoscale 6 (2014) 2603-2607.
https://doi.org/10.1039/C3NR05578G

[135] J. Shen, Y. Zhu, X. Yang, C. Li, Graphene quantum dots: emergent nanolights for bioimaging, sensors, catalysis and photovoltaic devices, Chem. Commun. 48 (2012) 3686e3699. https://doi.org/10.1039/C2CC00110A

[136] R. Tjandra, W. Liu, M. Zhang, A. Yu, All-carbon flexible supercapacitors based on electrophoretic deposition of graphene quantum dots on carbon cloth, J. Power Sources 438(2019) 227009. https://doi.org/10.1016/j.jpowsour.2019.227009

[137] S. Zhang, Y. Li, H. Song, X. Chen, J. Zhou, S. Hong. M. Huang,Graphene quantum dots as the electrolyte for solid state supercapacitors, Sci. Rep.6(2016) 19292. https://doi.org/10.1038/srep19292

[138] A.B. Ganganboina, A. D. Chowdhury, R. Doong, New avenue for appendage of graphene quantum dots on halloysite nanotubes as anode materials for high performance supercapacitors, ACS Sustain. Chem. Eng. 5 (2017) 4930–4940. https://doi.org/10.1021/acssuschemeng.7b00329

[139] H. Xia, C. Hong, B. Li, B. Zhao, Z. Lin, M. Zheng, S.V. Savilov, S.M. Aldoshin, Facile synthesis of hematite quantum dot/functionalized grapheme sheet composites as advanced anode materials for asymmetric supercapacitors. Adv. Funct. Mater. 25(2015) 627-635. https://doi.org/10.1002/adfm.201403554

Materials Research Forum LLC
https://doi.org/10.21741/9781644901250-10

Chapter 10

Role of Quantum Dots in Separation Processes

Neelam Verma[1,2]*, Rajni Sharma[2,3], Mohsen Asadnia[3]

[1]Chemistry and Division of Research and Development, Lovely Professional University, Phagwara-144401, India

[2]Biosensor Technology Laboratory, Department of Biotechnology, Punjabi University, Patiala-147002, India

[3]School of Engineering, Macquarie University, Sydney, New South Wales, 2109, Australia

*verma.neelam2@gmail.com

Abstract

Quantum dots (QDs), the fluorescent nanoparticles with multiplexing competency are applicable in broad range of fields. The application of QDs in separation processes is a relatively new approach, still presenting the spectacular advancement and wider future scope. The unique features of QDs endorse their use in wastewater treatment, chromatographic separation and heavy metal remediation. QDs also assist the separation of biomarkers, pathogens and tumor cells for biomedical applications. These tiny particles possess tremendous potential to deal with bigger global issues such as water desalination and early cancer diagnosis. To the best of our knowledge, it is the first most report summarizing the QDs uses for multiple separation processes.

Keywords

Quantum Dots, Magnetic Separation, Wastewater, Cancer, Membrane

Contents

1.   Introduction

Quantum dots (QDs) are the fluorescent nanoparticles possessing high surface-to-volume ratio, multiplexing competency, quantum confinement and tunable optoelectronic properties in accordance with their size and composition [1,2]. In general, the term 'Quantum dots' has most frequently been used for only semiconductor QDs such as lead sulphide (PbS) QDs, cadmium selenide (CdSe) QDs owing to their earlier discovery in 1980s. However, according to Cayuela et al. [3], it is a little bit confusing and must include semiconductors plus other types as well due to later discovery of organic QDs, for instance, carbon QDs (in 2004) and graphene QDs (in 2006). Carbon/graphene QDs (CQD/GQDs) differ dramatically from semiconductor QDs in terms of nature, properties and crystalline structure [3]. Semiconductor QDs are spherical elemental crystalline structures synthesized by bottom-up approach, typically <6 nm in size with broad absorption spectra (UV-Vis region), excitation independent photoluminescence, potential cytotoxicity (usually high with heavy metal QDs) and hydrophobic nature. On the other hand, carbon QDs (quasi-spherical, <10 nm size) and graphene QDs (one atom thick disc, <20 nm) are organic carbonaceous material, superior to conventional QDs owing to their water solubility and non-toxicity. These are superb chemically inert material with easy surface functionalization, excitation dependent photoluminescence, inexpensive scale up as well as synthesis by both bottom-up and top-down approaches [3,4]. Now-a-days, various strategies are put forward to obtain even semiconductor QDs with hydrophilic and nontoxic nature via surface modification with amphiphilic polymers which, on the whole, broadens the areas of application for QDs [5].

Materials Research Forum LLC
https://doi.org/10.21741/9781644901250-10

In addition to separation processes, QDs have been extensively applied in environmental remediation, sensors fabrication, drug delivery, super capacitors and cancer diagnostics [2,6,7]. The application of QDs in separation processes is a relatively new approach, despite that, there is spectacular advancement in this field and are legions of possibilities in the future. The ultra-small dimensions, chemical stability, hydrophilic nature and active functional groups of CQD/GQDs endorse their use in separation membranes and columns [8,9]. Further, the fluorescent nature of ZnS/CdSe QDs assists the cellular and molecular separation following specific recognition via combination with magnetic components. Magnetic quantum dots (MagDots), the synergism of QDs and magnetic particles presented the advanced research for separation of biomarkers, microbial and tumor cells which opened a new avenue for clinical diagnosis and biomedical applications [10–12]. Henceforth, the use of ultra-small size QDs in separation processes has an excellent potential to deal with big global issues such as desalination of water and early clinical diagnosis of cancer. In this chapter, we outlined the work related to the applications of QDs in various separation techniques. To the best of our knowledge, it is the first report summarizing the use of organic and inorganic QDs for multiple separation processes.

2. Quantum dots (QDs) in separation membranes

Despite three fourth fraction of water on earth, the increasing scarcity of fresh drinking water is a considerable issue of global concern. Since the major fraction of water resides in the oceans, the advancement in desalination techniques represents a possible solution for the purification of available water [13,14]. Here, QDs play a significant role in desalination, water purification and wastewater treatment by facilitating the waste/salt separation in the membrane systems. Many researchers reported the integration of carbon or graphene QDs (CQDs/GQDs) into various state-of-art membranes like ultrafiltration, nanofiltration, reverse osmosis, forward osmosis and membrane distillation to improve the water flux and waste filtration [9]. The favorable features like ultra-small dimensions, superb chemical stability, biocompatibility, antifouling properties, variable hydrophilicity and functional groups rich surface environment in addition to their easy and economical production endorsed their use in membrane systems. Further, the simple hydraulic cleaning and anti-fouling properties of QDs-incorporated membranes make them highly practical to use in separation process.

Technically, the membrane system can be customized with QDs in three different ways (Fig. 1 [9]). First, thin film nanocomposite (TFN) membranes where QDs are incorporated only in dense-selective layer via interfacial polymerization process. Secondly, QDs are assembled on top of membranes surface using coating agent and

Quantum Dots – Properties and Applications   Materials Research Forum LLC
Materials Research Foundations **96** (2021) 251-279   https://doi.org/10.21741/9781644901250-10

QD/polymer composite membranes in which QDs are integrated throughout the membrane homogeneously [9].

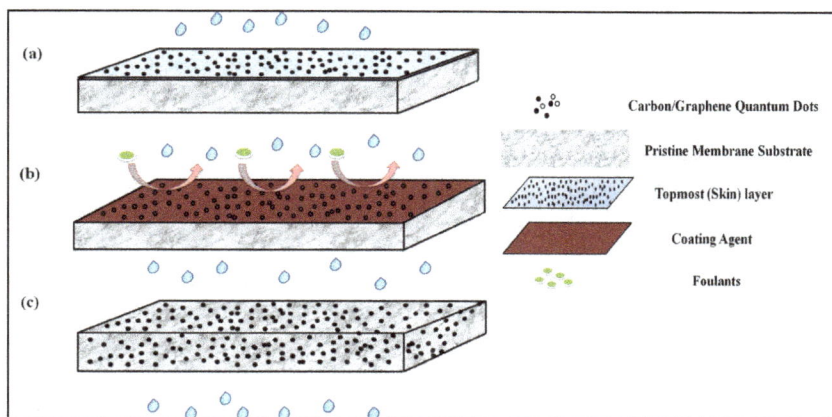

Fig. 1 Quantum dots in separation membranes (a) Thin-film nanocomposite (TFN) membranes (b) QDs on top of membranes surface with coating agent (c)QD/polymer composite membranes [9].

2.1   Thin-film nanocomposite (TFN) membranes

QDs tailored TFN membranes are highly beneficial in water purification attributing to an excellent robustness with stable water flux, salt/waste rejection and antifouling properties [15,16]. For the preparation of TFN membranes, the aqueous or organic phase dispersed QDs are used during the interfacial polymerization. Song et al. [17] explained the interfacial polymerization reaction between trimesoyl chloride and m-phenylenediamine/CQDs in the polyamide layer on polysulfone substrate. This strategy supports to cut off the membrane cost by maximizing the performance and reducing the material usage since QDs are assembled only in dense-selective layer [9].

Usually, the polymeric membranes comprising of polyvinylidene, polyethersulfone, polysulfone or polyacrylonitrile present weak mechanical strength and hydrophobicity. The incorporation of QDs overcomes these issues by altering the properties in different ways. Firstly, QDs modified TFN membranes demonstrated the chlorine resistance which is highly desirable in reverse osmosis membrane systems [14]. As the polyamide bonds in the matrix are more susceptible to break down by active chlorine released from the chlorine detergents which are used as disinfectants or cleaning agents. QDs shield the polyamide bonds either by making hydrogen bonds itself or excluding the contact of active chlorine (OCl⁻) being electron rich owing to the presence of COO⁻ groups. Secondly, steric hindrance induced by QDs leads to the smoother surface by retarding the rate of formation of selective layer during interfacial polymerization. The slow diffusion of monomers stimulated by QDs is due to their larger size as compared to other monomers (for instance, tannic acid and m-phenylenediamine) and higher aqueous concentration. Further, higher aqueous concentration enhance the surface hydrophilicity which in turn lead to enhance the water permeation flux without compromising the waste rejection. For example, GQDs customized TFN membrane presented 23.33 L/(m²·h·bar) water flux with high dye rejection efficiency of congo red (99.8%) and methylene blue (97.6%) which was 1.5 times compared to unaltered tannic acid TFC polyacrylonitrile membrane [18]. These QDs customized membranes increase the organic solvent nanofiltration selectivity by preventing the flow of non-polar solvents [19].

Recently, Bi and his team [15] developed a GQDs nano aggregated membrane presenting pure water flux up to 244.7 L/(m²·h·bar) with effective rejection of Alcian blue (92.9%) and Congo red (98.8%) dyes. The pore engineering of ultrafiltration membrane with GQDs gave rise to nanofiltration membrane which allowed the super-fast separation of molecules due to altered electronegativity and hydrophilicity. Inbiphasic process, the reaction of acyl chloride groups of trimesoyl chloride (organic) with hydroxyl or carboxyl groups of GQDs and amine groups of piperazine at the interphase brought about an *in-situ* polymerization. The interfacial polymerization within the voids of polyethersulfone substrate membrane reduced the pore size as compared to previous membrane which could be accustomed from 1.21 to 1.72 nm merely by changing the proportion of GQDs. The thermal treatment rendered the membrane more robust and chemically resistant by shrinking the bulk material of membrane permanently. GQDs embedded nanofiltration membrane with superior performance and long term operational stability proposed a path for advance molecular and ionic separations by manipulating the membrane pore structures via quantum dots or other nanomaterials. Similarly, Xu et al. [20] engineered a novel GQDs-embedded TFC membrane for forward osmosis process. The interfacial polymerization of GQDs/polyethyleneimine (aqueous)andtrimesoyl chloride (organic)was resulted in GQDs associated thin film (50 nm) formationin polyacrylonitrile matrix. The resulting membrane was found to have rough surface, hydrophilic nature and high stability due to covalent bonding between polyethyleneimine and GQDs. The developed membrane led to pure water flux of 12.9 L/(m²·h·bar) and salt rejection of 1.41 gm⁻²h⁻¹. Besides this, nitrogen-doped graphene oxide QDs (N-GOQD) were also used to formulate TFN polyamide reverse osmosis membrane by Fathizadeh and his team [14]. The addition of even a little amount (0.02 wt/v%) of N-GOQD increased the water permeability thrice without compromising the salt rejection efficiency leading to significant increase in surface area and hydrophilicity. The thermal stability of the membrane was also improved due to the chemical bonding of N-GOQD with the surface matrix. Hence, QDs coated TFN membranes present good potential for water reclamation and desalination processes.

2.2   QDs on top of membranes surface

In this approach, the upper surface of the membrane is covered with QDs using coating agents. Zeng and co-workers [21] immobilized the GOQDs on polyvinylidene difluoride (PVDF) membrane surface in three steps. Firstly, polyethylene glycol was attached on the membrane surface imparting surface hydroxylation. In the second step, hydroxylated surface was treated with a coating reagent, (3-aminopropyl) trimethoxysilane (APTMS) to create amine group. Eventually, covalent immobilization was made possible by reaction of carboxylic groups on GOQDs and amine groups on the membrane surface. The hydrophobic nature of PVDF membrane renders it highly susceptible to microbial contamination leading to membrane fouling [22]. However, GOQD coated PVDF Membrane revealed high performance in terms of bactericidal and anti-biofouling properties as compared to the earlier reported antifouling membranes. In addition, the water flux was drastically increased from 500 $L/(m^2 \cdot h \cdot bar)$ to more than 3800 $L/(m^2 \cdot h \cdot bar)$.

Zhao et al. [23] anchored the CQDs on hollow TFC poly(ethersulfone) membrane poly(ethersulfone) using polydopamine as coating agent. The carboxyl groups functionalized CQDs (~3.2 nm) were prepared via a simple process at large scale from inexpensive material, citric acid. The covalent bonding between QDs and poly(ethersulfone) substrate fabricated an excellent antifouling and antibacterial membrane which also exhibited the higher water recovery *i.e.* 94% as compared to 89% with pristine membrane. In addition, the developed membrane possessed the potential to generate green energy 'osmotic power' via pressure retarded osmosis (PRO) process *i.e.* the production of higher energy with power density 11.0 W/m^2 at 15 bar as compared to contamination-prone pristine membrane (8.8 W/m^2).

Although, the bacterial cells including *E. coli* and *S. aureus* were effectively inactivated by the membrane systems modified with QDs on the top of the surface, the exact mechanism for suppression of biofilm formation and antifouling property is not well understood. Three processes including charge repulsion, physical or oxidative stress explained the phenomenon up to some extent [24,25]. Zeng et al. [21] proposed the physical stress on the microbial cell surface induced by extremely small-sized and uniformly dispersed QDs with active edges which may penetrate or incise the microbial cells. Secondly, QD associated PVDF membrane presented three times more oxidative stress with respect to PVDF membrane linked with graphene oxide sheets. The oxidative stress provoked by uniform loading of GOQDs on the membrane surface reduced the biofilm formation which was proven by glutathione (GSH) oxidation *in vitro* tests. Third, CQD functionalized membranes possess various negatively charged carboxyl groups which are anticipated to electrostatically repel the negatively charged cell surface of bacteria due to the presence of peptidoglycan layer [23,26]. Inspite of these three possible explanations, there is the need for further study on the microbial interaction with QDs. The anti-fouling membranes usually slow down the biofilm formation rather than completely alleviating. Even though, the anti-fouling filtration membrane systems are highly significant as these prevent the deposition of biological layer which may reduce the water flux and lead to more power consumption. These membranes also help to trim down the maintenance cost due to less frequent need for cleaning and washing of the membrane surface [21].

2.3   QD/polymer composite membranes

In this technique, QDs are homogeneously distributed throughout the membrane matrix by spinning the QDs and polymer. The composite membranes have uniform distribution and mechanical strength which are attributed to the ultra-small size of QDs [27]. The QDs addition to polymeric matrix influences the membrane morphology, formation and dope rheology differently [28]. GQD/Polyvinyl alcohol composite membrane was formed by electrospinning which exhibited the efficiency to detect glucose and hydrogen peroxide [29]. Jafari and his workers [30] fabricated GQDs/polyvinylidene electrospun membranes for membrane distillation process which presented a denser structure, higher wetting resistance and more irregular surface on addition of GQDs as compared to the original membrane.

Some limitations restrict the wide use of QD/polymer composite membranes. There is a need to control the QDs distribution within the membranes and secondly, requirement of an appropriate solvent for both polymer as well as QDs [31]. Third, the leaching of QDs is the most vulnerable during filtration process of membrane system. Therefore, leaching must be prevented by making strong chemical bonding between polymeric matrix and QDs and assurance of long-standing stability of QDs within the membrane. Colburn et al. [27] proposed the fabrication of GQD/cellulose composite membranes for selective separation of small molecules (>300 Da) using ionic liquid *i.e.* 1-ethyl-3-methylimidazolium acetate) as the common solvent for both cellulose and QDs. The presence of GQDs (0.2 wt%, size 5 nm) in the membrane matrix endowed with more durable membrane due to the formation of hydrogen bonds between cellulose and functional hydroxyl and carboxyl groups on GQDs. Further, the formed composite membrane was more selective, hydrophilic and negatively charged which improved the separation processes. The stability and long-term permeability of membrane was presented with persistent water flux, dye rejection and no leaching of GQDs following convective flow.

Henceforth, the fabrication of efficient QDs functionalized separation membrane system confirmed the QDs suitability to improve water purification and wastewater treatment. The advanced research to prevent QD leaching and to make the highly durable membrane systems may help in the commercialization of the approach.

3. QDs in chromatographic separation column

The use of QDs is also recommended for the separation of gaseous components in addition to their application in liquid separation [9]. Zhang and his team [8] proposed the application of GQDs in separation of organic components on gas chromatographic (GC) column. The distinct features like adsorption affinities, high surface area, over and above the presence of many functional groups render GQDs as a superb stationary phase for gas chromatography. GQDs were compactly immobilized inside the fused silica capillary column with a coupling reagent, 3-aminopropyldiethoxymethyl silane (3-AMDS). The organic compounds including alkanes, ethylbenzene, dichlorobenzene, propylbenzene, styrene and xylene were separated efficiently on GQDs modified GC column on the basis of various interactions *i.e.* van der Waals forces, hydrogen bonding and/or π-π electrostatic interaction. Eventually, the analytes were eluted following the increasing order of their boiling points. The approach was extremely competent to separate linear and branched isomers and moreover, to distinguish isomers even with slightly different boiling points, for instance, m-dichlorobenzene (173°C) and p-dichlorobenzene (173.4°C). As compared to commercial capillary columns (HP-5 and HP-innowax), the GQDs modified column presented quick and high-resolution at lower temperature without temperature-programming. The GQDs based columns presented the high consistency with results. Hence, the use of QDs in the chromatographic column for the separation of volatile components is highly selective, reliable and reproducible approach which acquired a great potential for the scaling up of the process.

4. QDs in heavy metal remediation

The use of QDs for the on-site detection and separation of heavy metal ions was proposed by Gogoi and his group [6] using a CQD-embedded agarose hydrogel (Agr/CQD hydrogel) hybrid system. The electrostatic interaction between hydroxyl groups on

Materials Research Forum LLC
https://doi.org/10.21741/9781644901250-10

agarose and amino group of fluorescent CQDs attributed the formulation of hybrid sensing platform. The chitosan derived CQDs exhibited the production of colored chitosan-metal chelates on interaction with transition heavy metals. Different colored chelates yellow, blue, brown, white and tan brown were observed with different heavy metals Cr^{6+}, Cu^{2+}, Fe^{3+}, Pb^{2+} and Mn^{2+} respectively. The heavy metal separation was achieved by the filtration of chitosan-metal chelates by filtration membranes. The detection of five heavy metals in the analytic solutions was accomplished within 5-10 seconds by colorimetric change in the sensing platform. The platform was so sensitive to detect as low as 1 pM concentration. Therefore, the CQDs based fabricated system can be applied for quick on-site, sensitive and colorimetric detection as well as feasible separation of heavy metals (Fig. 2).

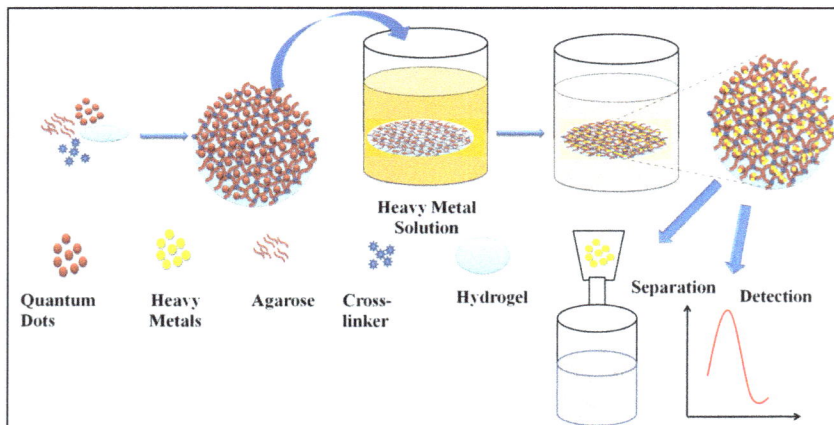

Fig. 2 Assembly of agarose/CQDs hydrogel film and its application for heavy metal sensing and separation.

5. Magnetic quantum dots (MagDots) for cellular/molecular separation

QDs have been appeared as promising tool for biomedical imaging owing to their strong fluorescence [2]. These are competent to exhibit high fluorescence with different color. In fact, the quantum size effect depicts the inverse correlation between size and energy band, as the QD size decreases, the energy gap between the excited and ground state increases leading to light emission with lower wavelength and different color [32,33]. The broad range of excitation and absorption spectra, narrow emission spectra and

Materials Research Forum LLC
https://doi.org/10.21741/9781644901250-10

quantum size effect allow the unique identification of the molecules [2]. Their applications and benefits can further be enhanced by conjugating QDs with magnetic nanoparticles such as super paramagnetic iron oxides, Fe_3O_4 or Fe_2O_3. Since the magnetic particles by themselves possess several characteristics: low surface to volume ratio, inducible magnetic character, no aggregation in solution and no hysteresis loss, the nanocomposites with QDs and magnetic particles attain multifunctional properties [34]. The nanocomposites with fluorescent quantum dots for detection/visualization and magnetic nanoparticles for magnetic separation are renowned as Magnetic Quantum Dots (MagDots) which can be prepared by both top-down and bottom-up approach (Fig. 3) [33]. Further, the conjugation of antibodies/aptamers specific to biomarkers or antigen further widens their area of applications.

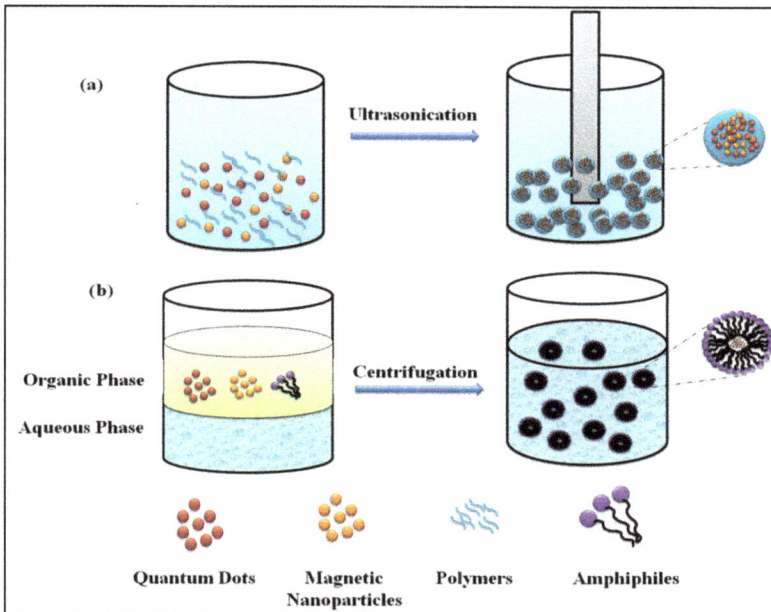

Fig. 3 Fabrication of magnetic quantum dots (MagDots) (a) Top-down approach e.g. emulsification (b) Bottom-up approach e.g. micellar assembly.

Table 1 Applications of quantum dots in cellular and molecular separation.

Quantum Dots	Target Cell/Molecule	Ligand	Application	References
ZnS/Mn^{2+}	CTCs	EpCAM	Multiple tumor cells	[7]
CdSe/ZnS	Jurkat (Leukemia) LNCaP (Prostate cancer)	CD3 PSMA	Leukemia, prostate cancer	[10]
CdSe/ZnS	MRSA and *Salmonella* DT104	Pardaxin peptide	Superbug detection	[11]
CdSe/ZnS	HeLa MCF-7	Folic acid receptor	Cervical, Breast cancer	[12]
CdSe/ZnS	DLD-1	EpCAM	Colorectal Cancer	[38]
QD-605	HCC	Specific to aptamer LY-1	Hepatocellular Carcinoma	[39]
GOQDs	HCC	Glypican-3	Hepatocellular Carcinoma	[40]
CQDs	Fluoride ions	Replacement of CQDs by F$^-$ ions	Ionic detection	[41]
CdSe/ZnS	*E. coli* O157:H7	Protein G antibody	Pathogen detection	[42]
GOQD	MCF-7	EpCAM	Breast Cancer	[43]
CdSe/ZnS	MCF-7	Cyclin	Breast cancer	[44]
CdTe	MDA-MB-435S	EGFR	Breast cancer	[45]
CdSe/ZnS	HeLa	Phosphatidyl serine	Apoptosis	[46]
CdSe/ZnS	HepG2 NIH3T3 4T1	Cell membrane	Live cells	[47]
CdTe	HeLa	CEACAM8	Cervical cancer	[48]
CdTe QDs	Molt-4 K562	EGFR	Leukemia	[49]

The traditional immunoassays were designed for the identification or quantitative measurement of biological molecules via antigen-antibody interactions using radioactive iodine which were later replaced by chemiluminescence or fluorescent dyes [35,36]. Further, QDs based immunoassays came into limelight which presented 1000 times higher sensitivity than that of conventional immune absorbent assays and around 100 times more than organic dyes based immunoassays [37]. Now-a-days, various reports (Table 1 [7,10–12,38–49]) are available for the use of fluorescent-magnetic nanocomposites (MagDots) in detection and isolation of a broad range of biomolecules (DNA/RNA/proteins), biomarkers, environmental inorganic/organic contaminants, microbial cells and tumor cells [12,50–52].

5.1   Separation of surface bound and soluble biomarkers

The biomarkers are the indicators for the existence and progression of diseases [53]. Sometimes, these molecules give the idea of mode of treatment and prognostic profile. For instance, estrogen receptors on cell-surface are considerable biomarker in breast cancer and quantitative measurements of biomarker give the estimate of disease vigorousness and envisage the accurate therapy viz. chemotherapy with low level of expression and endocrine therapy for high level of expression [12]. Therefore, it is also mandatory to find out the level of expression in addition to recognize cells as expressing and non-expressing the particular biomarker.

The precise detection and separation of biomarkers is very critical due to the presence of very low concentration or a few number of cells in the analytic samples. Moreover, the volume of analytic samples from the patients cannot be enhanced to concentrate the biomarkers. Here, the absorption based measurement techniques are incompetent to detect such a low concentrations in solutions. Also, the minute but significant biochemical changes like misfolded proteins and mismatched bases cannot be differentiated by absorbance measurement. In the current scenario, the detection by nanocomposites (MagDots) presents the superb alternate for accurate and specific diagnosis at nanoscale. MagDots displayed their ability to detect as low as femto molar (10^{-15} M) concentration using microlitres of samples and to differentiate DNA/RNA molecules with merely a single nucleotide mismatch [51]. Furthermore, the separation of cells at different stage [46] or multiple varieties of cells [10,44,49,54] can be achieved with small volume of analytic samples by MagDots whereas even the most commonly employed approach for cell separation i.e. fluorescence activated cell sorting (FACS) needs large volume samples.

Usually, the accuracy of results for biomarker detection is affected by multiple steps: Isolation, Purification, Labeling and Characterization [55]. Therefore, a one-step process was suggested for separation and quantification of biomarkers by Mahajan and his team [12]. They described a MagDots rooted lab-on-a-chip strategy on magnetic nano conveyors for the recognition and isolation of free as well as surface-bound biomarkers.

Materials Research Forum LLC

https://doi.org/10.21741/9781644901250-10

The soluble biomarkers are usually captured by sandwich assay whereas surface-bound biomarkers via direct binding (Fig. 4).

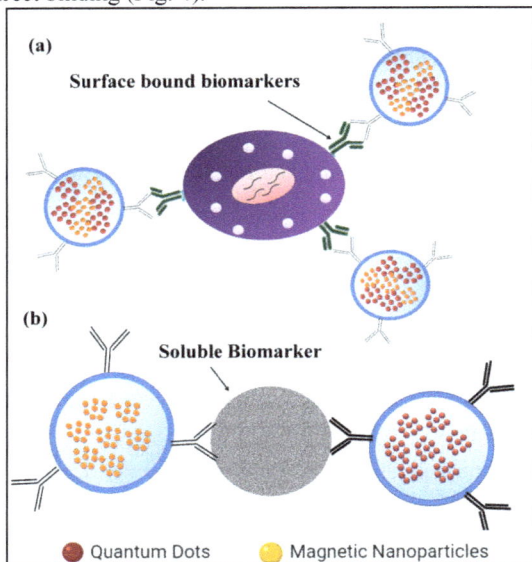

Fig. 4 Separation of (a) Cell-surface bound biomarker and (b) Soluble biomarkers by magnetic quantum dots.

The whole cells must be recognized and separated to detect the surface-bound targets. As shown in Fig.4, MagDots were synthesized by encapsulating both hydrophobic QDs and Fe_3O_4 magnetic nanoparticles in the core of polymer micelles (10-70 nm) with hydrophilic surface exposed to aqueous environment. The antibodies specific to cell surface-bound biomarkers (2–10 nm) are conjugated on micellar surface to capture the specific cells which are then separated using external magnetic field [12,56–58].The number of biomarkers on the cell surface can easily be estimated from the number of bound MagDots attributed to their size similarity. Assuming the negligible dimerization, the quantitative measurement is done either by fluorescent intensity associated with QDs or by the magnitude of magnetic force applied. QDs fluorescence provides spatial distribution and visualization of biomarkers in addition to quantification. The dual detection based assay is highly sensitive for the *in situ* separation and quantification of biomarkers among heterogeneous population of cells. The developed method is advantageous over FACS due to the requirement of small sample volume, strong and stable fluorescence of QDs over fluorescent dyes and accurate quantification despite the approximation from population average and furthermore, fluorescence intensity can be adjusted by QDs loading in the micelles [12].

On the other hand, detection of soluble biomarkers is dramatically challenging as compared to surface bound biomarkers due to very small size. Henceforth, Mahajan et al. [12] designed the more sensitive sandwich approach for soluble biomarkers in the dimensions of nanometers. In this assay, QDs and magnetic nanoparticles are separately encapsulated in polymer micelles and each micelle is conjugated with recognition probe specific to biomarker. Only in the presence of soluble biomarkers, two separated fluorescent and magnetic units combine together to form biomarker sandwich viz. QD-Biomarker-Magnetic nanoparticle. Again, the biomarker quantification can be done by either method i.e. fluorescent intensity or magnitude of magnetic force. The assay was miniaturized and so sensitive to analyze concentration of 10^{-15} M merely using 5 μL of analytic sample. The adjustable emission spectra on the basis of QD size allowed the multiplexing efficiency and the assay was applicable to detect and isolate DNA and protein targets on the same platform. The multiple biomarker detection efficiency for free or bound biomarkers imparts the foundation of path for use of MagDots assays in real clinical analysis.

5.2   QDs in tumor cells separation

The failure of available diagnostic techniques to detect metastasis and tumor cells at early stage is major reason of low survival rate of cancer patients [39]. Circulating tumor cells (CTCs) are the indicators of metastatic cancer stage and possess wealthy information for deadly disease which may be helpful in diagnostic and prognostic studies. Hence, their diagnosis in the blood samples may have huge advantage in the early cancer detection and improvement in survival rate of cancer patients. However, the detection of 1-100 CTCs among the billions (1.0×10^9) of blood cells in cancer patients is a big challenge [59,60]. The extremely low number and phenotypic heterogeneity demand the advanced techniques to detect and isolate these cells to get their actual benefit in clinical applications [61,62].The characterization of CTCs by physical properties like size and density is non-specific due to size similarity of blood cells [63,64]. However, the identification via specific cell surface proteins provides the more specific method [65]. The US Food and Drug Administration (FDA) approved medical equipment, named CellSearch system, for CTCs testing which is impractical for wider application owing to high cost, slow process and antibody-staining dependency.

In the current scenario, QDs attached with specific antibody or aptamers play a very significant role for the selective identification and fluorescent imaging of rare tumor cells. In spite of conventional imaging methods including fluorescent proteins or organic dyes, QDs are highly preferable for *in-vitro* as well as *in-vivo* due to their low cytotoxicity, biocompatiblility, stable photoluminescence and photobleaching resistancy. Further, the combination of magnetic nanoparticles with fluorescent QDs facilitates the identification and isolation of CTCs simultaneously.

Wang et al. [44] reported the detection and separation of MCF-7 breast cancer cells from the serum samples using anticycline E antibodies conjugated fluorescent/magnetic nanocomposite particles. The aqueous nanocomposites were comprised of polymer coating fluorescent CdSe/ZnS QDs and Fe_2O_3 magnetic beads. Usually, the breast cancer cells express protein cyclin on the surface which assisted the specific antigen-antibody interaction based detection of tumor cells in complex mixture. QDs layer was attached on thiol-functionalized magnetic beads surface via thiol-metal bond formation. The presence of carboxylic groups on the nanocomposites imparted the water solubility, fluorescence allowed the detection by conventional fluorescence microscope and magnetic component facilitated the separation by magnet. Similarly, Chu et al. [45] also explained the identification and isolation of breast cancer MDA-MB-435S cells using CdTe-QDs-labeled magnetic polystyrene nanospheres. The nanospheres were produced by electrostatic deposition of QDs on the magnetic component which were further cross linked with epidermal growth factor (EGF) using 1-ethyl-3(3-dimethylamino propyl)-carbodiimide (EDC). The EGF receptor over-expressing estrogen receptor negative breast cancer cells were easily detected and separated by EGF conjugated nanospheres.

The QDs associated magnetic nanoparticles were also used to recognize and separate apoptotic cells at various stages that can help to study the molecular changes during apoptosis or apoptosis associated disease like tumor and AIDS [46]. Usually, cellular surface of apoptotic cells expresses phosphatidylserine which can promptly be recognized by annexin V [66,67]. To analyze apoptotic cells, analytic samples were pre-incubated with biotinylated annexin V which were in turn recognized by avidin coupled MagDots via biotin-avidin interaction. Since a fixed amount of avidin coupled to nanocomposites, the reaction could be used for quantitative measurement. Further, the exposure to ultraviolet light or nuclear staining with propidium iodide offered the analysis at different stages of apoptosis due to fluorescence, and the magnetic property of nanocomposites allowed the separation of apoptotic cells. The efficiency of multifunctional nanospheres to quickly identify and sorting multiple cells renders their excellent potential for clinical diagnosis, drug development and other biomedical application.

The appropriate selection of biomarkers and corresponding aptamers or antibodies would help for the detection and separation at different stages for multiple cells including macrophages, bacteria and cancerous cells. Based on the same background, Song and his team [10] designed highly sensitive and specific monoclonal antibodies (mAb) coupled multifunctional nanobioprobes for detection and separation of multiple tumor cells. The CdSe/ZnS (core/shell) QDs and Fe_2O_3 magnetic nanoparticles were encapsulated in styrene/acrylamide nanospheres to prepare nanocomposites which are conjugated with mAbs specific to biomarkers, CD3 (cluster of differentiation3) and PSMA (prostate-specific membrane antigen) for leukemia cells and prostate cancer cells recognition respectively (Fig. 5). The developed method was quite sensitive and specific to detect and separate a few number of spiked leukemia cells and prostate cancer cells (0.01%) from a complex mixture containing prostate cancer cells (LNCaP), leukemia (Jurkat T) cells, human lung fibroblasts (MRC-5) and red blood cells within 25 min without sample pretreatment. The leukemia and prostate cancer cells were separated efficiently with capture efficiency of 96% and 97% respectively using ordinary magnet and fluorescence microscope.

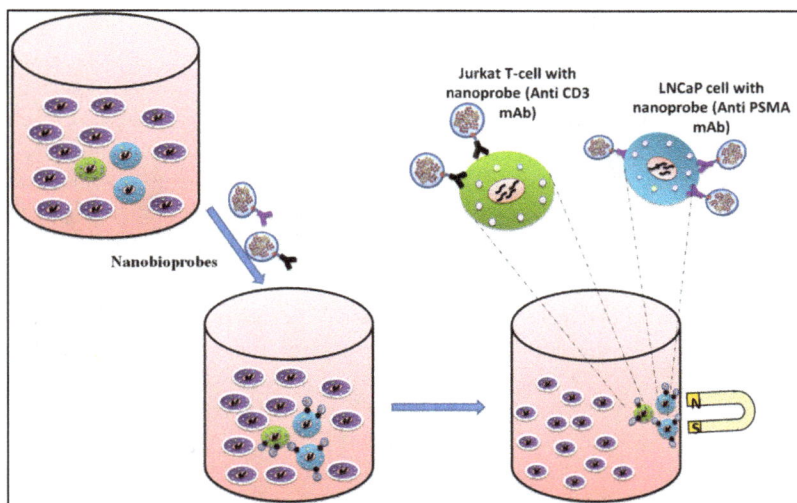

Fig. 5 Identification and Separation of Jurkat T-cells (Leukemia) and LNCaP cells (Prostrate cancer cells) using fluorescent magnetic nanoprobes.

Kale et al. [49] explained the selective separation of leukemic Molt-4 cells using fluorescent CdTe QD-magnetic nanocomposites conjugated with anti-EGFR antibody. EGFR, a glycoprotein is over-expressed on different tumor cells such as human breast cancer cells as well as T-cell leukemia. The developed nanocomposites were competent to differentiate EGFR expressing (+) T-cell leukemia (Molt-4 cells) from EGFR nonexpressing (-) myelogenous leukemic (K562cells) from their mixed solution.

The epithelial cell adhesion molecule (EpCAM) is the most frequently studied tumor-associated biomarker which is overexpressed in various types of cancer such as liver cancer, breast cancer, and colorectal cancer [59,68–70]. The aptamers specific for EpCAM receptors are better approach for CTCs analysis than the most frequently used antibodies due to their stability, biocompatibility and easy synthesis. Gazouli et al. [38] designed a simple immunoassay protocol for the detection of circulating human colorectal cancer cell using CdSe quantum-dots labeled monoclonal antibody and magnetic beads. In the first step, epithelial cell adhesion molecule (EpCAM) expressing human colon adenocarcinoma (DLD-1) cells were captured using EpCAM antibody linked magnetic beads and separated using normal magnet. The captured cells were then visualized and quantified by coupling with fluorescent QD-labeled mAb cytokeratin 19 (CK19). Gazouli and his team explained that the fluorescence was measured with spectrofluorometer, however the results were validated with fluorescence activated cell sorting (FACS) and Real Time PCR (RT-PCR) analysis as well. The limit of detection of the immunoassay was observed 10 DLD-1 cells/mL. Therefore, the described method is highly sensitive and comparatively inexpensive using spectrofluorometer and simple magnet for detection and separation of circulating colorectal cancer cells without sample pretreatment. Hence, it further confirms the use of QDs for the separation of circulating tumor cells (CTCs) in magnetic immunoassays by providing the highly specific detection and visualization.

In 2013, Wang and his team [39] demonstrated the production and utilization of molecular probe for the identification and isolation of metastatic hepatocellular carcinoma (HCC) cells. The molecular probe was the fluorescent QDs linked aptamer LY-1 which was highly specific to metastatic cell lines and synthesized via a whole cell-SELEX approach. One HCC cell line i.e. HCCLM9 cells with high metastatic potential, was used as target cells and other HCC cell line i.e. MHCC97-L cells with low metastatic potential, as subtractive cell to obtain highly specific aptamer. The aptamer was found competent to recognize metastatic HCC cells in cell cultures, animal models as well as in clinical HCC samples. The conjugation of aptamer with magnetic nanoparticles presented the easy separation of HCC cells by applying external magnetic field. The developed molecular probes exhibited the high commercialization possibility for early detection of metastatic HCC cells which might improve the survival rate of cancer patients.

A magnetic platform conjugated with GOQDs was designed by Shi and his team [40] for highly competent capturing (91%) and high luminescence imaging of Hep G2 liver CTCs. The antibodies specific to Glypican-3 (GPC3) particularly expressed on hepatocellular carcinoma CTCs were attached to GOQDs for specific detection in infected blood samples. The magnetic component assisted the separation of selective cells over the non-targeted GPC3 negative SK-BR-3 breast cancer cells. The proposed biocompatible multifunction designs are highly suitable for noninvasive and real-time analysis of rare cancerous cells with two photon luminescence (TPL) signal imaging.

Another magnetic fluorescent sensing system was designed for sensitive and quick identification and isolation of CTCs on the basis of nanosurface energy transfer (NSET) assay for EpCAM. The linear range achieved for the detection of EpCAM was 2–64 nM with limit of detection of 1.19 nM. In that assay, magnetic fluorescent nanocomposites (MFN) comprised of the aptamers specific to EpCAM receptors, nitrogen and sulfur doped blue luminescent GQDs and Fe_3O_4 nanoparticles were used as receptor whereas the molybdenum disulfide nanosheets used as quenchers The system was extremely specific to label and capture EpCAM expressing CTCs (low expression or overexpressed) in infected blood samples even as low as 10 out of billions of EpCAM non-expressing cells just in 15 min. The system presented high potential significance for early analysis and prognosis of tumor cells [43]. The enriched and captured cells can be used for imaging and further study as the rapid capturing of functional CTCs is highly helpful for clinical tumor therapy. Further, another one step detection and separation of CTCs based on anti-EpCAM antibodies was explained by Cui and his team [7] using MagDots containing fluorescent $ZnS:Mn^{2+}$ QDs and Fe_3O_4/SiO_2 magnetic nanoparticles. The magnetic fluorescent nanocomposites were encapsulated with biocompatible silica sphere via reverse microemulsion which remained stable in the whole blood samples. The tumor-specific biotinylated anti-EpCAM antibodies were then conjugated on silica coated nanocomposites via streptavidin grafting. The multifunctional nanocomposites exhibited high efficiency (90.8%) for CTCs capturing from infected blood samples within few minutes and visualization by yellowish orange fluorescent light. The capturing efficiency for SW480 and MCF-7 cell lines used as experiment models was recorded 90.8%. These sensitive and convenient methods for separation of multiple biomarkers and tumor cells enlighten a new avenue for advancement in cancer detection and treatment.

5.3   QDs in microbial cell separation

According to WHO report, foodborne pathogens cause illness in approximately 600 million people and lead to 4.2 million deaths every year worldwide [71,72]. *Campylobacter jejuni*, *Escherichia coli* O157:H7, *Listeria monocytogenes* and *Salmonella* spp. are the most frequently occurring foodborne pathogens. Fast screening of contaminated food and isolation of pathogenic bacteria are mandatory to alleviate the food borne diseases ranging from diarrhea to cancers [71,72]. Xue et al. [42] fabricated an extremely sensitive fluorescent biosensor for the detection and separation of food borne pathogen *Escherichia coli* O157:H7 using CdSe/ZnS quantum dots and high gradient magnetic separation. Magnetic nanoparticles (MNPs) conjugated with protein G antibodies were captured uniformly into twofold-layered channel in the presence of high gradient magnetic field (HGMF). The injection of contaminated food sample followed by target specific monoclonal antibodies (MAbs) resulted in the formation of MNP-MAb-bacteria complexes and background impurities were washed out of the channel. Further, MNP-MAb-bacteria-PAb-QDs complexes were formed on the application of biotinylated anti-*E. coli* polyclonal antibodies (PAbs) conjugated with quantum dots which were then streamed out in the absence of HGMF and measured on the basis of fluorescence intensity. The developed biosensor was competent to detect and quantitatively measure as low as 14 CFU/mL of *E. coli* and the separation efficiency of 86% was achieved with multiple rings of HGMF.

Further, there is also report for super bugs separation with the help of QDs. The drug resistant superbugs take a toll of 7.0 million people every year as predicted by the Center for Global Health [11,73]. It is astonishing to know WHO report that the number of deaths caused by superbugs may reach up to 10 million by 2050 which would be more than those caused by cancer if the conditions remain as present [74]. The inefficiency of traditional clinical methods to diagnose the severe cases in early stages is the major cause of deaths due to sepsis and hence, an ultrasensitive recognition and separation of superbugs are highly recommended. A novel multipurpose carbon dots based system was reported by Pramanik and others [11] for precise identification and separation of superbugs such as *Salmonella* DT104 and Methicillin resistant *Staphylococcus aureus* (MRSA) from infected blood samples. The nanosystem based on multicolor multifunctional carbon quantum dots exhibited the separation and exact identification of *Salmonella* DT104 and MRSA superbugs via multicolor fluorescence imaging. Further, the eradication of drug resistant superbugs was achieved with a specific system formed by combination of magnetic nanoparticles, pardaxin antimicrobial peptides and multicolor fluorescent carbon dots. Due to the resistance against mostly available antibiotics, the use of antimicrobial peptide was suggested to kill multidrug-resistant

(MDR) superbugs. The application of present system can also be extended for other superbugs using target specific antibodies as recognition probe which reveals its high potential for clinical analysis.

Despite the bacterial cell separation, QDs based assays are also reported for the detection and separation of virus antigens. The analysis of two types of virus: equine influenza virus (EIV) and equine infectious anemia virus (EIAV) was reported by Wang and his team [50]. Sensitive multiplex immunoassay for the analysis of virus antigens was suggested based on CdTe QDs and fluorescent magnetic nanocomposites. The nanocomposites were covalently conjugated with virus antigens (Ag) and the corresponding antibodies (Ab) with QDs on amine-functionalized biocompatible silica shell. Two different colored fluorescent nanocomposites (CN-1 and CN-2) were produced by changing ratio of nanocomposites and CdTe QD i.e. 5:7 for CN-1 and 3:1 for CN-2. In competitive immunoassay, the QD-labeled EIV Ab/EIV Ag/CN-1 and QD-labeled EIAV Ab/EIAV Ag/CN-2 complexes were formed which can be easily detected and separated. The developed assay was quick, highly sensitive (EIAV Ag, 1.2 ng/mL and EIV Ag, 1.3 ng/mL) and competent to multiplex analysis in vivo. Eventually, it is clear that QDs possess an excellent potential for detection and separation of tumor cells, bacterial cells, viruses or macrophages. The applications can be further extended by merely conjugating the specific molecular probes for the recognition of target.

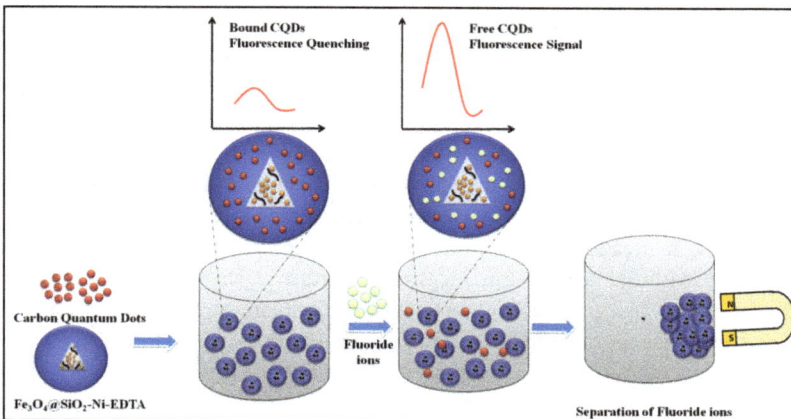

Fig. 6 Detection and separation of fluoride ions with carbon quantum dots.

5.4   QDs in ionic separation

CQDs were also used to design a simple, reusable and biocompatible nanosensor for fluoride ion detection, separation and bio-imaging [41]. The nanocomposites used in the process were comprised of fluorescent CQDs as fluorophore and water soluble silica coated nickel-EDTA-Fe_3O_4 magnetic complex as fluoride ions receptor. The detection strategy is based on competitive exchange of fluorescent CQDs on magnetic receptor in the presence of F^- ions which brought about the increase of fluorescence due to free CQDs, directly proportional to the concentration of F^- ions (Fig. 6). The developed nanosensor was competent to detect F^- with very selectivity and sensitivity. The linear range of detection was found to be 1.0 to 20 μM with a limit of detection of 0.06 μM. Later on, the magnetic receptor bound F^- ions can be separated easily by applying external magnetic field. The utility of nanosensor in aqueous solution and intracellular environment proposes the detection and separation of F^- ions in drinking water as well as in the living systems, for example in HT29 cells as mentioned by Sahu and his team [41].

Conclusion and future perspectives

The multi-featured quantum dots are emerged as highly astonishing tool for different fields. The application of QDs for the separation processes such as wastewater treatment, desalination, gaseous separation, heavy metal remediation, cellular and molecular separation using MagDots is comparatively new approach. Nevertheless, magnificent advancement has been achieved in this field and there are large possibilities for future advanced research and commercialization of the process. The ultra-small size, chemical inertness, hydrophilic nature and active functional groups of QDs endorsed their incorporation in previously reported separation membranes to improve the waste/salt filtration, water permeability and antifouling properties. However, the possibility of leaching of QDs in drinking water may hinder their scale up and demands the fabrication of superior and durable membrane systems. The strong bonding between QDs and matrices during fabrication, and sophisticated assessment may lower the risk of exposure of small QD particles in the drinking water. Further, QDs has been reported as a component of stationary phase in chromatographic column, however there is not much research work done in this direction. This area must be explored for inexpensive and convenient separation of volatile components. In addition, MagDots came into limelight as highly sensitive tool for the detection and capturing of rare CTCs in the blood samples. The appropriate selection of specific molecular probe to conjugate with MagDot for recognition and capturing of particular cell/biomarker promote their use in clinical diagnosis, treatment and other biomedical applications. Further, the risk of deposition of nanoparticles inside the body put a question mark on their use in *in-vivo* analysis. The use of non-toxic carbonaceous QDs instead of toxic inorganic semiconductor QDs and understanding the QD toxicity mechanism may be helpful to explore more opportunities. On a whole, quantum dots have huge potential and possibilities. However, there is a need of advanced research to discover the all possible benefits. The appropriate use of QDs would be highly beneficial to deal with major global concerns like desalination of water, detection of pathogens and lethal diseases.

References

[1]   K.J. McHugh, L. Jing, A.M. Behrens, S. Jayawardena, W. Tang, M. Gao, R. Langer, A. Jaklenec, Biocompatible semiconductor quantum dots as cancer imaging agents, Adv. Mater. 30 (2018) 1706356. https://doi.org/10.1002/adma.201706356

[2]   A.M. Wagner, J.M. Knipe, G. Orive, N.A. Peppas, Quantum dots in biomedical applications, Acta Biomater. 94 (2019) 44–63.

[3]   A. Cayuela, M.L. Soriano, C. Carrillo-Carrión, M. Valcárcel, Semiconductor and carbon-based fluorescent nanodots: The need for consistency, Chem. Commun. 52 (2016) 1311–1326. https://doi.org/10.1039/c5cc07754k

[4]   X. Wang, Y. Feng, P. Dong, J. Huang, A mini review on carbon quantum dots : preparation, properties and electrocatalytic application, Frontiers in chemistry. 7 (2019) 1–9. https://doi.org/10.3389/fchem.2019.00671

[5]   U. Kaiser, D. Jimenez De Aberasturi, M. Vázquez-González, C. Carrillo-Carrion, T. Niebling, W.J. Parak, W. Heimbrodt, Determining the exact number of dye molecules attached to colloidal CdSe/ZnS quantum dots in Förster resonant energy transfer assemblies, J. Appl. Phys. 117 (2015) 024701. https://doi.org/10.1063/1.4905025

[6]   N. Gogoi, M. Barooah, G. Majumdar, D. Chowdhury, Carbon dots rooted agarose hydrogel hybrid platform for optical detection and separation of heavy metal ions, ACS Appl. Mater. Interfaces. 7 (2015) 3058–3067. https://doi.org/10.1021/am506558d

[7]   H. Cui, R. Li, J. Du, Q.F. Meng, Y. Wang, Z.X. Wang, F.F. Chen, W.F. Dong, J. Cao, L.L. Yang, S.S. Guo, Rapid and efficient isolation and detection of circulating tumor cells based on ZnS:Mn^{2+} quantum dots and magnetic nanocomposites, Talanta. 202 (2019) 230–236. https://doi.org/10.1016/j.talanta.2019.05.001

[8]   X. Zhang, H. Ji, X. Zhang, Z. Wang, D. Xiao, Capillary column coated with graphene quantum dots for gas chromatographic separation of alkanes and aromatic isomers, Anal. Methods. 7 (2015) 3229–3237. https://doi.org/10.1039/c4ay03068k

[9]   D.L. Zhao, T.S. Chung, Applications of carbon quantum dots (CQDs) in membrane technologies: A review, Water Res. 147 (2018) 43–49. https://doi.org/10.1016/j.watres.2018.09.040

[10]   E. Song, J. Hu, C. Wen, Z. Tian, X. Yu, Z. Zhang, Y. Shi, D.-W. Pang, Fluorescent-magnetic-biotargeting multifunctional nanobioprobes for detecting and isolating multiple types of tumor cells, ACS Nano. 5 (2011) 761–770.

[11]   A. Pramanik, S. Jones, F. Pedraza, A. Vangara, C. Sweet, M.S. Williams, V. Ruppa-Kasani, S.E. Risher, D. Sardar, P.C. Ray, Fluorescent, magnetic multifunctional carbon dots for selective separation, identification, and eradication of drug-resistant superbugs, ACS Omega. 2 (2017) 554–562. https://doi.org/10.1021/acsomega.6b00518

[12]   K.D. Mahajan, G.B. Vieira, G. Ruan, B.L. Miller, B. Maryam, J.J. Chalmers, R. Sooryakumar, J.O. Winter, A MagDot-nanoconveyor assay detects and isolates molecular biomarkers, Chem Eng Prog. 2012. 108 (2012) 41–46.

[13]   A. Soroush, J. Barzin, M. Barikani, M. Fathizadeh, Interfacially polymerized polyamide thin film composite membranes: Preparation, characterization and performance evaluation, Desalination. 287 (2012) 310–316. https://doi.org/10.1016/j.desal.2011.07.048

[14]   M. Fathizadeh, H.N. Tien, K. Khivantsev, Z. Song, F. Zhou, M. Yu, Polyamide/nitrogen-doped graphene oxide quantum dots (N-GOQD) thin film nanocomposite reverse osmosis membranes for high flux desalination, Desalination. 451 (2019) 125–132. https://doi.org/10.1016/j.desal.2017.07.014

[15]   R. Bi, R. Zhang, J. Shen, Y. Liu, M. He, X. You, Y. Su, Z. Jiang, Graphene quantum dots engineered nanofiltration membrane for ultrafast molecular separation, J. Memb. Sci. 572 (2018) 504–511. https://doi.org/https://doi.org/10.1016/j.memsci.2018.11.044

[16]   M. Fathizadeh, A. Aroujalian, A. Raisi, Effect of lag time in interfacial polymerization on polyamide composite membrane with different hydrophilic sub layers, Desalination. 284 (2012) 32–41. https://doi.org/10.1016/j.desal.2011.08.034.

[17]   X. Song, Q. Zhou, T. Zhang, H. Xu, Z. Wang, Pressure-assisted preparation of graphene oxide quantum dot-incorporated reverse osmosis membranes: Antifouling and chlorine resistance potentials, J. Mater. Chem. A. 4 (2016) 16896–16905. https://doi.org/10.1039/c6ta06636d

Materials Research Forum LLC
https://doi.org/10.21741/9781644901250-10

[18]   C. Zhang, K. Wei, W. Zhang, Y. Bai, Y. Sun, J. Gu, Graphene oxide quantum dots incorporated into a thin film nanocomposite membrane with high flux and antifouling properties for low-pressure nanofiltration, ACS Appl. Mater. Interfaces. 9 (2017) 11082–11094. https://doi.org/10.1021/acsami.6b12826

[19]   Z. Yuan, X. Wu, Y. Jiang, Y. Li, J. Huang, L. Hao, J. Zhang, J. Wang, Carbon dots-incorporated composite membrane towards enhanced organic solvent nanofiltration performance, J. Memb. Sci. 549 (2018) 1–11. https://doi.org/10.1016/j.memsci.2017.11.051

[20]   S. Xu, F. Li, B. Su, M.Z. Hu, X. Gao, C. Gao, Novel graphene quantum dots (GQDs)-incorporated thin film composite (TFC) membranes for forward osmosis (FO) desalination, Desalination. 451 (2019) 219–230

[21]   Z. Zeng, D. Yu, Z. He, J. Liu, F.X. Xiao, Y. Zhang, R. Wang, D. Bhattacharyya, T.T.Y. Tan, Graphene oxide quantum dots covalently functionalized pvdf membrane with significantly-enhanced bactericidal and antibiofouling performances, Sci. Rep. 6 (2016) 1–11. https://doi.org/10.1038/srep20142

[22]   H. Shi, F. Liu, L. Xue, Fabrication and characterization of antibacterial PVDF hollow fibre membrane by doping Ag-loaded zeolites, J. Memb. Sci. 437 (2013) 205–215. https://doi.org/10.1016/j.memsci.2013.03.009

[23]   D.L. Zhao, S. Das, T.S. Chung, Carbon quantum dots grafted antifouling membranes for osmotic power generation via pressure-retarded osmosis process, Environ. Sci. Technol. 51 (2017) 14016–14023. https://doi.org/10.1021/acs.est.7b04190

[24]   L. Hui, J. Huang, G. Chen, Y. Zhu, L. Yang, Antibacterial property of graphene quantum dots (both source material and bacterial shape matter), ACS Appl. Mater. Interfaces. 8 (2016) 20–25. https://doi.org/10.1021/acsami.5b10132

[25]   F. Chen, W. Gao, X. Qiu, H. Zhang, L. Liu, P. Liao, W. Fu, Y. Luo, Graphene quantum dots in biomedical applications: Recent advances and future challenges, Front. Lab. Med. 1 (2017) 192–199. https://doi.org/10.1016/j.flm.2017.12.006

[26]   J. Zhu, J. Wang, J. Hou, Y. Zhang, J. Liu, B. Van der Bruggen, Graphene-based antimicrobial polymeric membranes: a review, J. Mater. Chem. A. 5 (2017) 6776–6793. https://doi.org/10.1039/c7ta00009j

[27]   A. Colburn, N. Wanninayake, D.Y. Kim, D. Bhattacharyya, Cellulose-graphene quantum dot composite membranes using ionic liquid, J. Memb. Sci. 556 (2018) 293–302. https://doi.org/10.1016/j.memsci.2018.04.009

[28]   B. Safaei, M. Youssefi, B. Rezaei, N. Irannejad, Synthesis and properties of photoluminescent carbon quantum dot/polyacrylonitrile composite nanofibers, Smart Sci. 6 (2017) 117–124. https://doi.org/10.1080/23080477.2017.1399318

[29]   Y. Zhang, Y.H. He, P.P. Cui, X.T. Feng, L. Chen, Y.Z. Yang, X.G. Liu, Water-soluble, nitrogen-doped fluorescent carbon dots for highly sensitive and selective detection of Hg^{2+} in aqueous solution, RSC Adv. 5 (2015) 40393–40401. https://doi.org/10.1039/c5ra04653j

[30]   A. Jafari, M.R.S. Kebria, A. Rahimpour, G. Bakeri, Graphene quantum dots modified polyvinylidenefluride (PVDF) nanofibrous membranes with enhanced performance for air Gap membrane distillation, Chem. Eng. Process. - Process Intensif. 126 (2018) 222–231. https://doi.org/10.1016/j.cep.2018.03.010

[31]   L.Y. Jiang, T.S. Chung, C. Cao, Z. Huang, S. Kulprathipanja, Fundamental understanding of nano-sized zeolite distribution in the formation of the mixed matrix single- and dual-layer asymmetric hollow fiber membranes, J. Memb. Sci. 252 (2005) 89–100. https://doi.org/10.1016/j.memsci.2004.12.004

[32]   U. Resch-Genger, M. Grabolle, S. Cavaliere-Jaricot, R. Nitschke, T. Nann, Quantum dots versus organic dyes as fluorescent labels, Nat. Methods. 5 (2008) 763–775. https://doi.org/10.1038/nmeth.1248

[33]   K.D. Mahajan, Q. Fan, J. Dorcéna, G. Ruan, J.O. Winter, Magnetic quantum dots in biotechnology - synthesis and applications, Biotechnol. J. 8 (2013) 1424–1434. https://doi.org/10.1002/biot.201300038

[34]   A.K. Gupta, M. Gupta, Synthesis and surface engineering of iron oxide nanoparticles for biomedical applications, Biomaterials. 26 (2005) 3995–4021. https://doi.org/10.1016/j.biomaterials.2004.10.012

[35]   A.K. Yetisen, M.S. Akram, C.R. Lowe, Paper-based microfluidic point-of-care diagnostic devices, Lab Chip. 13 (2013) 2210. https://doi.org/10.1039/c3lc50169h.

[36]   R.M. Lequin, Enzyme immunoassay (EIA)/enzyme-linked immunosorbent assay (ELISA), Clin. Chem. 51 (2005) 2415–2418. https://doi.org/10.1373/clinchem.2005.051532

[37]   A. Agrawal, T. Sathe, S. Nie, Single-bead immunoassays using magnetic microparticles and spectral-shifting quantum dots, J. Agric. Food Chem. 55 (2007) 3778–3782. https://doi.org/10.1021/jf0635006

[38]   M. Gazouli, A. Lyberopoulou, P. Pericleous, S. Rizos, G. Aravantinos, N. Nikiteas, N.P. Anagnou, E.P. Efstathopoulos, Development of a quantum-dot-labelled

magnetic immunoassay method for circulating colorectal cancer cell detection, World J. Gastroenterol. 18 (2012) 4419–4426. https://doi.org/10.3748/wjg.v18.i32.4419

[39]   F.B. Wang, Y. Rong, M. Fang, J.P. Yuan, C.W. Peng, S.P. Liu, Y. Li, Recognition and capture of metastatic hepatocellular carcinoma cells using aptamer-conjugated quantum dots and magnetic particles, Biomaterials. 34 (2013) 3816–3827. https://doi.org/10.1016/j.biomaterials.2013.02.018

[40]   Y. Shi, A. Pramanik, C. Tchounwou, F. Pedraza, R.A. Crouch, S.R. Chavva, A. Vangara, S.S. Sinha, S. Jones, D. Sardar, C. Hawker, P.C. Ray, multifunctional biocompatible graphene oxide quantum dots decorated magnetic nanoplatform for efficient capture and two-photon imaging of rare tumor cells, ACS Appl. Mater. Interfaces. 7 (2015) 10935–10943. https://doi.org/10.1021/acsami.5b02199

[41]   S. Sahu, S. Nayak, S.K. Ghosh, S. Mohapatra, Design of $Fe_3O_4@SiO_2@carbon$ quantum dot based nanostructure for fluorescence sensing, magnetic separation, and live cell imaging of fluoride ion, Langmuir. 31 (2015) 8111–8120. https://doi.org/10.1021/acs.langmuir.5b01513

[42]   L. Xue, L. Zheng, H. Zhang, X. Jin, J. Lin, An ultrasensitive fluorescent biosensor using high gradient magnetic separation and quantum dots for fast detection of foodborne pathogenic bacteria, Sensors Actuators, B Chem. 265 (2018) 318–325. https://doi.org/10.1016/j.snb.2018.03.014

[43]   F. Cui, J. Ji, J. Sun, J. Wang, H. Wang, Y. Zhang, H. Ding, Y. Lu, D. Xu, X. Sun, A novel magnetic fluorescent biosensor based on graphene quantum dots for rapid, efficient, and sensitive separation and detection of circulating tumor cells, Anal. Bioanal. Chem. 411 (2019) 985–995. https://doi.org/10.1007/s00216-018-1501-0

[44]   D. Wang, J. He, N. Rosenzweig, Z. Rosenzweig, Superparamagnetic Fe_2O_3 Beads – CdSe / ZnS quantum dots core – shell nanocomposite particles for cell, Nano Letters. 4 (2004) 3, 409-413.

[45]   M. Chu, X. Song, D. Cheng, S. Liu, J. Zhu, Preparation of quantum dot-coated magnetic polystyrene nanospheres for cancer cell labelling and separation, Nanotechnology. 17 (2006) 3268–3273. https://doi.org/10.1088/0957-4484/17/13/032

[46]   E.Q. Song, G.P. Wang, H.Y. Xie, Z.L. Zhang, J. Hu, J. Peng, D.C. Wu, Y.B. Shi, D.W. Pang, Visual recognition and efficient isolation of apoptotic cells with fluorescent-magnetic-biotargeting multifunctional nanospheres, Clin. Chem. 53 (2007) 2177–2185. https://doi.org/10.1373/clinchem.2007.092023

[47]   S.T. Selvan, P.K. Patra, C.Y. Ang, J.Y. Ying, Synthesis of silica-coated semiconductor and magnetic quantum dots and their use in the imaging of live cells, Angew. Chemie - Int. Ed. 119 (2007) 2500–2504. https://doi.org/10.1002/anie.200604245

[48]   K. Knop, R. Hoogenboom, D. Fischer, U.S. Schubert, Poly(ethylene glycol) in drug delivery: Pros and cons as well as potential alternatives, Angew. Chemie - Int. Ed. 49 (2010) 6288–6308. https://doi.org/10.1002/anie.200902672

[49]   A. Kale, S. Kale, P. Yadav, H. Gholap, R. Pasricha, J.P. Jog, B. Lefez, B. Hannoyer, P. Shastry, S. Ogale, Magnetite/CdTe magnetic-fluorescent composite nanosystem for magnetic separation and bio-imaging, Nanotechnology. 22 (2011) 225110. https://doi.org/10.1088/0957-4484/22/22/225101

[50]   G. Wang, Y. Gao, H. Huang, X. Su, Multiplex immunoassays of equine virus based on fluorescent encoded magnetic composite nanoparticles, Anal. Bioanal. Chem. 398 (2010) 805–813. https://doi.org/10.1007/s00216-010-4001-4

[51]   J.F. Rusling, C. V. Kumar, J.S. Gutkind, V. Patel, Measurement of biomarker proteins for point-of-care early detection and monitoring of cancer, Analyst. 135 (2010) 2496–2511. https://doi.org/10.1039/c0an00204f

[52]   A. Son, D. Dosev, M. Nichkova, Z. Ma, Quantitative DNA hybridization in solution using magnetic/luminescent core–shell nanoparticles, Anal. Biochem. 370 (2007) 186–194.

[53]   L. Chen, R. Liu, Z.P. Liu, M. Li, K. Aihara, Detecting early-warning signals for sudden deterioration of complex diseases by dynamical network biomarkers, Sci. Rep. 2 (2012) 1–8. https://doi.org/10.1038/srep00342

[54]   J. Thomas, P. Malla, T. Vu. Miniaturized single cell imaging for developing immuno-oncology combinational therapies. In: Tan SL. (eds) Immuno-Oncology. Methods in Pharmacology and Toxicology. Humana, New York, NY, 2020, pp 157-165.

[55]   G. Ruan, G. Vieira, T. Henighan, A. Chen, D. Thakur, R. Sooryakumar, J.O. Winter, Simultaneous magnetic manipulation and fluorescent tracking of multiple individual hybrid nanostructures, Nano Lett. 10 (2010) 2220–2224. https://doi.org/10.1021/nl1011855

[56]   B. Dubertret, P. Skourides, D.J. Norris, V. Noireaux, A.H. Brivanlou, A. Libchaber, In vivo imaging of quantum dots encapsulated in phospholipid micelles, Science (80-.). 298 (2002) 1759–1762. https://doi.org/10.1126/science.1077194

[57] J.H. Park, G. Von Maltzahn, E. Ruoslahti, S.N. Bhatia, M.J. Sailor, Micellar hybrid nanoparticles for simultaneous magnetofluorescent imaging and drug delivery, Angew. Chemie - Int. Ed. 120 (2008) 7394 –7398. https://doi.org/10.1002/anie.200801810

[58] G. Ruan, D. Thakur, S. Deng, S. Hawkins, J.O. Winter, Fluorescent-magnetic nanoparticles for imaging and cell manipulation, Proc. Inst. Mech. Eng. Part N J. Nanoeng. Nanosyst. 223 (2010) 81–86. https://doi.org/10.1243/17403499JNN178

[59] W.J. Allard, J. Matera, M.C. Miller, M. Repollet, M.C. Connelly, C. Rao, A.G.J. Tibbe, J.W. Uhr, L.W.M.M. Terstappen, Tumor cells circulate in the peripheral blood of all major carcinomas but not in healthy subjects or patients with nonmalignant diseases, Clin. Cancer Res. 10 (2004) 6897– 6904. https://doi.org/10.1158/1078-0432.CCR-04-0378

[60] C.L. O'Connell, R. Nooney, C. McDonagh, Cyanine5-doped silica nanoparticles as ultra-bright immunospecific labels for model circulating tumour cells in flow cytometry and microscopy, Biosens. Bioelectron. 91 (2017) 190–198. https://doi.org/10.1016/j.bios.2016.12.023

[61] X. Wu, T. Xiao, Z. Luo, R. He, Y. Cao, Z. Guo, W. Zhang, Y. Chen, A micro-/nano-chip and quantum dots-based 3D cytosensor for quantitative analysis of circulating tumor cells, J. Nanobiotechnology. 16 (2018) 1–9. https://doi.org/10.1186/s12951-018-0390-x

[62] Q.-Q. Huang, X.-X. Chen, W. Jiang, S.-L. Jin, X.-Y. Wang, W. Liu, S.-S. Guo, J.-C. Guo, X.-Z. Zhao, Sensitive and specific detection of circulating tumor cells promotes precision medicine for cancer, J. Cancer Metastasis Treat. 5 (2019) 1–18. https://doi.org/10.20517/2394-4722.2018.94

[63] P. Augustsson, C. Magnusson, M. Nordin, H. Lilja, T. Laurell, Microfluidic, label-free enrichment of prostate cancer cells in blood based on acoustophoresis, Anal. Chem. 84 (2012) 7954–7962. https://doi.org/10.1021/ac301723s

[64] Z. Chen, S.B. Cheng, P. Cao, Q.F. Qiu, Y. Chen, M. Xie, Y. Xu, W.H. Huang, Detection of exosomes by ZnO nanowires coated three-dimensional scaffold chip device, Biosens. Bioelectron. 122 (2018) 211–216. https://doi.org/10.1016/j.bios.2018.09.033

[65] M. Munz, P.A. Baeuerle, O. Gires, The emerging role of EpCAM in cancer and stem cell signaling, Cancer Res. 69 (2009) 5627–5630. https://doi.org/10.1158/0008-5472.CAN-09-0654

[66]   G.A.F. Van Tilborg, W.J.M. Mulder, N. Deckers, G. Storm, C.P.M. Reutelingsperger, G.J. Strijkers, K. Nicolay, Annexin A5-functionalized bimodal lipid-based contrast agents for the detection of apoptosis, Bioconjug. Chem. 17 (2006) 741–749. https://doi.org/10.1021/bc0600259

[67]   L. Prinzen, R.J.J.H.M. Miserus, A. Dirksen, T.M. Hackeng, N. Deckers, N.J. Bitsch, R.T.A. Megens, K. Douma, J.W. Heemskerk, M.E. Kooi, P.M. Frederik, D.W. Slaaf, M.A.M.J. Van Zandvoort, C.P.M. Reutelingsperger, Optical and magnetic resonance imaging of cell death and platelet activation using annexin A5-functionalized quantum dots, Nano Lett. 7 (2007) 93–100. https://doi.org/10.1021/nl062226r

[68]   M. Varga, P. Obrist, S. Schneeberger, G. Mühlmann, C. Felgel-Farnholz, D. Fong, M. Zitt, T. Brunhuber, G. Schäfer, G. Gastl, G. Spizzo, Overexpression of epithelial cell adhesion molecule antigen in gallbladder carcinoma is an independent marker for poor survival, Clin. Cancer Res. 10 (2004) 3131–3136,. https://doi.org/10.1158/1078-0432.CCR-03-0528

[69]   D. Fong, M. Steurer, P. Obrist, V. Barbieri, R. Margreiter, A. Amberger, K. Laimer, G. Gastl, A. Tzankov, G. Spizzo, Ep-CAM expression in pancreatic and ampullary carcinomas: Frequency and prognostic relevance, J. Clin. Pathol. 61 (2008) 31–35. https://doi.org/10.1136/jcp.2006.037333

[70]   G. Gastl, G. Spizzo, P. Obrist, M. Dünser, G. Mikuz, Ep-CAM overexpression in breast cancer as a predictor of survival, Lancet. 356 (2000) 1981–1982. https://doi.org/10.1016/S0140-6736(00)03312-2

[71]   WHO, Food Safety, (2015). https://www.who.int/en/news-room/fact-sheets/detail/food-safety (accessed June 21, 2020).

[72]   E. Scallan, R.M. Hoekstra, F.J. Angulo, R. V. Tauxe, M.A. Widdowson, S.L. Roy, J.L. Jones, P.M. Griffin, Foodborne illness acquired in the United States-Major pathogens, Emerg. Infect. Dis. 17 (2011) 7–15. https://doi.org/10.3201/eid1701.P11101

[73]   L.-S. Wang, A. Gupta, V.M. Rotello, Nanomaterials for the treatment of bacterial biofilms Li-Sheng, ACS Infect Dis. 2 (2016) 1–4. https://doi.org/10.1021/acsinfecdis.5b00116

[74]   WHO, Antimicorbial resistence global report on surveillance, 1 (2014) 1–8. https://prezi.com/nihmdozcvwvo/unidad-didactica-funcion-de-relacion/ (accessed June 21, 2020).

Quantum Dots – Properties and Applications
Materials Research Foundations **96** (2021) 280-304

Materials Research Forum LLC
https://doi.org/10.21741/9781644901250-11

Chapter 11

Quantum Dots Based Materials for Water Treatment

Chetna Tewari[1], Sumit Kumar[2], Neema Pandey[1], Sandeep Pandey[1], Nanda Gopal Sahoo[1*]

[1]Prof. Rajendra Singh Nanoscience and Nanotechnology Centre, Department of Chemistry, D.S.B. Campus, Kumaun University, Nainital, Uttarakhand, India

[2]EWRE Division, Department of Civil Engineering, Indian Institute of Technology-Madras (IITM), Chennai, India

ngsahoo@yahoo.co.in

Abstract

Quantum dot is a new class of nanomaterials having size in nanometers (<10 nm). This material has excellent photo-catalytic activity towards dyes and pollutants with great absorbance and photoluminescence properties. It shows shifting of peak in UV-FL data which indicates the excitation dependent emission spectra means tunable properties in different wavelength and this property makes it a wonderful probe for sensing application for different heavy metals, pollutants present in water. In this chapter the synthesis, properties, types, application of quantum dots and focus on the research that has been done in field of water treatment with possible future outcomes is discussed.

Keywords

Metal Detection, Photo-Catalytic Activity, Quantum Dots, Sensor, Water Treatment

Contents

1.   Introduction

The present research era of nanoscale materials is dealing with various aspects of water treatment due to their nano scale sizes, excellent optical and physical properties, high adsorption capacity and sensing properties, etc. These materials exhibit tremendous properties mostly due to their size in nanometer ($\sim 10^{-9}$m) scale. In the recent years, researchers have focus on the synthesis and enhancement of nanomaterials such as graphene oxide (GO), carbon nanotubes (CNTs), fullerene, quantum dots (QDs) etc. to carry out their potential in various application in the fields of energy, drug delivery, polymer nanocomposites, photo-voltaic cells, bio-imaging, and water treatment etc. Similarly, the utilization of carbon nanomaterials (CNMs) mainly QDs is huge in water purification to membrane filtration and metal detection. Basically QDs have been widely used to combat against variety of diseases in the human body, that contains approximately ~70% of water and it is vital for transportation of ions in blood, which helps the body to combat against a variety of diseases. The present scenario has revealed that the modest amount of freshwater (drinking water) is contaminated due to the rapid growth in number of industries which is adversely affecting the large population living on earth as well as in the water bodies. Several literatures reported that the major categories of pollutants are heavy metals, inorganic substances and organic pollutants especially dyestuff, polycyclic aromatic hydrocarbons, pesticides and pharmaceuticals etc. Most of the categorized pollutants are present in micro levels of concentration but their hazards are very severe on living organisms because those are highly persistent in environment as well as in the conventional water treatment systems available.

The hazards caused by heavy metal ions have been found as an alarming threat to the environment. Heavy metals such as Pb, Ni, Cd, As, Cu, Mn, Cr, Co etc. are very noxious for human health even in trace concentrations and the prominent source of these heavy metals in living beings is consumption of contaminated water. These types of pollutants can be bio accumulated and frequent consumption of these heavy metals leads to several human health hazards like damage to liver, brain, and cancer etc. The removal of these heavy metals from water is the need of entire living species and fulfilling this objective is not served by the existing treatment technologies, therefore, quantum dots (QDs) are emerging and promising materials for water purification and related applications.

Carbon based QDs are a new-class of nanomaterials initially prepared of carbon and graphene quantum dots (GQDs)) within the size of 10 nm. The first QDs came to picture in the year 2004 when Xu et al. [1] was experimenting with the purification of single-walled carbon nanotubes and they discovered a new class of carbon-based nanomaterials termed as CQDs. In recent years, QDs and their composites have been widely investigated for water treatment due to their unique range of size (~10 nm). The potential applications of QDs are in photocatalysis and sensors due to their strong quantum confinement and edge effects that can render excellent optical properties.

Zhang et al. [2] has synthesized g-C$_3$N$_4$/TNA, a visible light driven photo-catalytic membrane using g-C$_3$N$_4$ QDs assembled into TiO$_2$ nanotube array (TNA). This membrane was used for removal of rhodamine B dye and the removal efficiency was ~60%. Saud et al. [3] reported CQDs/TiO$_2$ composite nano-fibers possessing high antibacterial properties against *Escherichia coli*. Tang et al. [4] prepared carbon nitride quantum dots (g-CNQDs), by using guanidine hydrochloride and EDTA. The prepared g-CNQDs were used for detection of free chlorine present in water. Yu & Kwak [5] studied CQDs embedded in mesoporous α-Fe$_2$O$_3$ and this composite showed excellent photo catalytic efficiency towards the degradation of organic compounds in aqueous media under visible light irradiation and it showed up to 97% of initial catalytic capacity even after three cycles.

Materials Research Forum LLC
https://doi.org/10.21741/9781644901250-11

It has been found in many literatures for detection and quantification of substances such as Fe^{3+}, Ag^+, Hg^{2+}, PO_4^{3-}, thrombin, nitrite, glucose, biothiol, DNA and Cu^{2+} [6-16]. CQDs are frequently being used nowadays and this could be possible by changing in fluorescence intensity under physical and chemical stimuli. Moreover, such CQDs possess excellent photoluminescence properties and can be used for distinct bioimaging. Recent development in CQDs enhanced the selectivity and sensitivity in sensors for detection of dopamine (DA) and Fe^{3+} [17] due to oxidation hydroquinone groups of CQDs by Fe^{3+} and this results in fluorescence quenching from CQDs and given shelter by DA. This phenomenon termed as "mix and detect" for DA, Fe^{3+} and their analogs detection in merely 10 minutes of time under one step operation. In addition to this, this new strategy eliminates complexity of QDs, reduces the chemical modifications and more importantly it reduces the cost involved.

This book chapter describes the latest applications of quantum dots in the field of water treatment, synthesis of quantum dots, development of quantum dots based composites, and offers suggestion for future success. Furthermore, the applications of quantum dots based fluorescence sensor for discriminatory detection of pollutants are also discussed.

2. Classification of quantum dots

Excitation comprises of an electron and a hole, the energy difference between the both is referred to as excitation Bohr's energy. It is noticed that, when the semiconductor crystal is much larger than material's excitation Bohr's radius, the energy levels are treated as continuous while when crystal size of semiconductor becomes small in comparison to excitation Bohr's radius the energy levels appears to be discrete having a finite separation. This succeeding condition is termed as quantum confinement and when the semiconductor material differs with the bulk properties then it is named as quantum dots (QDs) [18]. Based on the type of electron confinement, QDs can be categorized into three classes as discussed below.

2.1 Planar quantum dots

In such categories, metal gates are fabricated on the surface of 2D electron gas heterostructure (2DGH). The 2DGH is constructed by the movement of 2D electron gas on heterostructures of GaAs/AlsGaAs. Electrostatic potential is raised by applying a negative gate bias. At suitable condition of biasing the 2DGH separates forming a small planer region like a dot at the center [19]. These planer dots are sensitive to gate voltage, shape, height, and external perturbation. The main characteristics of PQDs: i) the flows of current in the 2DGH plane; ii) the height of tunnel barrier is controllable.

2.2   Vertical quantum dots

A combination of GaAs disk (zero dimensional) in between two AlGaAs is referred to vertical QDs. In vertical QDs the surrounding gate electrode is negatively biased. On biasing the pillars negatively results the depletion of electrons from pillar's outer region and consequentially shrinks the disk. This result indicates that the confinement of electron at a very narrow region in the center of pillar [19, 20]. In vertical QDs are less sensitive to external fluctuations because of large band gap material and high tunneling blockages in which the flow of current is perpendicular to 2DGH.

2.3   Self-assembled quantum dots

When a few mololayers of InAs are grown on a GaAs substrate, the heterogeneity in the materials leading to the split of a 2D-layered structure often results in a random distribution of self-assembled QDs. The confinement of SAQDs is much stronger than PQDs and VQDs. This leads to exhibits lens-shaped and pyramidal structures with approx. 100A size [21].

3.   Synthesis of quantum dots

Quantum dots have been synthesized by many methods and these methods can be generally categorized into two approaches i.e. "Bottom-up" and "Top-down" shown in Fig. 1. In the bottom-up approach, small materials are used as precursor to building quantum dots. While in the top-down method, it is required to reduce the size of materials for the formation of quantum dots [22]. Sometimes modifications are required for enhancing the properties and application of the quantum dots based materials. The modification of quantum dots can be done during preparation or post-treatment of QDs. But the main problems faced by researchers in the synthesis of QDs are carbonaceous aggregation, size control, uniformity, and surface properties. To optimize the size and uniformity, the post-treatments i.e. dialysis, gel electrophoresis, and centrifugation can be employed. The aggregation problem can be overcome by using the electrochemical synthesis, solution chemistry, confirmed pyrolysis techniques. In this section, we focus on the synthesis of carbon based quantum dots (CQDs) in details.

Quantum Dots – Properties and Applications   Materials Research Forum LLC
Materials Research Foundations **96** (2021) 280-304   https://doi.org/10.21741/9781644901250-11

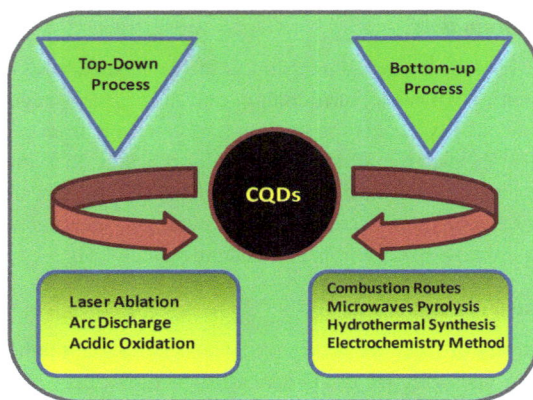

Fig. 1 The typical approaches for the synthesis of CQDs.

3.1   Top-down approach

3.1.1 Laser ablation

In laser ablation method, quantum dots are generated by the nucleation of precursor and growth of the vaporized spices in the background gases. The enormous fast reduction of vapor results in generation of highly pure quantum dots with a size of less than 10 nm. These nanoparticles show unique properties which cannot be observe in bulk materials [23]. The most critical properties of such nanoparticles (electrical, magnetic, optical etc.) are varied their size distributions [24]. In the laser ablation method, laser beam is bombarded on the surface of a solid target material either in gas or liquid medium and so the temperature of this spot is increased very rapidly and the vaporization of material in form of atoms and clusters starts. The excitation of the electron takes place by the collisions among the evaporated species and the present molecules. After the quenching of vapors, quantum dots are formed. In this method, the selection of the medium plays a critical role, so the ambient medium should be carefully selected because the vaporized particles are prone to react with the medium and formed complexes.

3.1.2 Arc discharge

Arc discharge is a top-down approach in which the quantum dots are formed by the decomposition of the bulk carbon precursor. In this method, carbon atoms are decomposed in acidic condition and electrode driven by the gas plasma generated in a sealed reactor system [10, 22]. The temperature of the reactor can reach as high as 3727°C under electric current, so that high energy plasma can produce and at the cathode the carbon vapors assembly to form CQDs.

3.1.3 Acidic oxidation

Acid oxidation method is used to decompose bulk carbon source and synthesize quantum dots. In this method, introduction of the hydrophilic groups (hydroxyl group, carboxyl group) in the material takes place during exfoliation and decomposition, which enhance the water solubility and fluorescence properties of the QDs. This enhances the application of synthesizing material in the field of life sciences. For carbonization of small organic molecules to carbonaceous material, this method requires strong oxidizing acids and further the obtained material can convert by quantum dots by controlled oxidation.

In 2014, Tang et al. [25] reported a large-scale synthesis of CQDs with doping of heteroatom using acid oxidation treatment, followed by hydrothermal reduction. In this synthesis carbon nanoparticles obtained from Chinese ink and oxidized by a solution of HNO_3, H_2SO_4, and $NaClO_3$. Then the doping of the nitrogen, sulfur, and selenium achieved using DMF, NaHS, NaHSe as sources for nitrogen, sulfur, and selenium in hydrothermal treatment. The doped quantum dots showed higher quantum yield, tunable PL performance, and longer fluorescence lifetime than the pure CQDs.

On the other hand, it has been seen from several literatures that the doped-CQDs showed the ability to interact with transition metal ions. The N-CQDs, S-CQDs, and Se-CQDs may also have the potential to absorb other metal ions, such as Co^{2+}, Fe^{3+}, and Ni^{2+}, to form the single-atom catalysts (SAC) [16, 22].

3.2   Bottom-up approach

3.2.1   Combustion/thermal routes

For preparation of smaller nanoparticles (< 10 nm), the most favorable approach is solution combustion. It has been found in the bottom-up approach, nano particle having high surface area can be produced in a single step. The method offers number of advantages over conventional methods such as time of synthesis, cost involved and energy intake. It is also offered in situ oxides doping method and the key aspects of this method are the molecular level homogeneous mixing, high-purity and highly crystallized final products formation due to high reaction temperature moreover growth that is hindered by the small reaction time.

3.2.2 Microwave pyrolysis

Microwave irradiation is a quick, environment-friendly and cost-effective approach of synthesizing CQDs with hydrophilic groups [26-29]. H. Zhu et al. [30] made a solution of poly (ethylene glycol) and a saccharide (glucose, fructose, etc.) in water to form a transparent solution and kept this solution in a microwave oven for CQDs synthesis. The obtained CQDs exhibited excitation-dependent PL properties.

Chen et al. [31] prepared highly fluorescent quantum dots by using microwave-mediated pyrolysis. For this, they used citric acid as a carbon source with various amine molecules. The amine molecules acted as passivating agents as well as doping source of nitrogen and enhanced the photoluminescence properties of the quantum dots. In this case, the quantum yield was proportional to the N content and showed quantum yield up to 30.2%. The CQDs synthesized was water-soluble, nontoxic and possess very high potential for life science applications.

3.2.3 Hydrothermal/solvothermal synthesis

Hydrothermal approach for synthesis of quantum dots is rather common because of size uniformity of the obtained product and simple setup [32, 33]. S. Zhu et al. [34] reported the highest quantum yield which was almost equal to fluorescent dye i.e. 80% using citric acid and amine (ethylene diamine). In this approach, the reaction precursor was formed by dissolving the small molecules (organic and/or polymers) in solvent i.e. water or any organic solvent. The obtained reaction precursor placed in the teflon-lined stainless steel autoclave. With the influence of high temperatures, the organic molecule merged together to form quantum dots of carbon-based materials. These particles have a size less than 10nm. The controllable doping and ease of synthesis process makes this approach very effective in designing and fabricating novel electro-catalysts. Niu et al. [35] prepared the nitrogen doped QDs by hydrothermal method and the TEM image of the obtained QDs in Fig. 2 [35] which indicates the size of QDs is less than 10 nm.

Quantum Dots – Properties and Applications
Materials Research Foundations **96** (2021) 280-304

Materials Research Forum LLC
https://doi.org/10.21741/9781644901250-11

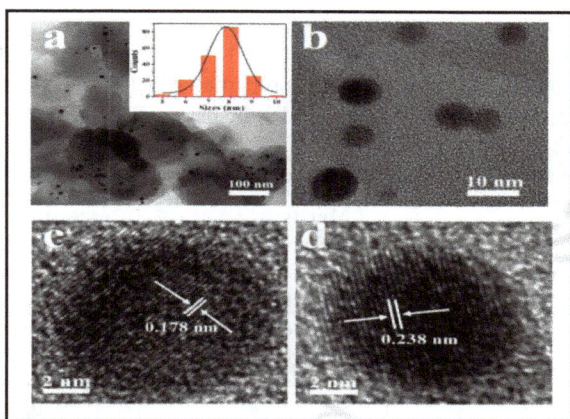

Fig. 2 LR- (a) and HR-TEM TEM (b, c, d) images of the as-synthesized N-doped carbon quantum dots (N-CQDs). The inset of (a) shows the particle size distribution histogram. Adapted with permission from Niu et al. [35]

3.2.4 Electrochemistry methods

This method is handy and conducive, which can be performed out in normal pressure and temperature conditions. The electrochemical method has its wide application in the synthesis of CQDS in order to articulate a quick PL performance of CQDS [36-38]. Luo et al. [39] performed an experiment engrossing the emission of blue CQD's having a particle size of ~2.4nm by carbonization of urea and sodium citrate in deionised (DI) water, which acts as a sensitive pathfinder for Hg^{2+} in waste water .

Electrochemical synthesis is a competent and efficacious method to fabricate an electro catalyst, but this CQD synthesis for the electro catalyst is infrequently reported. Thus, the amalgamation of CQD synthesis and electro catalyst development through one pot electrochemical method is captivating. Electrochemical soaking is a potential method to assemble CQD's using different bulk carbon ingredients as precursors [40-46]. But electrochemical carbonization of small molecules is rarely reported. Zang et al. [47] introduced the carbonization of low molecular weight alcohols for the preparation of CQD's. The experiment was performed using two Pt sheets acting as auxiliary electrode and a calomel electrode mounted on a freely adjustable stand. A reference electrode namely Luggin capillary was used. This carbonization resulted in the transformation of alcohols into CQD's under normal conditions. On increasing the applied potential, the size and graphitization degree also increased. This amorphous core CQD reflected exquisite excitation and size dependent PL properties which is as easy to proceed as a low hanging fruit [47]. The advantage and disadvantage of the above various synthesis approaches have been shown in Table 1.

Table 1   The advantage and disadvantage of the various synthesis approaches.

S.No.	Method	Merits	Demerits
1	Laser ablation	Effective and fast process, tunable surface states	It is difficult to control the size of obtained product. Modification needed
2	Arc Discharge	Production of high quality CQDs,	High temperature process, hard to control
3	Acidic Oxidation	Accessible	Harsh condition, strong acid requirement, Difficult to control the size
4	Combustion Routes	Single step process, less time require, cost effective	High temperature require
5	Microware irradiation	Eco-friendly , cost effective, fast process	It is difficult to control the size of obtained product.
6	Hydrothermal/solvothermal treatment	Cost effective, eco-friendly, non-toxic	It is difficult to control the size of obtained product.
7	Electrochemical carbonization	Products are controllable (nanostructure and size), stable, single step process	Precursors are limited in numbers due to their smaller molecular size (limited precursor)

4.   Properties of carbon based quantum dots

The interaction between the electrons and holes of the nanocrystal leads its optical properties. When a beam of light is irradiated onto the surface of material then material absorbs quanta and electron jump to conduction band from valence band. The handful electrons re-transition back to valence band and emits photons while the rest of the electrons fall into the electron trap. This leads quenching in form of non-radiation transition. Only a handful of photon electrons of the valence band absorb energy and then jump back to the guide belt. A deeper material for the electron trap semiconductor reduces luminous efficiency. [48]

Materials Research Forum LLC
https://doi.org/10.21741/9781644901250-11

4.1   Absorbance

The emission (photoluminescence) and absorbance of different wavelength of light are the major characteristics of any QDs and together give the optical behavior of QDs [18, 28]. It has been seen that the absorbance of QDs lies in UV-Visible range and QDs fluorescing correspondingly. The basic concept of illumination of different wavelength spectra is recombination of excited electrons with the holes in their ground state. In such recombination process, certain amount of symmetrical energy of narrow wavelength within UV region is released and showed color between red to blue. The variation of color emission is dependent on size of synthesized QDs, wavelength and intensity of energy released. Increase in size of QDs results towards blue to red spectrum [49] represents in Fig. 3.

Fig. 3 Decreasing size (left to right) and fluorescing under UV light.

4.2   Photoluminescence

Photoluminescence is the most mesmerizing features of CQDs in both fundamental research and practical application [50, 51]. The PL property of the CQDs is dependent of the emission wavelength and intensity. This may be due to the different emission traps on the surface or different size [16]. Pisanic et al. [49] studied the emission behaviors of dots at 470 nm wavelength with various concentrations. It was found that the PL strength of the CQDs solution first increased and then decreased as the concentration increased. Jaiswal et al. [52] synthesized quantum dot impregnated-chitosan film for heavy metal ion sensing and removal.

5.   Carbon quantum dots for water treatment

The huge industrial development along with global population burst has led the populous way of enrichment for nano technology based innovations and some of the main benefitted areas are biomedical, energy, and water purification along with sensor for detection of contaminants present in water samples [53-55]. The nanomaterials i.e. quantum dots, graphene, CNT etc. have huge potential to be utilized for various industries. In this section, we focus on carbon based QDs in the field of water treatment application especially as heavy metal ion detection sensors, filter papers and membranes developments etc. The high water demand in several fields of industries like mining, synthesis of organo chemical, energy related application and bio-medical are main sources of water pollution. Due to which water contaminates with the presence of heavy metal and their ions. Heavy metals can easily be bind with any cellular systems which can cause a serious tragedy to anyone [56, 57]. Some of the metals ions such as iron, copper, manganese, nickel play several important roles in various biological systems such as iron ion plays important role in catalysis, biotechnology, hemoglobin formation etc. Similarly, copper ion plays a crucial role in plants and various animals. The intake of exact amount of these ions is necessity for each biological system as their excess can cause harm [58]. Thus, the selective concentrations of each heavy metal ions have a crucial role for human to other animal health [59]. In this case, QDs have important role to verify their exact concentration and detection. Fig. 4 [52] represents the normalized absorbance of the QDs of ZnS and metals.

Fanet et al. [60] reported the CQDs aligned organic metal based framework for the more accurate detection of Fe^{3+} and Cu^{2+}. They developed the organometallic CQDs for their florescent sensing potential selectively for Fe^{3+} and Cu^{2+} ions. The developed system was reported to follow on–off strategy by insertion and removal of carbon quantum dots florescence from the system. This system worked as the probe for detection of iron and copper in water where the LOD value of CQDs based metal aligned oragnic framework possessed 1.3 and 2.3 ppb values for the sensing Cu^{2+} and Fe^{3+} ions, respectively. Simmaily, Dong et al. [61] reported the GQDs based monitoring system for the detection of residual and free chlorine ions in drinking water, which was based on the quenching of graphene based quantum dots i.e. GQDs. The developed system for the detection of chlorine ions exhibited many advantages over the existing sensors due to excellent response period, good selectivity along with good sensitivity. Here the responsive time range for the independent chlorine ions was found to be around 0.05 to 10 Mm. The lower value of detection limit was lower than the existing and mostly used N-N-diethyl-p-phenylenediamine colorimetric measurements methods. The reported sensing device was used to detect residual atoms and ions of chlorine also in the samples of tap water.

The mentioned colorimetric methodology suggested to have excellent candidature along with huge potential for the application of water sensing and purification as it has the advantage of greener and environment facile pathway.

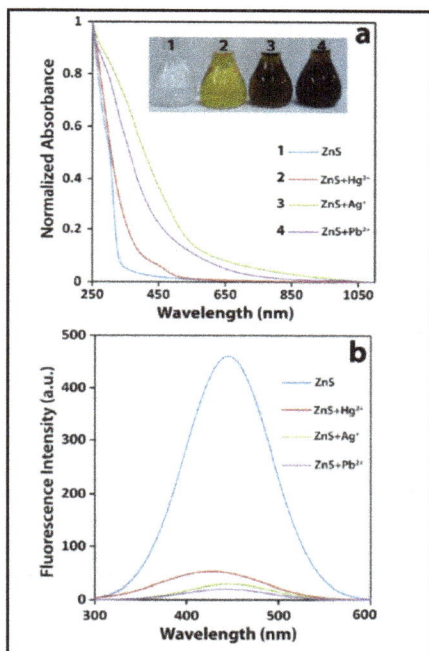

Fig. 4 (a) Normalized UV−vis-NIR spectra of ZnS Q-dots and that of ZnS^+Hg^{2+}, ZnS^+Ag^+, and ZnS^+Pb^{2+}, with inset showing the appearance (photographs) of the colloidal dispersions in visible light. (b) Fluorescence spectra of ZnS Q-dots and that of ZnS^+Hg^{2+}, ZnS^+Ag^+, and ZnS^+Pb^{2+} solution. Adapted with permission from Jaiswal et al. [52].

Song et al. [62] reported the multiple analytic based sensing system by the help of CQDs alignment to specifically target and detect iron ions by following the quenching approach. The basic principle of the process is the addition of iron micro molar concentrations which can specifically quench the florescence property of CQDs in both water and non-water system. Here, the detection limit was found to be around 0.25 mM for iron ions, whereas the reported system was mentioned as the photo-light stable and with least cytotoxicity.

Similarly, the garlic based CDs were synthesized to detect the iron ions by the controlled PL behavior of carbon dots [63]. The results exhibited the excellent correlation with the detection of iron by the turn on-off method. Yingshuai Liu et al. [64] reported the synthesis and utilizations of CQDs with excellent quantum yield. They prepared the polyethylenimine and CQDs composite by electrical deposition to detect the presence of copper ion by the alteration in its PL. The detection limit of 115 nm was reported with the dynamic range of 0.333 to 66.6 μM. The developed system tested by its direct utilization of river water samples for the successive detection of copper ions, which demonstrated the good selectivity.

Suryawanshi et al. [65] developed the bulk phase production of graphene quantum dots (GQDs) from agricultural waste for the detection of Ag+ ions. Agricultural waste based GQDs were found to have excellent photoluminescent property. The GQDs based system was modified by the addition of amine linkage to have good dispersibility. The system was reported to show the quenching of GQDs florescence by the Ag^+ ions and thus found to have excellent ability for their detection

In comparison to traditional organic composites, the QDs based systems exhibited higher sensitivity, excellent stability along with much appreciated selectivity for sensing applications [66]. They can also act as the excellent photoluminescent probes for detection of organic and inorganic ions [67]. The interactions between metals ions and QDs influence the florescence and photoluminescence properties of QDs [68]. The functionalized GQDs used for the detection of iron ions in the system with the detection limit of 7.22 μM [69]. Similarly, amino acid based graphene quantum dots were fabricated for the detection of Cu^{2+}Ion [70]. Here, probe possessed 6.9 Nm as detection limit due to the excellent interactions of metal ion to the surface of GQDs.

Jaiswal et al. [52] demonstrating the fabrication of ZnS Q-dots impregnated chitosan film along with the appearance of the drop cast of different metal ions onto the fabricated film under visible and UV light in Fig. 5 [52] and shows the synthesized material is detecting Hg^{2+}, Ag^+, Pb^{2+} [52]. Guo et al. [71] synthesized the functionalized GQDs for the tracking of iron ion in biological cell, where the orange-red emission at 580 nm showed an enhancement on the addition of iron ion. The detection of Pb^{2+} ion, Qian et al. [72] developed the functionalized GQDs with the low detection limit of corresponding ion. The QDs can also be utilized for the detection of various nonmetallic ions concentration to the system such as chloride ions, sulfide ions, nitrite ions detection [73]. Nitrite ion used as the regulator for the photoluminescence of nitrogen doped and phosphorous doped GQDs as they had excellent ionic interactions [74]. The detection time 2.5 nM was reported for the same composite. Liu et al. [16] reported a hydrothermal route for the synthesis of nitrogen doped and carbon rich polymer nanodots. The composite was found to have effective sensing and fluorescent property, which was further used for the selective detection of Cu(II) ions having the detection limit minimum as 1 nM. This composite showed its application in real water samples.

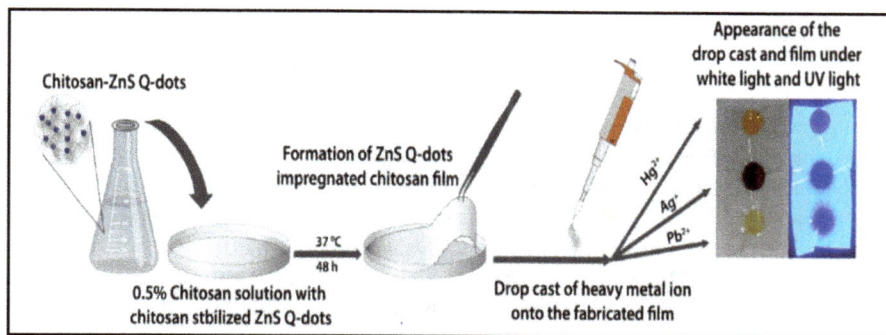

Fig. 5 Schematic demonstrating the fabrication of ZnS Q-dots impregnated chitosan film along with the appearance of the drop cast of different metal ions onto the fabricated film under visible and UV light. Adapted with permission from Jaiswal et al. [52].

Similarly, GQDs can be used for the detection and tracking of the several light weighted organic molecules i.e. amino acid, vitamins, carbohydrates etc. [67,75,76]. The hydrogen peroxide detection thorough GQDs was reported by Liu et al. [75] at the linear range of 3.33-500 μM and the detection time was 1.2 μM. Similarly, glucose can be sensed by PL on-off method, where GQDs were functionalized by 3-aminobenzeneboronic acid for the exact sensitivity and accurate selectivity of glucose [76]. Direct approaches for the detection of glucose had also been done by various researchers [77, 78]. The photoluminescence can be increase or decrease with the interaction between the metal ions and QDs, due to this property somehow QDs can be used for detection of the intensity and concentrations of the various organic as well as inorganic species [79, 80]. A facile, approach for the synthesis of CQDs by Konggang Qu et al. [81] was reported via economic and greener hydrothermal route. Thus, prepared nano particles were found to have 3.8 nm as average size. The photolytic emission spectra revealed in range of 380 nm to 525 nm in term of excitation wavelength. Due to excellent optical properties, the

QDs showed the favorable results for bio-imaging and bio-labeling applications. The QDs showed notable catechol groups alongside the surface, which helped to find specific response for the detection of iron ions. The obtained QDs were reported to have efficient probe for labeling and detection of iron ions along with dopamine having a detection limit as minimum as 0.32 nm for iron and 68 nm for dopamine. Further, the potential application of thus synthesized QDs confirmed in water samples with human urine. Li et al. [82] reported the carbon nanoparticles synthesis from carbon soot obtained by simple lighting of candle. The cost effective route of carbon nanoparticles synthesis was found to have amazing detection and sensing ability towards silver ions. The detection limit for silver ion was reported minimum 500 pM with excellent specific selectivity. Thus, obtained nanoparticles were also used for the practical detection of silver ions for various applications. The lake water was used as the sample to find out the exact and applied detection of silver ions via carbon based nano particle. The further evaluation showed the carbon nanoparticle probe withstands the existence of interference in the lake water. Lin [83] observed that the amplification of chemical luminescent property of peroxynitrous acid by incorporation of CQDs. Here the CQDs were reported as the energy acceptor for the investigations. This gives some new insights about the optical properties of CQDs. Author has mentioned the relationship between the variations of nitrous acids concertation and chemiluminescence behavior of QDs, that provided the methodology for the detection of nitrate ion into water system having the selectivity as high as 5.0×10^{-9} M. This method used for the detection of nitrate in tap water samples and the 98% recovery was reported.

Conclusions and future scope

Presently, many scientists are engaged to develop QDs based sensor for detection of different heavy metals and impurities present in water. But still limited research work is reported on QDs based membrane for water purification. The QDs mainly CQDs are hydrophilic in nature due to the presence of hydroxyl, carboxylic groups and water solubility of the CQDs can enhance their applications in the field of bio-imaging, sensing of metals in the water samples but limits their application as an adsorbent for the impurities present in water. Basically, metal doped QDs are hydrophobic in nature and shows great photocatalytic activity. It is reported Cu-doped SnO_2 QDs have great photocatalytic degradation capacity (~99%) towards methyl orange dye. For extending the reach of QDs in water purification, the passivation of surface is required, which can make CQDs hydrophobic in nature and thus it will further improve the absorptive nature of QDs. In this chapter, we discuss the synthesis of CQDs via different methods (including merits and demerits), classification, properties and the application in field of water treatment. We expect this chapter acts as an excellent modular to generate new ideas for expending the research area of "QDs based materials for water treatment".

Acknowledgements

The work is financially supported by Department of Science and Technology (Ref No.: DST/TM/WTI/WIC/2K17/82(G)), New Delhi, India and DST Inspire Division (IF150750).

References

[1] X. Xu, R. Ray, Y. Gu, H. J. Ploehn, L. Gearheart, K. Raker, W. A. Scrivens, Electrophoretic analysis and purification of fluorescent single-walled carbon nanotube fragments, J. Am. Chem. Soc. 126 (2004) 12736-12737. https://doi.org/10.1021/ja040082h

[2] Q. Zhang, X. Quan, H. Wang, S. Chen, Y. Su, Z. Li, Constructing a visible-light-driven photocatalytic membrane by gC_3N_4 quantum dots and TiO_2 nanotube array for enhanced water treatment, Sci. Rep. 7 (2017) 3128.

    https://doi.org/10.1038/s41598-017-03347-y.

[3] P.S. Saud, B. Pant, A.M. Alam, Z.K. Ghouri, M. Park, H.Y. Kim, Carbon quantum dots anchored TiO_2 nanofibers: Effective photocatalyst for waste water treatment, Ceram. Int. 41 (2015) 11953-11959. https://doi.org/10.1016/j.ceramint.2015.06.007

[4] Y. Tang, Y. Su, N. Yang, L. Zhang, Y. Lv, Carbon nitride quantum dots: a novel chemiluminescence system for selective detection of free chlorine in water, Anal. Chem. 86 (2014) 4528-4535. https://doi.org/10.1021/ac5005162

[5] B.Y. Yu, S.Y. Kwak, Carbon quantum dots embedded with mesoporous hematite nanospheres as efficient visible light-active photocatalysts, J. Mater. Chem. 22 (2012) 8345-8353. https://doi.org/10.1039/C2JM16931B

[6] S. Qu, H. Chen, X. Zheng, J. Cao, X. Liu, Ratiometric fluorescent nanosensor based on water soluble carbon nanodots with multiple sensing capacities. Nanoscale, 5 (2013), 5514-5518. https://doi.org/10.1039/C3NR00619K

[7] M.J. Krysmann, A. Kelarakis, P. Dallas, E.P. Giannelis, Formation mechanism of carbogenic nanoparticles with dual photoluminescence emission. J. Am. Chem. Soc. 134 (2011) 747-750. https://doi.org/10.1021/ja204661r

[8] H. Li, J. Zhai, X. Sun, Sensitive and selective detection of silver (I) ion in aqueous solution using carbon nanoparticles as a cheap, effective fluorescent sensing platform, Langmuir. 27 (2011) 4305-4308. https://doi.org/10.1021/la200052t

[9] J. Li, Q. Zhou, L.C. Campos, The application of GAC sandwich slow sand filtration to remove pharmaceutical and personal care products, Sci. Total Environ. 635 (2018) 1182-1190. https://doi.org/10.1016/j.scitotenv.2018.04.198

[10]　R. Li, Y. Ren, P. Zhao, J. Wang, J. Liu, Y. Zhang, Graphitic carbon nitride (g-C_3N_4) nanosheets functionalized composite membrane with self-cleaning and antibacterial performance, J. Hazard. Mater. 365 (2019) 606-614. https://doi.org/10.1016/j.jhazmat.2018.11.033

[11]　J. Liu, J. Li, Y. Jiang, S. Yang, W. Tan, R. Yang, Combination of π–π stacking and electrostatic repulsion between carboxylic carbon nanoparticles and fluorescent oligonucleotides for rapid and sensitive detection of thrombin, Chem. Commun. 47 (2011) 11321-11323. https://doi.org/10.1039/C1CC14445F

[12]　Z. Lin, X. Dou, H. Li, Y. Ma, J.M. Lin, Nitrite sensing based on the carbon dots-enhanced chemiluminescence from peroxynitrous acid and carbonate, Talanta. 132 (2015) 457-462. https://doi.org/10.1016/j.talanta.2014.09.046

[13]　W. Shi, Q. Wang, Y. Long, Z. Cheng, S. Chen, H. Zheng, Y. Huang, Carbon nanodots as peroxidase mimetics and their applications to glucose detection, Chem. Commun. 47 (2011) 6695-6697. https://doi.org/10.1039/C1CC11943E

[14]　D. Dey, T. Bhattacharya, B. Majumdar, S. Mandani, B. Sharma, T.K. Sarma, Carbon dot reduced palladium nanoparticles as active catalysts for carbon–carbon bond formation, Dalton Trans. 42 (2013) 13821-13825. https://doi.org/10.1039/C3DT51234G

[15]　H. Li, Y. Zhang, L. Wang, J. Tian, X. Sun, Nucleic acid detection using carbon nanoparticles as a fluorescent sensing platform, Chem. Commun. 47 (2011) 961-963. https://doi.org/10.1039/C0CC04326E

[16]　S. Liu, J. Tian, L. Wang, Y. Zhang, X. Qin, Y. Luo, A.M. Asiri, A.O. Al-Youbi, X. Sun, Hydrothermal treatment of grass: a low-cost, green route to nitrogen-doped, carbon-rich, photoluminescent polymer nanodots as an effective fluorescent sensing platform for label-free detection of Cu (II) ions, Adv. Mater. 24 (2012) 2037-2041. https://doi.org/10.1002/adma.201200164

[17]   K. Qu, J. Wang, J. Ren, X. Qu, Carbon dots prepared by hydrothermal treatment of dopamine as an effective fluorescent sensing platform for the label-free detection of iron (III) ions and dopamine, Chem. Eur. J. 19 (2013) 7243-7249. https://doi.org/10.1002/chem.201300042

[18]   F.W. Wise, Lead salt quantum dots: the limit of strong quantum confinement, Acc. Chem. Res. 33 (2000) 773-780. https://doi.org/10.1021/ar970220q

[19]   S. Nagaraja, P. Matagne, V.Y. Thean, J.P. Leburton, Y.H. Kim, R.M. Martin, Shell-filling effects and Coulomb degeneracy in planar quantum-dot structures, Phys. Rev. B. 56 (1997) 15752. https://doi.org/10.1103/PhysRevB.56.15752

[20]   P.V. Joglekar, P.V. Mandalkar , D.J. Nikam, M.A. Pande, N.S. Dubal, A review article on quantum dots: Synthesis, properties and application, International Journal of Research in Advent Technology, 7 (2019) 2321-9637. https://doi.org/10.32622/ijrat.712019113

[21]   K.L. Wang, D. Cha, J. Liu, C. Chen, Ge/Si self-assembled quantum dots and their optoelectronic device applications, P IEEE. 95 (2007) 1866-1883. https://doi.org/10.1109/JPROC.2007.900971

[22]   X. Wang, Y. Feng, P. Dong, J. Huang, A mini review on carbon quantum dots: preparation, properties and electrocatalytic application, Front. Chem. 7 (2019) 671. https://doi.org/10.3389/fchem.2019.00671

[23]   C. Buzea, I.I. Pacheco, K. Robbie, Nanomaterials and nanoparticles: sources and toxicity, Biointerphases. 2 (2007) 17-71. https://doi.org/10.1116/1.2815690

[24]   J.R. Kim, E. Kan, Heterogeneous photo-Fenton oxidation of methylene blue using CdS-carbon nanotube/TiO_2 under visible light, J. Ind. Eng. Chem. 21 (2015) 644-652. https://doi.org/10.1016/j.jiec.2014.03.032

[25]   S. Yang, J. Sun, X. Li, W. Zhou, Z. Wang, P. He, G. Ding, X. Xie, Z. Kang, M. Jiang, Large-scale fabrication of heavy doped carbon quantum dots with tunable-photoluminescence and sensitive fluorescence detection, J. Mater. Chem. A. 2 (2014) 8660-8667. https://doi.org/10.1039/C4TA00860J

[26]   A.M. Schwenke, S. Hoeppener, U.S. Schubert, Synthesis and modification of carbon nanomaterials utilizing microwave heating, Adv. Mater. 27 (2015) 4113-4141. https://doi.org/10.1002/adma.201500472

[27]   S. Rai, B.K. Singh, P. Bhartiya, A. Singh, H. Kumar, P.K. Dutta, G.K. Mehrotra,

Lignin derived reduced fluorescence carbon dots with theranostic approaches: Nano-drug-carrier and bioimaging, J. Lumin. 190 (2017)492-503. https://doi.org/10.1016/j.jlumin.2017.06.008

[28]   G. Zhao, C. Li, X. Wu, J. Yu, X. Jiang, W. Hu, F. Jiao, Reduced graphene oxide modified NiFe-calcinated layered double hydroxides for enhanced photocatalytic removal of methylene blue, Appl. Surf. Sci. 434 (2018) 251-259. https://doi.org/10.1016/j.apsusc.2017.10.181

[29]   Z. Shen, C. Zhang, X. Yu, J. Li, Z. Wang, Z. Zhang, B. Liu, Microwave-assisted synthesis of cyclen functional carbon dots to construct a ratiometric fluorescent probe for tetracycline detection, J. Mater. Chem. C. 6 (2018) 9636-9641. https://doi.org/10.1039/C8TC02982B

[30]   H. Zhu, X. Wang, Y. Li, Z. Wang, F. Yang, X. Yang, Microwave synthesis of fluorescent carbon nanoparticles with electro chemiluminescence properties, Chem. Commun. 34 (2009) 5118-5120. https://doi.org/10.1039/B907612C

[31]   S. Chen, Y. Wu, G. Li, J. Wu, G. Meng, X. Guo, Z. Liu, A novel strategy for preparation of an effective and stable heterogeneous photo-Fenton catalyst for the degradation of dye, Appl. Clay. Sci. 136 (2017) 103-111. https://doi.org/10.1016/j.clay.2016.11.016

[32]   J. Shen, Y. Zhu, X. Yang, J. Zong, J. Zhang, C. Li, One-pot hydrothermal synthesis of graphene quantum dots surface-passivated by polyethylene glycol and their photoelectric conversion under near-infrared light, New J. Chem. 36 (2012) 97-101. https://doi.org/10.1039/C1NJ20658C

[33]   T.L. Lai, Y.L. Lai, C.C. Lee, Y.Y. Shu, C.B. Wang, Microwave-assisted rapid fabrication of Co_3O_4 nanorods and application to the degradation of phenol, Catal. Today. 131 (2008) 105-110. https://doi.org/10.1016/j.cattod.2007.10.039

[34]   S. Zhu, Q. Meng, L. Wang, J. Zhang, Y. Song, H. Jin, K. Zhang, H. Sun, H. Wang, B. Yang, Highly photoluminescent carbon dots for multicolor patterning, sensors, and bioimaging, Angew. Chem. 52 (2013) 3953-3957. https://doi.org/10.1002/ange.201300519

[35]   W.J. Niu, Y. Li, R.H. Zhu, D. Shan, Y.R. Fan, X.J. Zhang, Ethylenediamine-assisted hydrothermal synthesis of nitrogen-doped carbon quantum dots as fluorescent probes for sensitive biosensing and bioimaging, Sens. Actuators B Chem. 218 (2015) 229-236. https://doi.org/10.1016/j.snb.2015.05.006

[36]   J. Deng, Q. Lu, N. Mi, H. Li, M. Liu, M. Xu, L. Tan, Q. Xie, Y. Zhang, S. Yao, Electrochemical synthesis of carbon nanodots directly from alcohols, Chem. Eur. J. 20 (17) 4993-4999. https://doi.org/10.1002/chem.201304869

[37]   S. Ahirwar, S. Mallick, D. Bahadur, Electrochemical method to prepare graphene quantum dots and graphene oxide quantum dots, ACS Omega. 2 (2017) 8343-8353. https://doi.org/10.1021/acsomega.7b01539

[38]   S. Anwar, H. Ding, M. Xu, X. Hu, Z. Li, J. Wang, L. Liu, L. Jiang, D. Wang, C. Dong, M. Yan, Recent Advances in Synthesis, Optical Properties and Biomedical Applications of Carbon Dots, ACS Appl. Bio Mater. 2 (2019) 2317-2338. https://doi.org/10.1021/acsabm.9b00112

[39]   X. Luo, Y. Han, X. Chen, W. Tang, T. Yue, Z. Li, Carbon dots derived fluorescent nanosensors as versatile tools for food quality and safety assessment: A review, Trends Food Sci. Tech. (2019). https://doi.org/10.1016/j.tifs.2019.11.017

[40]   J. Zhou, C. Booker, R. Li, X. Zhou, T. K. Sham, X. Sun, Z. Ding, An electrochemical avenue to blue luminescent nanocrystals from multiwalled carbon nanotubes (MWCNTs), J. Am. Chem. Soc. 129 (2007) 744-745. https://doi.org/10.1021/ja0669070

[41]   D. B. Shinde, V. K. Pillai, Electrochemical preparation of luminescent graphene quantum dots from multiwalled carbon nanotubes, Chem. Eur. J. 18 (2012) 12522-12528. https://doi.org/10.1002/chem.201201043

[42]   L. Bao, Z. L. Zhang, Z. Q. Tian, L. Zhang, C. Liu, Y. Lin, D. W. Pang, Electrochemical tuning of luminescent carbon nanodots: from preparation to luminescence mechanism, Adv. Mater. 23 (2011) 5801-5806. https://doi.org/10.1002/adma.201102866

[43]   L. Zheng, Y. Chi, Y. Dong, J. Lin, B. Wang, Electrochemiluminescence of water-soluble carbon nanocrystals released electrochemically from graphite, J. Am. Chem. Soc. 131 (2009) 4564-4565. https://doi.org/10.1021/ja809073f

[44]   H. Ming, Z. Ma, Y. Liu, K. Pan, H. Yu, F. Wang, Z. Kang, Large scale electrochemical synthesis of high quality carbon nanodots and their photocatalytic property, Dalton Trans. 41 (2012) 9526-9531. https://doi.org/10.1039/C2DT30985H

[45]   H. Li, X. He, Z. Kang, H. Huang, Y. Liu, J. Liu, S. T. Lee, Water-soluble fluorescent carbon quantum dots and photocatalyst design, Angew. Chem. 49 (2010)

4430-4434. https://doi.org/10.1002/ange.200906154

[46] J. Deng, Q. Lu, N. Mi, H. Li, M. Liu, M. Xu, S. Yao, Electrochemical synthesis of carbon nanodots directly from alcohols, Chem. Eur. J. 20 (2014) 4993-4999. https://doi.org/10.1002/chem.201304869

[47] Y., Wang, A. Hu, Carbon quantum dots: synthesis, properties and applications, J. Mater. Chem. 2 (2014) 6921-6939. https://doi.org/10.1039/C4TC00988F

[48] L. Zou, Z. Gu, M. Sun, Review of the application of quantum dots in the heavy-metal detection, Toxicol Environ. Chem. 97 (2015) 477-490. https://doi.org/10.1080/02772248.2015.1050201

[49] T.R. Pisanic Ii, Y. Zhang, T.H. Wang, Quantum dots in diagnostics and detection: principles and paradigms, Analyst. 139 (2014) 2968-2981. https://doi.org/10.1039/C4AN00294F

[50] H. Peng, J. Travas-Sejdic, Simple aqueous solution route to luminescent carbogenic dots from carbohydrates, Chem. Mater. 21 (2009) 5563-5565. https://doi.org/10.1021/cm901593y

[51] Z. Gan, X. Wu, G. Zhou, J. Shen, P.K. Chu, Is there real upconversion photoluminescence from graphene quantum dots?, Adv. Opt. Mater, 1 (2013) 554-558. https://doi.org/10.1002/adom.201300152

[52] A. Jaiswal, S.S. Ghsoh, A. Chattopadhyay, Quantum dot impregnated-chitosan film for heavy metal ion sensing and removal, Langmuir, 28 (2012) 15687-15696. https://doi.org/10.1021/la3027573

[53] Y. Wu, M. Xu, X. Chen, S. Yang, H. Wu, J. Pan, X. Xiong, CTAB-assisted synthesis of novel ultrathin $MoSe_2$ nanosheets perpendicular to graphene for the adsorption and photodegradation of organic dyes under visible light, Nanoscale. 8 (2016) 440-450. https://doi.org/10.1039/C5NR05748E

[54] M.P. Wei, H. Chai, Y.L. Cao, D.Z. Jia, Sulfonated graphene oxide as an adsorbent for removal of Pb^{2+} and methylene blue, J. Colloid Interface Sci. 524 (2018) 297-305. https://doi.org/10.1016/j.jcis.2018.03.094

[55] P. Tan, J. Sun, Y. Hu, Z. Fang, Q. Bi, Y. Chen, J. Cheng, Adsorption of Cu^{2+}, Cd^{2+} and Ni^{2+} from aqueous single metal solutions on graphene oxide membranes, J. Hazard. Mater. 297 (2015) 251-260. https://doi.org/10.1016/j.jhazmat.2015.04.068

[56] M. Li, R. Cao, A. Nilghaz, L. Guan, X. Zhang, W. Shen, "Periodic-table-style" paper device for monitoring heavy metals in water. Anal. Chem. 87 (2015) 2555-2559. https://doi.org/10.1021/acs.analchem.5b00040

[57] J.P. Devadhasan, J. Kim, A chemically functionalized paper-based microfluidic platform for multiplex heavy metal detection, Sens. Actuators B-Chem. 273 (2018) 18-24. https://doi.org/10.1016/j.snb.2018.06.005

[58] Y. Wu, G.P. Yang, X. Zhou, J. Li, Y. Ning, Y.Y. Wang, Three new luminescent Cd (II)-MOFs by regulating the tetracarboxylate and auxiliary co-ligands, displaying high sensitivity for Fe^{3+} in aqueous solution, Dalton Trans. 44 (2015) 10385-10391. https://doi.org/10.1039/C5DT00492F

[59] G. Zhang, H. Zhang, J. Zhang, W. Ding, J. Xu, Y. Wen, Highly selective fluorescent sensor based on electro synthesized oligo (1-pyreneboronic acid) enables ultra-trace analysis of Cu^{2+} in environment and agro-product samples, Sens. Actuators B-Chem. 253 (2017) 224-230. https://doi.org/10.1016/j.snb.2017.06.144

[60] L. Fan, Y. Wang, L. Li, J. Zhou, Carbon quantum dots activated metal organic frameworks for selective detection of Cu (II) and Fe (III), Colloids Surf. A Physicochem. Eng. Asp. 588 (2020) 124378. https://doi.org/10.1016/j.colsurfa.2019.124378

[61] Y. Dong, G. Li, N. Zhou, R. Wang, Y. Chi, G. Chen, Graphene quantum dot as a green and facile sensor for free chlorine in drinking water, Anal. Chem. 84 (2012) 8378-8382. https://doi.org/10.1021/ac301945z

[62] P. Song, L. Zhang, H. Long, M. Meng, T. Liu, Y. Yin, R. Xi, A multianalyte fluorescent carbon dots sensing system constructed based on specific recognition of Fe (III) ions, RSC Advances 7 (2017) 28637-28646. https://doi.org/10.1039/C7RA04122E

[63] Y. Chen, Y. Wu, B. Weng, B. Wang, C. Li, Facile synthesis of nitrogen and sulfur co-doped carbon dots and application for Fe (III) ions detection and cell imaging, Sens. Actuators B-Chem. 223 (2016) 689-696. https://doi.org/10.1016/j.snb.2015.09.081

[64] Y. Liu, Y. Zhao, Y. Zhang, One-step green synthesized fluorescent carbon nanodots from bamboo leaves for copper (II) ion detection, Sens. Actuators B-Chem. 196 (2014) 647-652. https://doi.org/10.1016/j.snb.2014.02.053

[65]   A. Suryawanshi, M. Biswal, D. Mhamane, R. Gokhale, S. Patil, D. Guin, S. Ogale, Large scale synthesis of graphene quantum dots (GQDs) from waste biomass and their use as an efficient and selective photoluminescence on–off–on probe for Ag^+ ions, Nanoscale. 6 (2014) 11664-11670. https://doi.org/10.1039/C4NR02494J

[66]   X. Hai, J. Feng, X. Chen, J. Wang, Tuning the optical properties of graphene quantum dots for biosensing and bioimaging, J. Mater. Chem. 6 (2018) 3219-3234. https://doi.org/10.1039/C8TB00428E

[67]   R. Liu, J. Zhao, Z. Huang, L. Zhang, M. Zou, B. Shi, S. Zhao, Nitrogen and phosphorus co-doped graphene quantum dots as a nano-sensor for highly sensitive and selective imaging detection of nitrite in live cell, Sens. Actuators B-Chem. 240 (2017) 604-612. https://doi.org/10.1016/j.snb.2016.09.008

[68]   M. Liu, T. Liu, Y. Li, H. Xu, B. Zheng, D. Wang, J. Du, D. Xiao, A FRET chemsensor based on graphene quantum dots for detecting and intracellular imaging of Hg^{2+}, Talanta. 143 (2015) 442-449. https://doi.org/10.1016/j.talanta.2015.05.023

[69]   A. Ananthanarayanan, X. Wang, P. Routh, B. Sana, S. Lim, D.H. Kim, K.H. Lim, J. Li, P. Chen, Facile synthesis of graphene quantum dots from 3D graphene and their application for Fe^{3+} sensing, Adv. Funct. Mater. 24 (2014) 3021-3026. https://doi.org/10.1002/adfm.201303441

[70]   H. Sun, N. Gao, L. Wu, J. Ren, W. Wei, X. Qu, Highly photoluminescent amino-functionalized graphene quantum dots used for sensing copper ions, Chem. Eur. J. 19 (2013) 13362-13368. https://doi.org/10.1002/chem.201302268

[71]   C. Sun, Y. Zhang, P. Wang, Y. Yang, Y. Wang, J. Xu, Y. Wang, W.Y. William, Synthesis of nitrogen and sulfur co-doped carbon dots from garlic for selective detection of Fe^{3+} , Nanoscale Res. Lett. 11 (2016) 110. https://doi.org/10.1186/s11671-016-1326-8

[72]   R. Guo, S. Zhou, Y. Li, X. Li, L. Fan, N.H. Voelcker, Rhodamine-functionalized graphene quantum dots for detection of Fe^{3+} in cancer stem cells, ACS Appl. Mater. Interfaces. 7 (2015) 23958-23966. https://doi.org/10.1021/acsami.5b06523

[73]   Z.S. Qian,, X.Y. Shan, L.J. Chai, J.R. Chen, H. Feng, A fluorescent nanosensor based on graphene quantum dots–aptamer probe and graphene oxide platform for detection of lead (II) ion, Biosens. Bioelectron. 68 (2015) 225-231. https://doi.org/10.1016/j.bios.2014.12.057

[74]   L. Lin, X. Song, Y. Chen, M. Rong, T. Zhao, Y. Jiang, Y. Wang, X. Chen, One-pot synthesis of highly greenish-yellow fluorescent nitrogen-doped graphene quantum

dots for pyrophosphate sensing via competitive coordination with Eu^{3+} ions, Nanoscale. 7 (2015) 15427-15433. https://doi.org/10.1039/C5NR04005A

[75] Q. Zhang, C. Song, T. Zhao, H.W. Fu, H.Z. Wang, Y.J. Wang, D.M. Kong, Photoluminescent sensing for acidic amino acids based on the disruption of graphene quantum dots/europium ions aggregates, Biosens. Bioelectron. 65, (2015) 204-210. https://doi.org/10.1016/j.bios.2014.10.043

[76] H. Liu, W. Na, Z. Liu, X. Chen, X. Su, A novel turn-on fluorescent strategy for sensing ascorbic acid using graphene quantum dots as fluorescent probe, Biosens. Bioelectron. 92 (2017) 229-233. https://doi.org/10.1016/j.bios.2017.02.005

[77] Z.B. Qu, X. Zhou, L. Gu, R. Lan, D. Sun, D. Yu, G. Shi, Boronic acid functionalized graphene quantum dots as a fluorescent probe for selective and sensitive glucose determination in microdialysate, Chem. Commun. 49 (2013) 9830-9832. https://doi.org/10.1039/C3CC44393K

[78] L. Wang, W. Li, B. Wu, Z. Li, S. Wang, Y. Liu, D. Pan, M. Wu, Facile synthesis of fluorescent graphene quantum dots from coffee grounds for bioimaging and sensing, Chem. Eng. J. 300 (2016) 75-82. https://doi.org/10.1016/j.cej.2016.04.123

[79] Y. Zhang, H.Y. Shen, X. Hai, X.W. Chen, J.H. Wang, Polyhedral oligomeric silsesquioxane polymer-caged silver nanoparticle as a smart colorimetric probe for the detection of hydrogen sulfide, Anal. Chem. 89 (2016) 1346-1352. https://doi.org/10.1021/acs.analchem.6b04407

[80] D. Pan, L. Guo, J. Zhang, C. Xi, Q. Xue, H. Huang, J. Li, Z. Zhang, W. Yu, Z. Chen, Z. Li, Cutting sp^2 clusters in graphene sheets into colloidal graphene quantum dots with strong green fluorescence, J. Mater. Chem. 22 (2012) 3314-3318. https://doi.org/10.1039/C2JM16005F

[81] K. Qu, J. Wang, J. Ren, X. Qu, Carbon dots prepared by hydrothermal treatment of dopamine as an effective fluorescent sensing platform for the label-free detection of iron (III) ions and dopamine, Chem. Eur. J 19 (2013) 7243-7249. https://doi.org/10.1002/chem.201300042

[82] H. Li, J. Zhai, X. Sun, Sensitive and selective detection of silver (I) ion in aqueous solution using carbon nanoparticles as a cheap, effective fluorescent sensing platform, Langmuir, 27 (2011) 4305-4308. https://doi.org/10.1021/la200052t

[83] Z. Lin, X. Dou, H. Li, Y. Ma, J.M. Lin, Nitrite sensing based on the carbon dots-enhanced chemiluminescence from peroxynitrous acid and carbonate, Talanta, 132 (2015) 457-462. https://doi.org/10.1016/j.talanta.2014.09.046

Materials Research Forum LLC
https://doi.org/10.21741/9781644901250-12

Chapter 12

Semiconductor Quantum Dots

N.B. Singh[1,2]*, Richa Tomar[2]

[1]RTDC, Sharda University, Greater Noida-201310, India

[2]Department of Chemistry and Biochemistry, Sharda University, Greater Noida-201310, India

nbsingh43@gmail.com

Abstract

Semiconductor particles in the range of 2-10 nm are known as quantum dots (QDs) and nano-crystals where in all the three spatial dimensions, excitons are confined. Because of very small size and special electronic properties, QDs are expected to be building blocks of many electronic and optoelectronic devices. These particles possess tunable quantum efficiency, continuous absorption spectra, narrow emission and long term photostability. These are important for various biomedical applications. In this chapter definition of semiconductor QDs, their methods of preparation and characterization along with their properties and applications have been discussed.

Keywords

Semiconductor, Quantum Dots, Band Gap, Photodetectors, Photovoltaics

Contents

Materials Research Forum LLC
https://doi.org/10.21741/9781644901250-12

1.   Introduction

Semiconductors are solids having electrical conductivity (σ) between conductors and insulators. Contrary to metals, the electrical conductivity of semiconductors increases with increase of temperature and obey Eq. (1).

$$\sigma = \sigma_o \ \exp{(E_g/2k_B\,T)} \tag{1}$$

Where σ_o is a constant, E_g is activation energy for conduction, T is temperature in Kelvin. Semiconductors of $(2-10\ nm)$ (normally spherical), with excitons confined in all the three dimensions are known as QDs [1]. They have unique electronic, optical and structural properties [2] and the properties lie between discrete molecules and bulk materials (Fig. 1) [3].

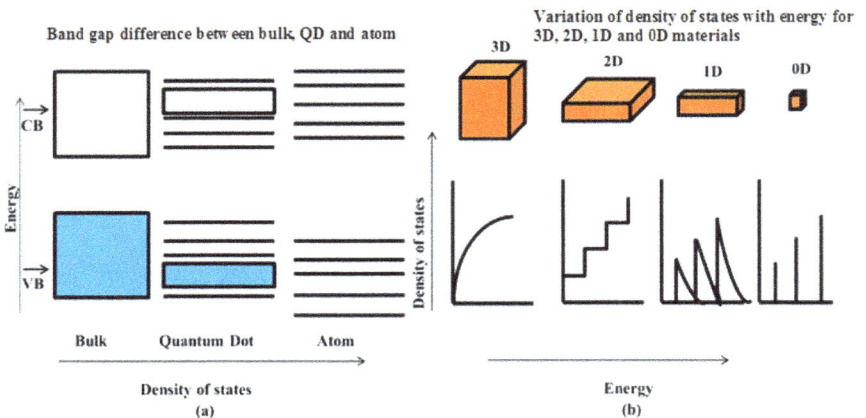

Fig. 1 (a) band gap and (b) Energy of states for 3D, 2D, 1D and 0D materials [3].

Quantum dots have narrow emission and broad absorption spectra. To understand the behavior of QDs the first approach is the investigation of their optical properties. The optical phenomenon of QDs was observed due to the unique phenomenon of quantum confinement. The electronic excitations shift to higher energy in QDs. QDs absorb white light and then re-emit a specific color a few nanoseconds afterward, depending on the band gap of the material. In this chapter, different type of QDs, their preparation, characterization, properties and applications have been discussed.

2. Types of quantum dots

The quantum dots because of varying size show different colors. They can be classified into different types depending on their composition and structure (Table 1) [1].

Table 1 Types of semiconductor quantum dots [1].

Type	QDs
II-VI	CdSe, CdTe, CdS, ZnSe, ZnS, HgS, ZnTe, PbS, PbSe, HgSe
III-V	InGaAs, GaAs, InAs, InP
IV	Si, Ge
IV-VI	PbSe, ZnO

3. Preparation of quantum dots

QDs are generally synthesized by top-down and bottom-up approaches. The top-down approach takes the help of initial macroscopic structures but it has certain drawbacks. The bottom-up approach uses number of wet chemical methods [3]. Some of the methods have been discussed below.

3.1 Colloidal method

Colloidal method depends on existence of a three-component. It is used to produce multilayered QDs. It is a simple process and can be used for mass production. Colloidal methods have been used to produce number of QDs such as lead selenide, lead sulfide, cadmium selenide, etc. Redl synthesized PbSe and Fe_2O_3 QDs by employing colloidal route [4].

3.2 Emulsion method

This method uses oil-in-water or water-in-oil. However, in some cases polar solvents like alcohol has also been used. Using surfactants, water droplets of nanosize can be obtained in solution. Micelles formed in the oil medium, act as 'nanoreactors. Mandal et al. [5] has prepared some QDs by using this method.

3.3 Sol-gel method

It is a chemical method for the preparation of QDs. Different steps involved in the synthesis of QDs by sol-gel method are schematically represented in Fig. 2 [3].

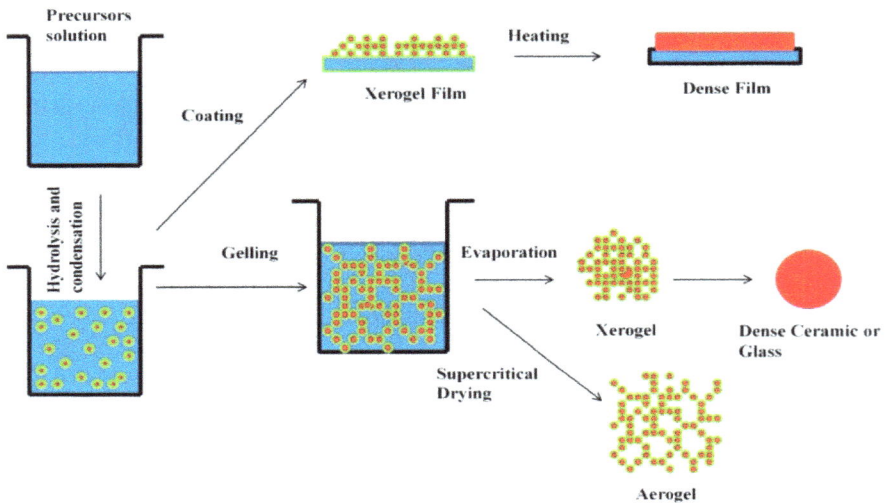

Fig. 2 Sol-gel method for preparation of QDs [3].

3.4 Co-precipitation

Co-precipitation method is used to have precipitates of two substances under controlled pH. Fig. 3 [3] shows a schematic diagram for the preparation of ZnX and CdX (X=S, Te, Se) QDs using co-precipitation technique.

3.5 QDs of metal chalcogenide by hot injection method

The fig. 4 [3] shows the apparatus for QD synthesis by hot-injection method. In this method powder of X (S, Se, Te) in a solvent is prepared (solution I) and metal oxide (e.g. CdO) in another solvent (solution II) containing surfactant are mixed and heated to 250-270 °C, different size of QDs are formed at different time interval.

309

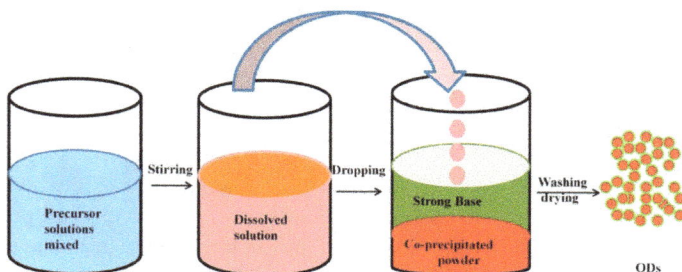

Fig. 3 Co precipitation method for the preparation of QDs [3].

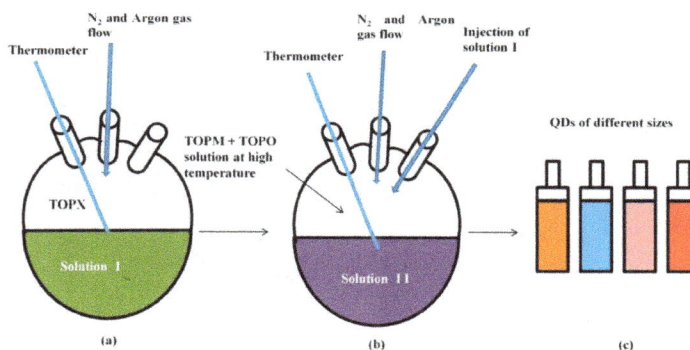

Fig. 4 Hot-injection method for preparation of QD [3].

3.6 Aqueous and hydrothermal synthesis

This synthesis route involves water. As the synthesized QDs have compatibility with water, they are found of biological applications. The quantum yield obtained via this route is comparatively low and large size distribution is observed. The post treatment of QDs may lead to improvement in QDs quality but adds on to the final cost. Thiols, thioglycolic acid and 3-mercaptopropionic acid acts as stabilizers during the synthesis. In most of the cases two of the precursors containing respective anion and cation are made to react at high temperature (around 160°C) in an autoclave. Finally on centrifugation of product dissolved in n-hexane and ethanol, desired nanocrystals are obtained [6]. Ya-Dong Li et. al [7] developed a different strategy in which water –oil heterogeneous system is used for QDs synthesis. QDs of CdTe, CdSe, $Zn_xCd_{1-x}Se$, ZnSe, CdS, PbS,

ZnS, CdTe (core)/CdS(Shell), CdTe/CdSe and CdTeS alloyed system have been synthesised from this route.

3.7   Solvothermal synthesis

This is an extensively used method for quantum dot synthesis. Earlier the nanocrystals obtained are not in the QDs as no quantum confinement is shown by the product. Later organic solvents are used for the synthesis. It starts with the mixing of all precursors at low temperature only and then raised to desired growth temperature in an autoclve, which generates the conditions for the growth of nanocrystals. Using this method PbSe, CdS, $CuInS_2$, CdSe and CdSe(core)/ CdS (Shell), InP are prepared. The QDs obtained via this route show very less size distribution.

3.8   Microwave assisted synthesis

In microwave assisted synthesis, the precursor solution was kept in a microwave oven at 100°C for 1 minute. The QDs synthesised by this approach, possesses excellent photostability and high photoluminescence quantum yield [8]. CdSe, CdTe, CdTe/CdS/ZnS, ZnSe(S) alloyed, CdSe- CdS are the QDs synthesized successfully using this method.

3.9   Direct adsorption method

The literature reveals that this is the most accepted method for QDs preparation supported on semiconductors. QDs and semiconductor prepared separately and linked together via adsorption. During adsorption van der Waal forces helps in binding of QDs on the surface of semiconductor. Some of the composites need to be annealed at varying temperatures from 160 to 600°C. The QDs so obtained are dispersed in ethanol, water, chloroform or dichloromethane. These are mixd with powdered semiconductor nanocrystals. Few of the QDs obtained using direct adsorption method are reported in Table 2 [9,10].

Table 2 Semiconductor quantum dots obtained using direct adsorption method [9-10].

QDs	Size of QDs (nm)	QDs combined with semiconductor NPs via	Size of Semiconductor	Ref.
$CuInS_2/TiO_2$	3.4	Mixing- Annealing	Approx. 25 nm	[9]
SnO_2/TiO_2	2.7	Mixing	Nanoparticles	[10]

3.10   Linker assisted adsorption method

To accomplish more effective coupling, bifunctional linker molecules like thioglycolic acid (TGA), mercaptopropionic acid (MPA), aminopropyltrimethoxysilane (APTMS), chitosan, thiolactic acid (TA), 3-aminopropyltrethoxysilane (APTES) were used to stabilize and disperse the QDs. As bifunctional linker molecules are used this route is named as linker assisted adsorption method. Bifunctional linker molecules help to attain the stable heterojunctions by tethering the QDs on to the surface of semiconductors [11].

The quantum dot sensitized solarcells are generally obtained by in situ or ex situ method. There is more attention on the ex situ method. In ex situ method, prepared QDs are attached to WBGS-films. More details are displayed in Fig. 5 [12]. A few of the QDs obtained using linker assisted adsorption method are enlisted in Table 3 [13,14].

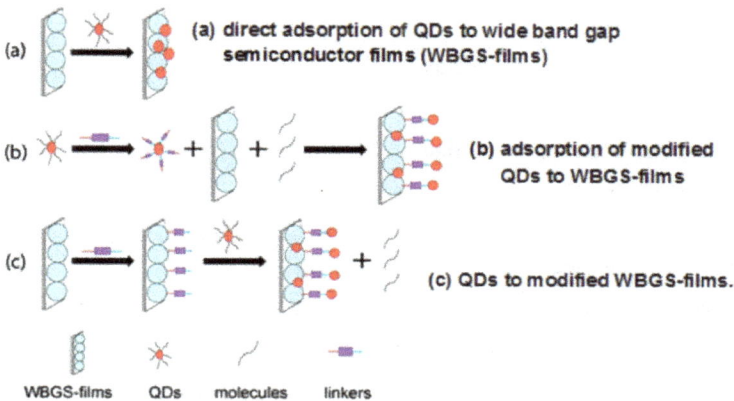

Fig. 5 Ex situ method and linker assisted QDs [12].

Table 3 Quantum dots obtained using linker assisted adsorption method [13-14].

Composite QDs	Size (nm)	Tethering of quantum dots on semiconductors via	Linker Molecule	Ref.
CdS/TiO$_2$	8	Stirring	Thiolactic acid	[13]
CdS-ZnO/TiO$_2$	6	Heating	3-Mercaptopropionic Acid	[14]

3.11 Green method of synthesis of CdS QD

Green routes are eco-friendly, economical and and of medicinal importance [15]. *O. ficus-indica* fruit extract was mixed with freshly prepared 0.5 M $CdCl_2 \cdot H_2O$ solution and 0.5 M of $Na_2S \cdot H_2O$ was added drop by drop till orange-yellow color, showing the formation of CdS was obtained. CdS nanoparticles when centrifuged, washed with water and dried at 110°C for 4 h, spherical QD of 3–5 nm size was obtained.

3.12 Biological method

This method is economical and efficient but comparatively less studied. E. coli, on incubaton with $CdCl_2$ and Na_2S gives CdS in wurtzite crystal phase with a size distribution of 2–5 nm [16].

4. Characterization

The various characterization techniques of semiconductor QDs are discussed as given below.

4.1 Optical characterization methods

The QDs show following characteristic optical properties.

  i. By controlling the composition and size, the wavelength of emitted radiation by QDs can be adjusted.

 ii. The symmetrical and narrow emission peaks with small overlap are obtained.

iii. With a desired emission wavelength, QDs can be prepared.

 iv. The fluorescence intensity is normally stronger for QDs than ordinary dyes.

Fig. 6 [17] shows some characteristics of QDs.

4.2 Electrical characterization

Electrical characterization techniques are used for knowing the electronic structure of QDs. The electrical properties of QDs can be investigated on thin films and ordered arrays. Difficulty appears in single-particle experiments, in accomplishing the controlled electrical contacts to the SQD as the size variation is of few nanometers. Electrodes and QDs when joined together generate an electrical response which depends on the properties of semiconducor QDs and also on the Fermi-level alignment and the contact resistances between the metal electrodes and semiconductor QD and also on electrical charging energies (Coulomb blockade).

The above problem can be overcome by establishing the electrical contacts for single QDs via scanning probe techniques. For electrical scanning probe experiments, semiconductor QDs are spread on a conductive surface which act as the ground electrode and scanning tip serves as the second electrode. The adjustment between the tip–QD distance controls the tunneling resistance. The tip and QD resistance should be much higher than the substrate-QD resistance. This confirms that the voltage almost completely drops on the QD-probe tip tunneling junction. The single QDs for planar electrode devices are generally characterized by fabrication of electrode pairs and the electrodes should be at a distance of less than the diameter of QDs. Ordered arrays and thin films can be examined electrically after deposition of the film on a planar Si/SiO_2 substrate.

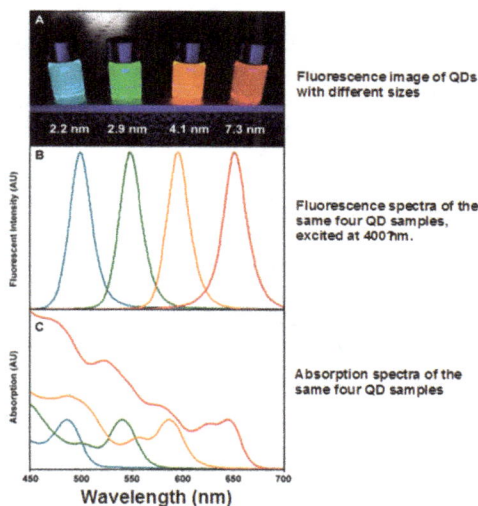

Fig. 6 Optical properties of CdSe QDs in chloroform [17].

4.3   TEM studies

TEM (Transmission electron microscopy) imaging involves the interaction of a high-energy electron beam (≥ 80) keV with a solid specimen. The beam is focused on the sample by electromagnetic lenses and the transmitted or diffracted electrons from the specimen form an image and helps in examining the morphology and structure of the sample down to nanometric scale. Besides, the secondary signals (Fig. 7) [2] emerging from the specimen can be helpful in obtaining information about the composition of specimen down to the nanometer scale.

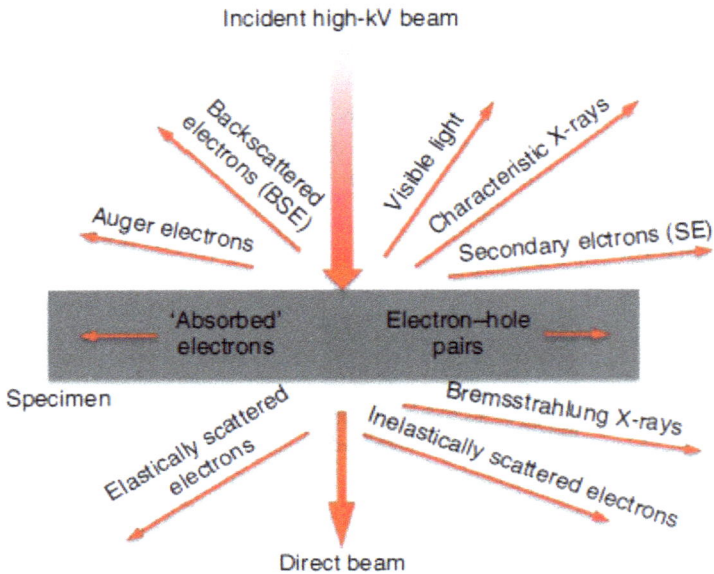

Fig.7 Secondary signals produced by the interaction of electron beam with specimen in an electron microscope [2].

The conventional TEM (CTEM) instrument can be used for capturing the image in dark or bright-field mode. The phase-contrast image can be taken via HRTEM (high-resolution TEM) and STEM (both in bright field and dark field). STEM can also be used for capturing Z-contrast images. The elemental and chemical mapping can be done using EDX (energy-dispersive X-ray spectroscopy) or EELS (electron energy-loss spectroscopy).

If the QDs are matrixed in solid material, a very thin specimen of thickness not more than 100 nm is required. The uniform distribution of particles in specimen can be obtained by sonicating for sometime and then placing the drop on the TEM grid and drying it. Dimpling is another procedure to obtain the thickness of 10 mm in the central region of square or central disk. One more known procedure is ion milling in which the sample is glued on a grid and ion milling is carried out. In ion milling, bombardment in the specimen's central zone is carried out with high energy ions till it is thin enough that can be imaged under TEM. TEM images of some QDs are given in Fig. 8 [2].

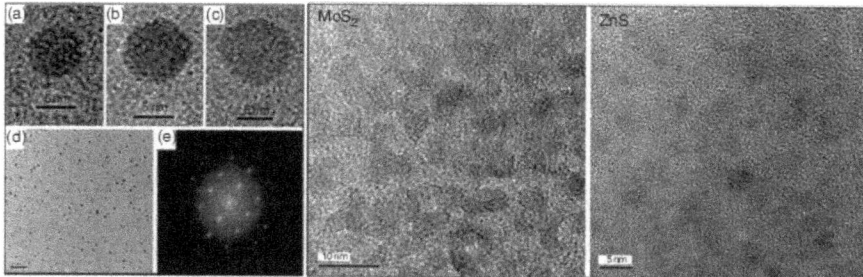

Fig. 8 TEM images of PbS nanocrystals (a-d) and (e) electron diffraction (spraying technique), nanocrystals of MoS$_2$ and ZnS (spraying technique) [2].

EDX determinations have some limitations in accuracy and therefore EELS analysis is preferred. Most of the electrons lose part of their energy due to plasmons or core excitation done on thin samples and therefore bears lesser energy than the incident electrons. This technique is helpful in getting information about the type of elements, chemical environment in the sample and their oxidation state.

4.4 Raman spectroscopy

This spectroscopy is sensitive to the inelastic scattering processes. The three steps mentioned below helps in understanding the resonant Raman scattering on collective excitation i.e. (1) the light incident on the sample produces an exciton; (2) scattering of exciton and generation of phonon and then (3) the radiative recombination of the excitons. In case of phonons, the anionic - cationic lattice planes oscillate giving an electric field oscillation. The resulting field interacts with the dipole field of the exciton via a mechanism known as Froehlich interaction. As the crystal momentum is related to diameter of QDs inversely, their finite size results in the transfer of crystal momentum. The longitudinal optical phonon frequency of semiconductor QDs is red shifted due to negative dispersion of the longitudinal optical phonon and the phonon peak width broadens as seen in CdSe [18]. The energy of phonon is inversely proportional to the QDs diameter in case of confined acoustical phonons as observed in CdS QDs. Expansions and contractions of semiconductor QDs are the characteristic oscillations of acoustical phonons and this mode is called as breathing mode.

4.5   X-ray-related techniques for structural/compositional studies

The powder X-ray diffraction pattern of semiconductor QDs bears the basic information and the diffraction peaks show the crystalline phase. The smaller the size of nanoparticles more important is the study of surface-related effects seen in diffraction pattern.

5.   Properties

5.1   Photoactivity

The UV/Visible light induced excitation mechanism of semiconductor QDs depends on the position of valence band and conduction band and the band gaps. Thus, these can be distinguished into three categories [19]:

(i) Semiconductors with wide band gap embedded by narrow band gap QDs;

(ii) Semiconductors with narrow band gap embedded by QDs of narrow band gap;

(iii) Wide and Narrow band gap semiconductors embedded by carbon QDs.

When visible light is absorbed by QDs, the electrons are excited from valence band to the conduction band of the QDs, leaving holes behind in the valence band (Fig. 9) [19]. Generally there are three types of photocatalytic excitation of semiconductors with narrow band gap embedded by narrow band gap QDs. It depends on the energy level position of QDs and semiconductor nanoparticles and the transference of photoexcited electrons (Fig. 10) [19].

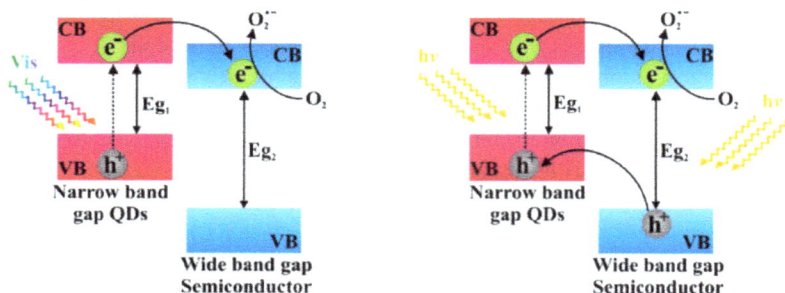

Fig.9 Mechanism of photogenerated electron transfer for a wide band gap semiconductor embedded by narrow band gap QDs in (a) visible and (b) UV-V light irradiation [19].

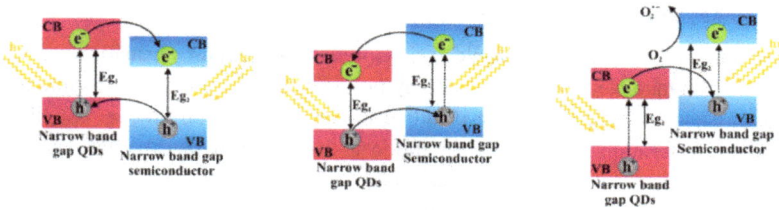

Fig.10 Type I, II, III mechanism of photogenerated electron transfer in semiconductors with narrow band gap embedded by narrow band gap QDs [19].

5.2 Controllable emission light

The semiconductor QDs show an important property of emission of light which depends on the size, composition and shape of the nanostructures. The intensity and wavelength of emission can be adjusted most of the time by altering the conditions in which reaction is carried out or changing the size or composition of the nanostructures [20]. A single light source can be used to excite the semiconductor QDs of different size and different compositions giving distinctive emission spectrum with very slight overlapping. This property makes Semiconductor QDs an attractive material for multiplexed cell imaging. At present QDs also show emission spectra in the near IR region.

5.3 Good light intensity and stability

In biological analysis, different organic dyes as molecular probe are well studied. But with the passage of time, more flexibility is in demand as the traditionally used dyes are not able to set up on expectations. In this context SQDs have played a superior part over the traditionally used organic dyes on various aspects such as stability (less photo-bleaching) and brightness because of the high extinction co-efficient along with almost similar quantum yield as that of fluorescent dyes [21].

Semiconductor QDs are very stable and show very high light intensity which is many times better than traditional fluorescent dyes [22]. Therefore, they can be used to study the long term interactions between biomolecules in living system and also for long term tracking. They act as good tracer for clinical diagnosis. It allows the collection of data related to many, one after the other focal plane images and the best part is that they can be reconstructed into 3D image of high-resolution.

5.4   Stokes shift and fluorescence lifetime

Stokes shift is related to vibronic i.e. vibrational-electronic transitions. Electronic transitions can take place by various energy providing mechanisms, but the interaction of electronic states with the vibrational lattice modes i.e. electron-phonon coupling takes place due to vibrational energy consumption and a red shift of the photoluminescence peak in comparison to fundamental transition is observed. This shift is named as stokes shift. Observation of large stokes shift in semiconductor QDs is the most important quantity which determines the optical behaviour of QDs. With the increase in the radius, there is a decrease in the shift which disappear beyond a limit. This stokes shift mechanism can be understood by the energy calculation of excitonic states [23]. As the QDs have fluorescence lifetime of around 50 ns, biological structures show the auto fluorescence of just 10 ns and due to this fact semiconductor QDs found wide applications in biological systems.

5.5   Active surface for good biocompatibility and functionalization

Semiconductor QDs along with excellent properties discussed above also possesses active surface for functionalization and good biocompatibility. However, during the last two decades much efforts were made to overcome some of the key limitations of semiconductor QDs by reducing the cytotoxicity and improving the biocompatibility. For functionalization, various biomolecules like antibodies, peptides, DNA and enzymes are used to facilitate the multipurpose biological analysis.

In biomedical applications, the basic requirement is high water solubility and dispersibility of QDs. The solubility depends on the hydrophobic and hydrophilic behavior of the layer. CdTe in water is well dispersed but SQDs of CdSe if synthesized in organic solvents are not dispersed in water. Electrostatically, using small charged ligands like charged surfactants that intercalate with the hydrophobic ligands, mercaptopropionic acid or cystamine, water-dispersibility to such materials can be provided [24].

5.6   Quantum Confinement Effect

The zero-dimensional semiconductor nanomaterials have been explored greatly during the past few decades, taking care of novel aspects for controlling and tuning of their electro-optical properties. The confinement of material in at least one dimension (1D) display well-defined change of optical or electronic properties, the structure is referred as quantum-confined structure. Best part in it is the zero-dimensional (0D) materials i.e. semiconductor QDs, which are confined in all 3 dimensions and exhibits quantized levels with varying density of states in comparison to bulk materials. Accordingly, the optical behavior of semiconductor QDs, differ a lot from the bulk materials. This effect describes

Materials Research Forum LLC
https://doi.org/10.21741/9781644901250-12

electrons in terms of valence bands, conduction bands, electron energy band gaps, energy levels and potential wells. Obviously, the confinement of hole and an electron in semiconductor QDs very much depends on the material properties like Bohr radius. The bandgap increases due to the quantum size effect in comparison to the bulk semiconductor (Fig. 11) [19] and ultimately leads to various fluorescent colors pertaining to small differences in the particle size.

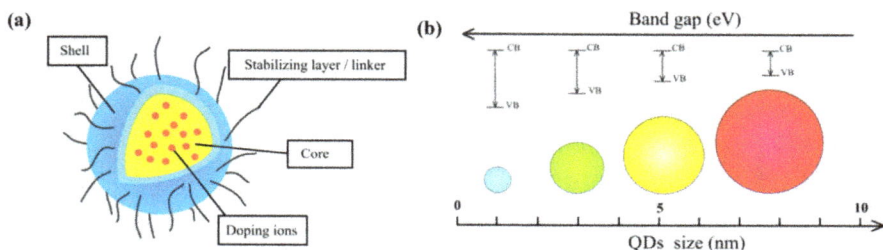

Fig. 11 Pictorial representation of (a) structure of QDs (b) Increase in the band gap with decrease in size of quantum dots i.e. the quantum confinement effect [19].

5.7 In cancer therapy and drug delivery

Capability to target the living cells i.e. human organs like lungs without affecting other organs of the body is the impressive avenue in favour of the use of semiconductor QDs. The nano-conjugate semiconductor QDs of doxorubicin (Dox) and CdSe/CdS/ZnS have been designed to spot cells, alveolar macrophages play a crucial role in the pathogenesis for inflammatory lung injuries. It has been observed via confocal imaging that doxorubicin releases from the QDs-Dox conjugates. It is clearly seen from its aggregation in the nucleus of the cell and induction of apoptosis, portraying that the bioactivity of drug is retained after coupling with nanoparticle. Paramount discovery done in last few years shows that the molecular identity of cancerous cells is different from original cells and it keeps on changing with their growth. Therefore, any kind of tumor may probably contain variety of cells having different growth cycle, genetic make up and moreover this behavior makes them susceptible to any kind of drug therapies. Here hybrid semiconductor QDs with linked antibodies play an important role in detecting the proteins associated with cancer cells, which can be used for the mapping of heterogeneous behaviour of molecules in tumors diagnosed in cancer patients.

5.8   DNA Labelling

To understand the complications in biological molecules, it is necessary to monitor and anticipate the interaction of DNA or proteins sequences of cells. Quantitative imaging of biochemical contents in organisms is a favorable strategy for the same. Using confocal micro-spectrophotometer, diverse receptors (ligands) present on cells are focused and analyzed simultaneously that will help in providing an approach for identification of DNA sequencing which further can correlate in the presence of certain diseases. Already available approaches to visualize and label the DNA and multiple protein molecules depend on the light-emitting properties of chemical dyes, radioactive elements, and proteins. Several drawbacks are there for these labelling techniques. Organic dyes bear few fixed colours and fade away quickly and radioactive markers have short lifetime. Therefore, there is great demand of reliable and powerful labelling fluorophores for biomedical research in terms of quantitative determination and real-time imaging of different types of molecule available in the tissues or cells. The functional limitations with organic dyes and normal chemicals can be overcome by using luminescent QDs. Since semiconductor QDs have narrow symmetric emission spectra, they are stable against photo bleaching. By changing the particle size or composition of QDs, their emission wavelength can be continuously tuned. Therefore, a single light source may be used for excitation of SQDs of different color. Above mentioned optical properties of semiconductor QDs opens their area of application as multicolour and multiplexing sensitive ideal fluorophores. So, semiconductor QDs have tremendous future in the field of fluorescence imaging and their analysis. With the increase in temperature, the fluorescence intensity of DNA-CdS diminishes and therefore, can be used without fail for medical diagnostic analysis.

5.9   Photodynamic therapy (PDT)

An ample range of medical worries like cancers, muscular degeneration with age are treated using Photodynamic therapy. The three key components of PDT are (a) light source (b) photosensitiser and (c) tissue's oxygen in the form of atomic oxygen. On merging of these three components, chemical annihilation of tumor tissues can be done. Semiconductor QDs produces atomic oxygen, which are mostly consumed by the tumor or cancer cells. Thus, specifically the cancer cells were demolished in presence of semiconductor QDs on exposure to laser radiation. On comparing the potential of photosensitization of QDs in PDT with the organic dyes used conventionally it was found that latter produces the reactive cytotoxic oxygen species whereas semiconductor QDs along with their conjugates have proven to be novel and effective materials for therapeutic applications. Iridium complexes embedded QDs have potential applications in cancer treatment via PDT due to sufficient emission from CdSe/ZnS QDs. Moreover, iridium complexes have the tendency to sensitize oxygen molecules for the generation of singlet oxygen on exposure of light. The application of QDs in PDT was noted by the interaction between CdSe QDs with PDT photosensitizer - silicon phthalocyanine [25].

6.   Applications of quantum dots

6.1   Luminescent probes for inorganic-trace analysis

For detecting metals {Ag(I), Hg(II), Pb(II), Cu(II)}, metalloids, non metals and organometals; semiconductor QDs assays were developed [26]. During studies it was found that the low concentration of metal like Ag(I) enhances the QDs fluorescence and high concentration decreases it. Here particle size is an important factor, as in case of small size particle, trapping defect originates on the surface, the passiveness of which may enhance the fluorescence. For Cu(II) detection, quenching of QDs luminescence method has been used. Researchers have successfully combined the unique behavior of CdTe QDs and inhibition of enzymatic activities for Cu(II) ion determination. The alcohol oxidase activity inhibition and decreased quenching of QDs fluorescence have marked the selective detection of Cu(II) ions. Because of the superior fluorescence property of semiconductor QDs, the low limit of detection, around 0.176 ng/mL was observed [27]. The approach for lead detection was based on fluorescence quenching. The drawback of this method is the reduced selectivity of the CdTe-QD/AuNP assembly.

6.2   Biomonitoring and bioimaging

The single particle tracking (SPT) is successfully used for tracking of single bio molecule with high resolution. Actual molecular dynamics can be observed by using the brighter probes like semiconductor QDs for SPT. The conventional semiconductor QDs because of the presence of Cd and Pb suffer from toxicity but the nanostructures like Carbon QDs shows bright and colourful fluorescence emission and used for biomonitoring [28]. The photophysical properties of semi conductor QDs mainly depends on the composition, shape and size i.e. why the synthesis of carbon QDs is done by taking care of their biocompatibility and low toxicity. The synthesized carbon QDs are capped with hydrophobic molecules and solubilized in aqueous solution. For solubilization either encapsulation with ampipathic polymers or water soluble silica shells can be used. Advancement in research have successfully eliminated the solubilization step. More interesting direction is the synthesis of carbon QDs inside the living organisms like yeast or E.coli cells. The only requirement is the insertion of semiconductor QDs precursors

Materials Research Forum LLC
https://doi.org/10.21741/9781644901250-12

into the cells [29]. There were reports on the synthesis of CdTe QDs inside earthworm [30].

6.3   Biosensing

The unique photochemical and photophysical properties of semiconductor QDs in comparison to organic dyes and protein makes them a suitable candidate of fluorescence probes. The fluorescence probes are used for sensitive optical biosensing applications namely immunoassays, biomolecules sensing, catalysis monitoring and nucleic acid detection [31]. Immunoassay is a very useful and helpful tool for biomedical applications and clinical tests. The field of semiconductor QDs have led to the development of competitive fluoroimmunoassay with the help of which targets ranging from very small molecules to micro organisms like viruses and bacteria are detected. DNA and RNA segments are conjugated on QDs surface, as they act as recognition moieties to form fluorescent probes for genetic target analysis. On the basis of high specificity among the DNA probes-multicolored QD and the complementary sequence of target strand. Similarly, once the DNA sequence probe gets coupled, the different emission colors of QDs were used for multiplexed detection of the complementary sequences.

6.4   Photodetectors or photoelectrochemical sensors

In photoelectrochemical sensors the semiconductor QDs are made immobilized using an organic linker layer onto an electrode in most of the cases. On applying the appropriate potential and on excitation of the semiconductor QDs the electrons tunnel to the valence band of the semiconductor QDs from the electrode, electrons from the conduction band of the semiconductor QDs tunnel to oxidant molecules in the solution. Thus, with the increase in the concentration of the oxidant present in solution, the monotonically generated cathodic photocurrent increases. In contrary, if electron donors exist in solution then tunneling of electrons to the valence band of the semiconductor QDs from the solution phase occurs and of electrons from the conduction band to the electrode takes place. The amplitude and the direction of the resulting photocurrent depends on the bias potential applied to the electrode and concentration of molecules to be detected. Semiconductor QD-based photoelectrochemical sensors bear following advantages: (1) photoelectrochemical sensors have a good sensitivity and fast response and if coupled with biocatalytic reactions, these sensors have the tendency to detect certain substances which are not detected via other analytical methods. (2) the QDs act as photoactivators of the sensor. As the QDs have wide absorption spectra, a common white light source can be used to excite their photoelectrochemical sensor systems. This opens the ways to design simple, cheap and portable sensor systems.

6.5   Photovoltaics (solar cells)

The tunable energy band gap and multiexciton generation have made semiconductor quantum dot-sensitized solar cells (SQDSSCs) well known in market. They are considered as ideal candidates for energy devices and the next generation solar cells because of their defined qualities like (i) their capability to harvest sunlight, helps in generation of multiple electron–hole pairs, (ii) its low cost and simplicity in fabrication. Although because of charge recombination and narrow absorption ranges the power conversion efficiency (η) rates of a number of SQDSSCs are lower in comparison to dye-sensitized solar cells [32].

For the improvement of power conversion efficiency, choice of semiconductor with appropriate band gap is the essential criteria. Because of their high absorption spectra these materials show high applicability for photosensitization and the absorption and emission spectra depends the size of the QDs. Few such materials include CdS, ZnSe, CdSe, Ag_2S, PbS, InP, CdHgTe and CdTe. Among the mentioned, CdS and CdSe are very much used. $CuInS_2$ is a direct band gap semiconductor compound with band gap of almost 1.5 eV having many advantageous features like excellent stability, non-toxicity and the higher absorption coefficient of 10^5 cm^{-1} [33].

6.6   Light emitting diodes

Semiconductor QDs in the form of light-emitting diodes (QLEDs) are very extensively explored for two important applications, lighting and displays, brightness, spectral tunability and improved colour saturation. Advancement in the field have satisfied the demand of high external quantum efficiency (EQE) but still have relatively low luminance range (<2,000 cdm^{-2}) for red QLEDs, far below the lightning threshold value [34]. In 2018, for green QLEDs Xiang Li et. al recorded a high luminance of 460,000 cdm^{-2} but EQE of only 6% [35]. In case of blue QLEDs, although the EQE has been improved to >10% but its a challenge to get good Luminescence which is extremely low of the order of 10^2 cdm^{-2}. So, the main challenge is in achieving the simultaneously high luminescence and high EQE for these three primary colours. Huaibin Shen, very recently in 2019, demonstrated devices for blue, green and red colours with maximum EQE of 8.05%, 22.9% and 21.6% and brightness of 10,100 cd m^{-2}, 52,500 cd m^{-2} and 13,300 cd m^{-2} respectively peak luminance of 62,600 cd m^{-2}, 614,000 cd m^{-2} and 356,000 cd m^{-2} respectively. These attributes are found in semiconductor QD light emitting diodes of ZnSe/CdSe shell/core structures. The use of Se throughout the core/shell regions can be the reason for high performance and also due to the presence of alloyed bridging layers at the shell/core interfaces [36].

Materials Research Forum LLC
https://doi.org/10.21741/9781644901250-12

6.7 Electrochemiluminescence

Electrochemiluminescence (ECL) is the conversion of electrical energy into radiative energy. On applying voltage, species produced at electrodes undergo high-energy electron-transfer reactions and excited states are produced that emit measurable luminescent signals. The materials like transition metal complexes ($Ru(bpy)_3^{2+}$) and clusters, organic molecules (luminol) are known to produce ECL.

The semiconductor QDs have recently attracted much attention as they emerged a new group of ECL luminophores, because they bears the good ECL activities and have the advantage of easy labeling. Therefore, these materials have applications in cell-imaging and bio-sensing etc. As metal-based semiconductor QDs namely CdTe and CdSe have issues related to biocompatibility, stability and environmental toxicity, alternatives are being explored. Carbon QDs (CQDs) and graphene QDs (GQDs) show photoluminescence, chemiluminescence, especially the ECL. Due to superiorities such as excellent chemical inertness, facile synthesis, low toxicity and easy labeling of CQDs and GQDs, they have been projected as emerging ECL luminophores.

6.8 Bio conjugation and surface functionalization of QDs for biological applications

Semiconductor QDs are highly fluorescent and their emission spectra is highly size dependent. The exceptionally good photophysical properties of these nanoscale crystals can be utilized for drug delivery, imaging and cell targeting and makes them a reliable alternative to fluorescent proteins and organic fluorescent dyes. For biomedical applications, semiconductor QDs needs to be chemically stable in aqueous solutions and can be tagged with recognition of molecules or drugs. Conjugates of Semiconductor QDs with peptides, proteins, small molecules, antibodies and oligonucleotides are utilized for fluorescent targeting, imaging and tracking both in vivo and in vitro. Semiconductor QDs have found applications in diagnostic systems, ultrasensitive detection, drug delivery approaches, imaging and delivery in a single assay and accurate targeting. The highly bright fluorescence of semiconductor QDs is due to high molar adsorption coefficients (many times higher than the fluorescence of proteins and dyes) in combination with a high quantum yield. Semiconductor QDs bioconjugates in comparison to phycoerythrin, fluorescein isothiocyanate and AlexaFluor488 bioconjugates have been found to be 2600-fold, 4200-fold and 420-fold more resistant to photobleaching respectively. Medical and biological applications require that semiconductor QDs must be biocompatible, functionalized with drugs and/or biomolecules and highly water-soluble. Ample number of proposals on functionalization and solubilization strategies have been proposed. Applications of semiconductor QDs in conjugation with biomolecules can be used for single molecule tracking, cell labeling, flow cytometry, biosensing, targeted drug delivery, photodynamic therapy and deep-tissue and tumor imaging. The two major approaches for the surface modification of semiconductor QDs are (i) hydrophobic interactions of amphiphilic molecules with quantum dots and (ii) the ligand exchange strategy i.e. the interaction of polar groups of the coated molecules with the QDs surface.

Conclusions

In recent years semiconductor QDs have attracted the interest of researchers for various applications. Preparation of semiconductor QDs by using colloidal, emulsion, sol–gel, co-precipitation, hot injection, hydrothermal, solvothermal, green method, biological methods, etc have been discussed. The chapter also focuses on the optical, electrical, X-ray diffraction and microscopic techniques used to characterize semiconductor QDs. Properties of semiconductor QDs have been studied to get in depth knowledge about these materials and the results were discussed. The importance of semiconductor QDs particularly in luminescent probes for inorganic-trace analysis, biomonitoring and bioimaging, biosensing, photodetectors or photoelectrochemical sensors, photo voltaics (solar cells), light emitting diodes, electrochemiluminescence, and surface functionalization for biological applications have been elaborated in this chapter. Applications of semiconductor QDs in conjugation with biomolecules can be used for single molecule tracking, cell labeling, flow cytometry, biosensing, targeted drug delivery, photodynamic therapy and deep-tissue and tumor imaging. In general semiconductor QDs have enormous applications in different sectors.

References

[1] J. Xu, J. Zheng, Nano-inspired biosensors for protein assay with clinical applications, quantum dots and nanoclusters, Elsevier, 2019, pp. 67-90. https//doi.org/10.1016/C2016-0-01779-5

[2] D. Dorfs, R. Krahne, A. Falqui, L. Manna, C. Giannini, D. Zanchet, Quantum Dots: Synthesis and characterization, comprehensive nanoscience and nanotechnology (Second Edition), Elsevier, 2011, pp. 17-60. https//doi.org/10.1016/b978-0-12-374396-1.00028-3

[3] S. Chand, N. Thakur, S. C. Katyal, P. B. Barman, Vineet Sharma, Pankaj Sharma, Recent developments on the synthesis, structural and optical properties of chalcogenide quantum dots, Sol. Energy Mater. Sol. Cells 168 (2017) 183–200. https//doi.org/10.1016/j.solmat.2017.04.033

[4] F. X. Redl, K.-S. Cho, C. B. Murray and S. O'Brien, Three Dimensional Binary superlattices of magnetic nanocrystals and semiconductor quantum dots, Nature 423 (2003) 968-971. https//doi.org/10.1038/nature01702

[5] S. K. Mandal, N. Lequeux, B.Rotenberg, M. Tramier, J. Fattaccioli, J. Bibette, B. Dubertret, Encapsulation of magnetic and flourescent nanoparticles in emulsion droplets, Langmuir 21(2005) 4175-4179. https://doi.org/10.1021/la047025m

[6] H. Zhang, L. Wang, H. Xiong, L. Hu, B. Yang, W. Li, Hydrothermal synthesis for high-quality CdTe nanocrystals, Adv. Mater. 15 (2003) 1712–1715. https//doi.org/10.1002/adma.200305653

[7] J.-P. Ge, S. Xu, J. Zhuang, X. Wang, Q. Peng, Y.-D. Li, Synthesis of CdSe, ZnSe, and $Zn_xCd_{1-x}Se$ nanocrystals and their silica sheathed core/shell structures, Inorg. Chem. 45 (2006) 4922–4927. https//doi.org/10.1021/ic051598k

[8] Y. He, H.-T. Lu, L.-M. Sai, Y.-Y. Su, M. Hu, C.-H. Fan, W. Huang, L.-H. Wang, Microwave synthesis of water-dispersed CdTe/CdS/ZnS core shell-shell quantum dots with excellent photostability and biocompatibility, Adv. Mater. 20 (2008) 3416–3421. https//doi.org/10.1002/adma.200701166

[9] F. Shen, W. Que, Y. Liao, X. Yin, Photocatalytic activity of TiO_2 nanoparticles sensitized by $CuInS_2$ quantum dots, Ind. Eng. Chem. 50 (2011) 9131–9137. https//doi.org/10.1021/ie2007467

[10] K.T. Lee , C.H. Lin, S.Y. Lu, SnO_2 quantum dots synthesized with a carrier solvent assisted interfacial reaction for band-structure engineering of TiO_2 photocatalysts. J. Phys. Chem. 118 (2014) 14457–14463. https//doi.org/10.1021/jp5045749

[11] S. Qian, C. Wang, W. Liu, Y. Zhu, W. Yao, X. Lu, An enhanced CdS/TiO_2 photocatalyst with high stability and activity: effect of mesoporous substrate and bifunctional linking molecule, J. Mater. Chem. 21 (2011) 4945–4952. https//doi.org/10.1039/c0jm03508d

[12] D. Zhang, P. Ma, S. Wang, M. Xia, S. Zhang, D. Xie, X Zhou, Y. Lin, The in situ ligand exchange linker-assisted assembly of oil-soluble CdSe quantum dots to TiO_2 films, Appl. Surf. Sci. 475 (2019) 813–819. https//doi.org/10.1016/j.apsusc.2018.12.289

[13] Q. Zhou, M.L. Fu, B.L. Yuan, H.J. Cui, J.W. Shi, Assembly, characterization, and photocatalytic activities of TiO_2 nanotubes/CdS quantum dots nanocomposites, J. Nanopart. Res. 13 (2011) 6661–6672. https//doi.org/10.1007/s11051-011-0573-y

[14] M.A. Mumin, G. Moula, P. A. Charpentier, Supercritical CO_2 synthesized TiO_2 nanowires covalently linked with core-shell CdS-ZnS quantum dots: enhanced photocatalysis and stability, RSC Adv. 5 (2015) 67767–67779. https//doi.org/10.1039/c5ra08914j

[15] K. Kandasamy, M. Venkatesh, Y. A. Syed Khadar, Paramasivan Rajasingh, One-pot green synthesis of CdS quantum dots using Opuntia ficus-indica fruit sap, Materials today Proc. 2020. https//doi.org/10.1016/j.matpr.2019.06.003

16] B. T Dubertret, P. Skourides, D. J. Norris,V. Noireaux, A. H. Brivanlou, A. Libchaber, In vivo imaging of quantum dots encapsulated in phospholipids micelles, Science 298 (2002) 1759-1762. https//doi.org/10.1126/science.1077194

[17] B. Liu, B. Jiang, Z. Zheng, T. Liu, Semiconductor quantum dots in tumor research, J. Lumin. 209 (2019) 61-68. https//doi.org/10.1016/j.jlumin.2019.01.011

[18] C. Trallero-Giner, A. Debernardi, and M. Cardona, E. Menendez-Proupın, A.I. Ekimov, Optical vibrons in CdSe dots and dispersion relation of the bulk material, Phys. Rev. B 57 (1998) 4664–4669. https//doi.org/10.1103/physrevb.57.4664

[19] B. Bajorowicz, M.P. Kobylański, A. Gołąbiewska, J. Nadolna, D. Zaleska-Medynska, A. Malankowska, Quantum dot-decorated semiconductor micro- and nanoparticles: A review of their synthesis, characterization and application in photocatalysis, Adv. Colloid Interface Sci. 256 (2018) 352–372. https//doi.org/10.1016/j.cis.2018.02.003

[20] I. L. Medintz, H. T. Uyeda, E. R. Goldman, H. Mattoussi, Quantum dot bioconjugates for imaging, labelling and sensing, Nat. Mater. 4 (2005) 435-446. https//doi.org/10.1038/nmat1390

[21] C. Delerue, M. Lannoo, Nanostructures: Theory and modelling, nanoscience and nanotechnology, Springer, 2004, pp. 47-80. https//doi.org/10.1007/978-3-662-08903-3

[22] A. R. Clapp, I. L. Medintz, J. M. Mauro, B. R. Fisher, M. G. Bawendi, H. Mattoussi, Fluorescence resonance energy transfer between quantum dot donors and dye-labeled protein acceptors, J. Am. Chem. Soc. 126 (2004) 301-310. https//doi.org/10.1021/ja037088b

[23] A. Bagga, P. K. Chattopadhyay, S. Ghosh, Stokes shift in quantum dots: Origin of dark exciton, International Workshop on Physics of Semiconductor Devices 2007. https//doi.org/10.1109/iwpsd.2007.4472661

[24] W.C.W. Chan, S.Nie, Quantum dot bioconjugates for ultrasensitive non isotopic detection, Science 281 (1998) 2016–2018. https//doi.org/10.1126/science.281.5385.2016

[25] A. C. Samia, X. Chen , C. Burda , Semiconductor quantum dots for photodynamic theory, J. Am. Chem. Soc. 125 (2003) 15736-15737. https//doi.org/10.1021/ja0386905

[26] I. Costas-Mora, V. Romero, I. Lavilla, C. Bendicho, An overview of recent advances in the application of quantum dots as luminescent probes to inorganic-trace analysis, Trend Anal. Chem. 57 (2014) 64–72. https//doi.org/10.1016/j.trac.2014.02.004

[27] C. Guo, J. Wang, J. Cheng, Z. Dai, Determination of trace copper ions with ultrahigh sensitivity and selectivity utilizing CdTe quantum dots coupled with enzyme inhibition, Biosens. Bioelectro. 36 (2012) 69–74. https//doi.org/10.1016/j.bios.2012.03.040

[28] P. G. Luo, F. Yang, S.T. Yang, S.K. Sonkar, L. Yang, J.J. Broglie, Y. Liu, Y.P. Sun, Carbon-based quantum dots for fluorescence imaging of cells and tissues, RSC Adv. 4 (2014) 10791–10807. https//doi.org/10.1039/c3ra47683a

[29] R. Cui, H.H. Liu, H.Y. Xie, Z.L. Zhang, Y.R. Yang, D.W. Pang, Z.X. Xie, B. Bei Chen, B. Hu, P. Shen, Living yeast cells as a controllable biosynthesizer for fluorescent quantum dots, Adv. Funct. Mater. 19 (2009) 2359–2364. https//doi.org/10.1002/adfm.200801492

[30] S. R. Sturzenbaum, M. Hockner, A. Panneerselvam, J. Levitt, J-S. Bouillard, S. Taniguchi, L-A. Dailey, R. Ahmad Khanbeigi, E. V. Rosca, M. Thanou, K. Suhling, A. V. Zayats, M. Green, Biosynthesis of luminescent quantum dots in an earthworm, Nat. Nanotechnol. 8 (2013) 57–60. https//doi.org/10.1038/nnano.2012.232

[31] Y. Wang, R. Hu, G. Lin, I. Roy, K-T Yong, Functionalized quantum dots for biosensing and bioimaging and concerns on toxicity, ACS Appl. Mater. Inter. 5 (2013) 2786–2799. https//doi.org/10.1021/am302030a

[32] M. Kouhnavard, S. Ikeda, NA Ludin, NB Ahmad Khairudin, BV Ghaffari, MA Mat-Teridi, M. A. Ibrahim, S. Sepeai, K. Sopian, A review of semiconductor materials as sensitizers for quantum dot sensitized solar cells, Renew. Sust. Energ. Rev. 37 (2014) 397–407. https//doi.org/10.1016/j.rser.2014.05.023

[33] B. Fu, C. Deng and L. Yang, Efficiency enhancement of solid-state cuins$_2$ quantum dot-sensitized solar cells by improving the charge recombination, Nanoscale Res. Lett. 14 (2019) 1981-1988. https//doi.org/10.1186/s11671-019-2998-7

[34] X. Dai, Z. Zhang, Y. Jin, Y. Niu, H. Cao, X. Liang, L. Chen, J. Weng, X. Peng, Solution-processed, high-performance light-emitting diodes based on quantum dots, Nature 515 (2014) 96–99. https//doi.org/10.1117/12.591299

[35] X. Li, Y.-B. Zhao, F. Fan, L. Levina, M. Liu, R. Quintero-Bermudez, X. Gong, L. N. Quan, J. Fan, Z. Yang, S. Hoogland, O. Voznyy, Z.-H. Lu, E. H. Sargent, Bright colloidal quantum dot light-emitting diodes enabled by efficient chlorination, Nat. Photon. 12 (2018) 159-164. https//doi.org/10.1038/s41566-018-0105-8

[36] H. Shen, Q. Gao, Y. Zhang, Y. Lin, Q. Lin, Z. Li, L. Chen, Z. Zeng, X. Li, Y. Jia, S. Wang, Z. Du, L. Song Li, Z. Zhang, Visible quantum dot light-emitting diodes with simultaneous high brightness and efficiency, Nat. Photon. 13 (2019) 192–197. https//doi.org/10.1038/s41566-019-0364-z

Materials Research Forum LLC
https://doi.org/10.21741/9781644901250-13

Chapter 13

Quantum Dots: Properties and Applications

Amal I. Hassan, Hosam M. Saleh*

Radioisotope Department, Nuclear Research Center, Atomic Energy Authority, Dokki 12311, Giza, Egypt

*hosam.saleh@eaea.org.eg, hosamsaleh70@yahoo.com

Abstract

Quantum dots (QDs) are very small nanoparticles and are composed of hundreds to thousands of atoms. These semiconducting materials can be made from an element, such as silicon or germanium, or compounds such as cadmium sulphide (CdS) or cadmium selenide (CdSe). The colour of these small particles does not depend on the type of semiconducting material from which the dots are made, but rather on its diameter. Besides, ODs attract the most attention because of their unique visual properties. Therefore, these are used in all kinds of applications where precise control of coloured light is important. As these dots are of great importance in chemical, biological and medical applications, they can be designed to deliver anti-cancer drugs and direct them to specific areas of the body. Therefore, with this technique, the harmful side effects of chemical treatments can be reduced. It is possible to examine and study the properties of these nanomaterials and make sure they are analyzed using some scientific devices and techniques, the most important of which are: transmittance electron microscopy (TEM), scanning electron microscopy (SEM), atomic forces microscopy (AFM) with dielectrics, and X-ray diffraction (XRD). This chapter opens horizons towards knowing what quantum dots are and their unique properties, as well as methods of preparation and then placing our hands on the chemical, and biological applications of these dots.

Keywords

Quantum Dots, Nanotechnology, Cadmium Sulphide, Cadmium Selenide, Semiconducting Materials

Contents

1.   Introduction

As a field, nanotechnology endeavors to explore, understand and utilize the unique physical and chemical properties of materials that appear with a reduced volume based on scales up to 100 nanometres or less[1]. Nanotechnology (NT), in particular, having optical and electrical phenomena associated with semiconductors (SC) on the nanometre scale [2]. Nanoscale semiconductor crystals, often referred to as "quantum dots" (QDs), exhibit strange optical and electrical behaviors that are not found in their bulk analogs, including high luminosity, extinction factors, and light. Quantum dots are small particles or nanocrystals of a semiconducting material with diameters ranging from 2-10 nanometres (10-50 atoms) [3], were first discovered in 1980 [4]. These dots are characterized by certain optoelectronic properties, which are moderate between the properties of massive SC and separate molecules, partly due to the very large surface area of these particles [5]. The conspicuous consequence of this is fluorescence where nanocrystals can emit distinct colours defined by the size of molecules [6]. These characteristics have aroused great interest in fields ranging from quantum computing and solar cells to biomarker marking and high sensitivity in laboratory diagnostics [6]. Here we summarize the basic physical realities accompanied by QD and exemplary examples of how these scientific peculiarities are easily exploited in multiple benefits in nanotechnology with a particular attraction on their implementation in biomedical investigation. Quantum dots are now being used in place of organic pigments in a variety of experiments using optoelectronics to track how drugs and other molecules move in the body, and these points may have a new supporting role in health care. Scientists have designed nanoparticles of quantum dots that produce chemicals that can make bacteria weaker in front of antibiotics, a step forward against drug-resistant infections, such as superbugs and the infection they cause [7].

The antibiotics in this study were fortified with experimental quantum points, which resulted in a thousand times greater effectiveness in fighting bacteria than antibiotics alone [8]. The quantum dots were thickened by a thread of deoxyribonucleic acid (DNA) - 3 nanometres in diameter - and made from cadmium telluride, a stable crystalline compound usually used in photoelectric products [9]. Quantum dot electronics interact with the green light at a certain frequency, causing them to bond with the oxygen molecules in the body, creating superoxide [9]. Also, bacteria that absorb this substance are unable to resist antibiotics, due to a chemical imbalance inside them. Various quantities of quantum points were mixed with varying concentrations of each of the five antibiotics to produce a spectrum of test samples [10]. They then added these samples to five strains of drug-resistant bacteria, including streptococcus anti-methicillin (MRSE) and salmonella. Of the 480 tests for different combinations of quantum dots, antibiotics, and bacteria, more than 75% of samples with quantum dots managed to stop bacterial growth or permanently kill them with low doses of antibiotics.

2. Optical properties

In the seventies, it was also possible to manufacture the first quantum well in two dimensions with a single atomic thickness, followed by manufacturing quantum dots with a zero dimension that matured with their applications nowadays. Commonly, the smaller size of the crystal, has a greater band-gap, and the grander the energy discrepancy between the greatest valence band and the lowest conduction beam it transforms more energy are needed to excite the dots, more energy is emitted when the crystal returns to its stable state. For example, in the pigment of neon applications (fluorescent), this corresponds to the higher frequencies of light emitted after excitation because of the size of the crystal becoming smaller, resulting in a colour shift in the light emitted from red to blue. Besides such harmony, a major advantage of quantum dots is that because of the extraordinary permissible control of the dimension of the crystal produced, there can be strict control of the productive qualities of the substance (Fig. 1) [11]. Also, quantum dots of various sizes can be grouped into a graded, multilayer nanofilm.

Likely implementations of QDs include single-electron transistors, solar cells, a light-emitting diode lasers, mono photon sources [12], the 2nd symmetric generation [13], and Q-medicalization imaging. Its small scale permits to be placed easily in a solution, which may trigger their utilization in the printing and rotary coating methods [14,15].

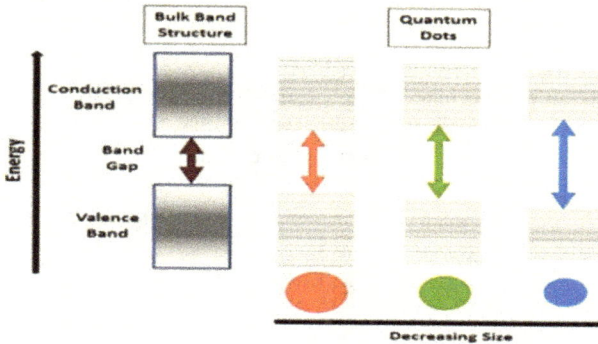

Fig. 1 Dissociation of energy levels in quantum points [11].

There are many methods for restricting and limiting the excitons in SC, leading to many methods for manufacturing quantum dots. The quantum wires, quantum energy bundles, and quantum points grow through advanced metallurgical methods in nanocrystals generated by chemical processes or by the ions cultivation process, or in nanostructures produced by art-based lithography methods [16].

3.   Manufacturing of quantum dots

Cadmium is principally utilized in building nanoparticles in nanotechnology such as QDs, which are semi-metal semiconductor constructions of metal SC [17]. Because of their tiny size, QDs have novel photoelectric and optical features that give nanoparticles great tunable stability "fluorescence". The large surface area also allows for rapid QD quality operation with targeting links to site-oriented leverage based on these characteristics, they have the capability for renewal in cancer apprehension and medication, including biological imaging at the cellular level [18]. However, a keen interest in the QD is moderately mitigated because QDs include large quantities of cadmium interactively, whereas the knowledge of the health hazards of vulnerability to cadmium nanoparticles is scanty [19].

Scientists made CdSe QDs in the eighties of the last century, by using the top-down process to manufacture nanomaterials. Notwithstanding, divergences in the dimension, unique visual features, crystal shortages, and insufficient transcription of these QDs have represented them inadmissible for progressed applications [20]. These codons were highly hydrophobic because the nanocrystals were covered with non-polar surfactant particles, and these non-polar aliphatic groups were present on the surface of the QDs [21]. The prevalent technique of synthesis of QDs by injecting organic mineral antecedents into trioctylphosphine (TOP) and trioctylphosphine oxide (TOPO) surface at excrescent temperature (190 to 320 °C) [22]. The hydrophobically coated CDS, CdSe and CdTe can be processed through thermoplastic precursors of cadmium (dimethyl cadmium) and selenium in a coordinated solvent mixture consisting of TOP and TOPO[23]. In TOPO (mainly phosphine acids and alkyl phospholipids) may restrain the development of molecules. However, boosting a definite number of composites such as hexylphosphonic acid to the response will appear in QDs with homogeneous volume allocation whereas restraining their growth [23]. The Cd predecessors like myristate, acetylacetonate, and oxide were prepared [24]. Therefore, adjustable photoluminescence (PL) and better retention of the amount of colloidal QDs were attained by this approach, which allured variousinvestigators. The extra more old-fashioned is Ostwald, which brought about a progressive disintegration of minor QDs and the formation of larger crystals which were accomplished by isolating the unconstrained core approach from the remaining low sufficient to reinforce the growth of crystals. The concentricity of monomers is relatively moderate nanocrystalline development method. The main expediency of this strategy is that QDs can get its volume stabilized by selecting injection temperature and growth [25]. Because this technique includes an intricate strategy, it is less used.

Colloidal (C) semiconductor nanocrystals are composed of pioneering compounds dissolved in solutions and are very similar to conventional chemical procedures. Where the installation of CQDs is built on a three-part framework consisting of monomers, organic surface compounds, and solvents [26]. Temperature is a crucial agent in governing the optimal conditions for the growth of nanocrystals. They should be high sufficient to permit reorganization and softening of the atoms throughout the manufacturing operation while another influential agent that must be rigorously controlled during the growth of nanocrystals [26]. At high monomer concentrations, the decisive size; at which nanocrystals do not grow, and do not shrink, and is resulting in the growth of virtually all molecules. In this system, more diminutive molecules develop more agile than massive molecules, since large crystals need further atoms to develop grow than small crystals. This results in a concentrate on volume distribution, resulting in an almost improbable distribution of single-mass molecules [27]. The establishment in volume is appropriate when monomer concentration is maintained so that the median size of the nanocrystalline is perpetually insignificantly greater than the critical size (CZ). Additionally, the concentration of the monomer decreases, the CZ becomes greater than the mean, and the distribution "removes the concentration"[28]. The QDs may encompass as few as a hundred and up to 10000 atoms inside the QD size, with a diameter of 10 to 50 atoms. This size conforms between 2 to 10 nanometres, where about 3 million quantum dots in 10 nanometres can line up to all of them jointly, which are proportional to the width of the human thumb [29]. Great quantities of QDs can be performed through colloidal synthesis. Colloidal structures are a promising area for commercial applications. These methods are to be the least toxic of all other methods of installation or manufacture [29].

Plasma synthesis has stemmed to be one of the most trivial gaseous state methods for producing QDs, exclusively those with covalent bonds [30]. For example, the quantum dots are created of silicon (Si) and germanium (Ge) utilizing non-thermal plasma. The volume, form, and configuration of the dots can be controlled in non-thermal plasma [31]. Anabolic steroids that imply extraordinarily challenging to QDs have likewise been achieved in plasma synthesis. Quantum dots that are generated by plasma are frequently powdered, and that surface amendment may be achieved [32]. This can engender a distinguished scattering of quantum points in biological solvents [33] or water (that is, CQDs) [34].

The highly arranged beams of QDs may be self-amassed by electrochemical methods. Where a model was designed by inducing an ionic reaction at the interface of the metallic electrolyte which results in an unrestrained accumulation of nanostructures, inclusive of QDs, on the surface of the metal, which become a cover to engrave the mesa of these nanostructures on a selected substrate [35, 36].

On the quantum level, bandgap energy, which determines the energy of fluorescent light, is inversely proportional to the size of the QDs. Therefore, the larger QDs have higher energy levels, which are also more closely related to the distance between them. These critical structures can be called shells. Single sub-layer shells can also be powerful strategies for passivating QDs, which includes CbS single-layer shells PbS [37]. The nucleus of a self-assembled quantum dot under certain conditions during the period of the radial molecular beam and the extract of the metallic vapor phase, when a substance is grown on a substrate that the net does not match [38, 39].

The QDs energy spectrum can be created by controlling the geometric volume, and hardness of the confinement potential [40]. Likewise, in contrast to the atoms, it is relatively easy for quantum points to be connected by tunneling barriers for conduction behavior, which allow research and investigations to be carried out around them by applying scanning tunnel spectroscopy methods [41].

The attributes of QDs absorption correspond to the transitions between states of a particle in a separate three-dimensional electron and hole box (particle in a box), both bound by the same nanoscale box. Whereas, these discrete transformations are exciting for the atomic spectroscopy and result in the emergence of QDs called artificial atoms [42]. Adherence to quantum points may also be triggered by states of the electronegative potential (Generated by external electrodes, activation, stress, or impurities). A model was created by inducing an ionic reaction at the interface of metallic electrolysis, which results in an automatic gathering of nanostructures, including quantum dots, on the surface of the metal, which is then used as a cover to engrave the mesa of these nanostructures on a selected substrate [41].

The production of standard small QDs was based on a manner referred to "high-temperature binary injection" which is not practical in most nonscientific applications i.e. trade and that requires huge quantities of QDs [43]. One of the ways to increase and multiply, which can be used to manufacture larger quantities of quantum points of higher quality and more compatible, involves the production of nanoparticles of chemical isotopes in the presence of molecular mass collected under conditions and conditions through which the unit of molecular mass has been preserved in addition to playing a role ready seed template. Each molecule illustrates within the mass of the compiled particles as seed or nucleation point based on which the nanoparticles can appear to grow. In this procedure, the excessive temperature nucleation step does not become crucial to start the growth of the nanoparticle because the nucleation sites are ready by molecular masses. Also noteworthy is that this process is makeable extensible [43].

Another method of mass creation of colloidal QDs can be noticed in transferring the well-known method of thermal injection for synthesis to a mechanical perpetual flow system. Scientists can overcome differences from one batch to another through the mentioned method by using technical ingredients (TI) of mixing and growth, transport and temperature adjustments [44]. To produce CdSe-based semiconductor particles, this technique was investigated and adjusted to production quantities of kg per month. Considering the use of TI permits easy exchange concerning the maximum size and volume, it can be notably improved for tens or indeed hundreds of kilograms [45].

4. CdSe quantum dots

One of the most well-known types of quantum dots that have been extensively studied is quantum cadmium selenide [46]. These points are semiconducting nanocrystals consisting of a core of CdSe and ligand shell. The ligands play an important role in the stability and dissolution of these nanoparticles [47]. During the reaction, the ligands stabilize the crystal growth to prevent nanocrystals from aggregation or precipitation

[47]. These devices also affect electronic and optical properties by passivating surface electronic shells. Quantum CdSe dots have been used in a wide range of applications involving solar cells. Light-emitting diodes and fluorescent tagging, in addition to cadmium selenide-based materials, have applications related to biomagnification. There are some limitations to the applications related to the quantum dots of the semiconductor, foremost of which are related to toxicity, for the use of heavy metals in their preparation [48]. The toxicity of these substances appears despite being surrounded by crusts of non-toxic substances because it is difficult to guarantee that cadmium ions or other toxic metals will not leak.

The quantum dots are called cadmium-free quantum (CFQD). As it is spread in many parts of the world today a ban or restriction from the use of heavy metal heavy metals within the production of many household requirements, which means that the quantum points based on cadmium become unused within the applications of goods used in supplies for homes.

For commercial viability, many quantum dots were developed that are free of heavy metals and that show bright emissions in the visible and near-infrared spectroscopy. Besides, they have the same visual properties as the quantum dots that are used in their manufacture of cadmium selenide.

There are some documents of the use of such as phosphorous nanoparticles for targeting and bioimaging. Besides, the ZnS coating layer presents a lot of positive effects than different materials because it can minimize the toxicity of metal by limiting the dissolution of free ions as well as, stop the reaction of the CdSe core, finally increased photostability of QDs. At an equivalent time, the QD core size has not altered whereas the ZnS cover layer is directly developed on the core surface. Thus, the luminescence properties of the QDs square measure maintained principally, and solely a little shift (less than five nm) is detected within the most visible radiation wavelength. redoubled use of biocompatible purposeful teams or coatings will confer desired important activities on the core QDs-shell [49]. Fresh synthesis of quantitative points makes them unsuitable for biological use, due to the hydrophobic coverage on the surface of mineral cores throughout the fixation in organic solvents. Fresh factory-made QDs are useful or given secondary coatings to enhance water solubility, primary sturdiness, and suspension properties, creating them biologically compatible [49]. For instance, QD nuclei are often coated with water-resistant polyethylene glycol (PEG), QDs with sensible biocompatibility associate degreed dispersion in a liquid solution; they will even be plus bioactive compounds to focus on cellular structural options or specific biological issues [50]. Consequently, associations with different molecular entities will operate QD nuclei for specific therapeutic or diagnostic functions. These ways of operation usually embody static interactions, valence bonding, and multi-element heavy metal removal in mind for stability/durability and in vivo interaction with QD reactions [50].

5.   Carbon quantum dots (CDs)

Carbon dots, also called carbon nanoparticles, are a new type of material that has shown great application potential [51]. Accidentally carbon dots were discovered for the first time by Shaw and his colleagues in 2004 while purifying single-walled carbon nanotubes by electrophoresis. Sun and his colleagues were able to prepare carbon dots from graphite powder using a laser as a thermal source [51]. Carbon dots are usually divided into graphene carbon dots, quantum carbon dots, and polymeric dots. The main difference between these types of carbon materials is their structure. The quantum dots have a monolayer carbon core bound to the surface of chemical functional groups, whereas the polymeric dots are aggregated [52].

Unlike conventional quantum dots, this type of material is not prepared from SC, and instead, small organic molecules are condensed together to form nanoparticles smaller than 10 nanometres. Quantum carbon dots seem to have properties similar to the traditional SC in terms of high fluorescence and photoresist, despite their different chemical nature [53]. In addition to the above, the carbon points are characterized by their low toxicity biocompatibility, as well as the stability of their optical properties and the low cost of preparing them [53].

5.1   The mechanism of photoluminescence in quantum carbon dots

carbon core states

Luminescence is light emission from certain materials at a relatively low temperature that is, light emission is not a result of heat. It is a light that differs from the light emitted by incandescent or glowing objects resulting from the burning of wood or coal, or the light from molten iron or from wire through which an electric current pass through it [54]. Luminescence can be seen in neon or fluorescent lights, on television and in some organic materials such as luminol or luciferins, a chemical compound responsible for living things that glow in the deep sea, as well as some organic dyes used in billboards [55]. Light is emitted in all of these phenomena, not due to heat, and this is why the luminescent phenomenon is sometimes known as cold light. The idea of glowing materials lies in their ability to convert invisible energy forms into visible light [56].

Luminescence emission occurs when a suitable substance absorbs energy from a source such as ultraviolet light, X-ray, electron beam, or by chemical reactions, among others [56]. This energy causes the atoms of the matter to be excited by transferring them from the ground state level to the excited state energy level, and where the excited energy level is unstable, another transition occurs in which the excited atoms return to the ground level and the amount of energy is released in the form of light or heat or both. Excitation involves the electrons of the outermost orbit of an atom. The luminescence depends on how much the excitation energy converts into the light, and there are a few appropriate substances that have sufficient luminescence efficiency to be used in practical applications [57].

Several types of luminescence can be distinguished according to the excitation energy source. When the light energy comes from a chemical reaction, as in the case of slow oxidation of phosphorus at normal temperatures, this type is called chemical luminescence, chemiluminescence [58]. When a chemical reaction results in light that results in living systems, as in the glow of living creatures in the depths of the sea and oceans, light is called bioluminescence [58]. Other types of luminescence arise from the flow of some forms of energy from outside the body into it. Depending on the source of this stimulated energy, photoluminescence is described as cathodoluminescence if the energy comes from ejaculation, and radioluminescence is excited in the excitation state by X-ray or gamma-rays. If the excitement is by ultraviolet light, visible or infrared, it is called photoluminescence, and it is called electroluminescence if the excitation source is an electric field [59]. Fluorescence, phosphorescence, and photoluminescence occur when the sample is excited by photon absorption and then excitation photons are emitted when the excited atoms relax with time. The term fluorescence is used by chemists to absorb and emit atoms and molecules [59]. The phosphorescence process is very similar to the fluorescent process, but the time between absorption and emission is much longer than fluorescent. The process of photoluminescence is used by physicists to describe the process of absorption and emission of light by materials such as SC and carbon nanotubes [60]. Regardless of the term used, when the sample absorbs photons and then the emission process occurs at different wavelengths it is monitored by a spectrograph with a CCD chip and the information, we get gives us information about the sample.

6.   Chemical and biological applications of nanoparticle dots

Quantum dots play a very important role in optical applications, particularly because of their high damping or extinguishing coefficient [60]. The ability to adjust the quantum dot volumes plays a distinctive role, useful in many applications. Whereas, the larger QDs have a more significant spectral shift towards red compared to the smaller dots, besides that they show less clear quantum properties [61]. In contrast, smaller QDs in their particle sizes allow taking advantage of the most subtle quantum effects. Quantum dots have higher transitional and optical properties, as well as an opportunity to research them for use in the manufacture of a diode laser, amplifiers, and biometric sensors [62]. Quantum points can be triggered within the domestically supported electromagnetic domain generated by Au-NPs, which can be noticed from the echo of the surface of the plasmon in the spectral excitation spectrum of crystals (cadmium selenide) zinc sulfide nanoparticles [62]. High-quality QDs are appropriate for optical coding applications because of their multiple agitation features and narrow/symmetric emission spectra. This new generation of QDs provides a long-term possibility for studying intracellular processes at the individual molecular level, in addition to high-quality and high-quality cell imaging observed in the long-term (Vivo) biological viability of cellular passage, as well as targeting tumors and various diagnostic processes [63].

Many types of organic pigments are used in the field of modern biological analysis. Despite this, with each passing new year, such pigment is required more flexibility, and the traditional pigment is not able to meet these expectations [63]. In many respects, perhaps one of the most obvious currently is its luster (due to the quenching coefficient associated with the quantitative return compared to the fluorescent pigment in addition to the property of its stability and stability and it was assessed that the quantum points are 20 times more luminous and more stable 100 times than the traditional fluorescent counterpart [64]. To track a single particle, the irregular flash of quantum points represents a slight defect. This has been a major advance in the use of quantum dots in high-sensitivity cellular imaging over the past ten years. Another model for applications that take advantage of the unique optical stabilization feature of a quantum dot probe is tracking the actual time of particle movement and cellular activity over long periods. With the previous feature, the investigators were eligible to monitor the quantum points in the lymph nodes (LN) of small animals like mice for over 120 days. Also, quantum semiconductor points were used in the biological imaging (in the vital) process of cells before classification [65]. This ability to visualize individual cell immigration in real-time is anticipated to be especially significant in many fields of research including embryology, metastatic cancer, stem cell treatments, and lymphocyte immunity. Recent studies have proven that quantum points are more reliable than methods currently present in the siRNA connected the interlaced short RNA into the cell. The nanocrystals of cadmium selenide are extremely lethal to cells grown under ultraviolet (UV) [66]. The nanocrystals of cadmium selenide are acutely virulent to cells grown under UV enlightenment. The intensity from UV radiation is also approaching to that of the covalently of the cadmium selenide nanocrystals. As a proceeding, SC molecules can be dismissed in a refining referred to photolysis, to release noxious cadmium ions into the intermediate stream [66]. With the nonexistence of ultraviolet rays, nevertheless, quantum dots with stable polymer casings were observed to be substantially non-toxic.

Materials Research Forum LLC
https://doi.org/10.21741/9781644901250-13

Researchers have studied quantitative points in both solar cells, light-emitting diodes, and dyes used in pathological imaging, and scientists expect to apply them in the future in the field of quantum computing. Quantum nanoparticles can be negatively injected into cells, and are designed to bond and self-assemble the desired enzymes, and then activate these enzymes based on the orders sent by using certain wavelengths of light [67]. There is a recent study confirmed that theoretically the Hiezenberg model can be considered a good model in describing these dots, and it can use this model to identify the physical properties of these materials such as ground energy, excitation energies, and magnetization [68]. This study used the spin-wave theory in dealing with the Hiezenberg model, as it is characterized by the ease of mathematical manipulation and the speed of obtaining results. It is also an effective theory, especially in cases of ferromagnetic materials. The researchers studied an infinite number of quantum dots containing an unlimited number of electrons supporting the impact ofan external magnetic field by treating them with the Heisenberg model within the spindle wave theory [69].

Conclusions

Quantum dots have great importance, due to many of their unique properties, which include optical and electrical non-linear properties. The quantum dots are characterized by the phenomenon of quantitative restriction, where the movement of electrons is restricted in the three dimensions. Besides, these dots have the advantages and characteristics of nanomaterials. These properties are partly provoked by the unusual increment in the surface ratio of these particles compared to their size. The characteristics of these dots also, the possibility of emitting light when excited, the largest light emission energy is attributed to the smaller dots. This ability allows making dots that emit rainbow colors, allowing them to be used as biological sensors. These dots can be used in many areas, which are a few nanometers in size, have a behavior similar to molecules. As well as we can modify or change the shape, size, and the number of electrons of those materials. This means that we can modify their electrical or optical properties to accommodate many targets such as modern display technology, photovoltaics, photocatalysts, and biomedical applications.

References

[1]   A. Aboulaich, D. Billaud, M. Abyan, L. Balan, J.-J. Gaumet, G. Medjadhi, J. Ghanbaja, R. Schneider, One-pot noninjection route to CdS quantum dots via hydrothermal synthesis, ACS Appl. Mater. Interfaces. 4 (2012) 2561–2569. https://doi.org/10.1021/am300232z

[2]   F.C. Adams, C. Barbante, Nanoscience, nanotechnology and spectrometry, Spectrochim. Acta Part B At. Spectrosc. 86 (2013) 3–13. https://doi.org/10.1016/j.sab.2013.04.008

[3]   C. Adlhart, J. Verran, N.F. Azevedo, H. Olmez, M.M. Keinänen-Toivola, I. Gouveia, L.F. Melo, F. Crijns, Surface modifications for antimicrobial effects in the healthcare setting: A critical overview, J. Hosp. Infect. 99 (2018) 239–249. https://doi.org/10.1016/j.jhin.2018.01.018

[4]   Á. Andrade-Eiroa, M. Canle, V. Cerdá, Environmental applications of excitation-emission spectrofluorimetry: an in-depth review I, Appl. Spectrosc. Rev. 48 (2013) 1–49. https://doi.org/10.1080/05704928.2012.692105

[5]   M. Bawendi, K.F. Jensen, B.O. Dabbousi, J. Rodriguez-Viejo, F.V. Mikulec, Highly luminescent color selective nanocrystalline materials, (2005).

[6]   D. Bera, L. Qian, T.-K. Tseng, P.H. Holloway, Quantum dots and their multimodal applications: a review, Materials (Basel). 3 (2010) 2260–2345. https://doi.org/10.3390/ma3042260

[7]   S. Berardi, S. Drouet, L. Francas, C. Gimbert-Suriñach, M. Guttentag, C. Richmond, T. Stoll, A. Llobet, Molecular artificial photosynthesis, Chem. Soc. Rev. 43 (2014) 7501–7519. https://doi.org/10.1039/C3CS60405E

[8]   V. Biju, T. Itoh, A. Anas, A. Sujith, M. Ishikawa, Semiconductor quantum dots and metal nanoparticles: syntheses, optical properties, and biological applications, Anal. Bioanal. Chem. 391 (2008) 2469–2495. https://doi.org/10.1007/s00216-008-2185-7

[9]   J.B. Blanco-Canosa, M. Wu, K. Susumu, E. Petryayeva, T.L. Jennings, P.E. Dawson, W.R. Algar, I.L. Medintz, Recent progress in the bioconjugation of quantum dots, Coord. Chem. Rev. 263 (2014) 101–137. https://doi.org/10.1016/j.ccr.2013.08.030

[10]   G. Blasse, B.C. Grabmaier, A general introduction to luminescent materials, in: Lumin. Mater., Springer, 1994: pp. 1–9. https://doi.org/10.1007/978-3-642-79017-1_1

[11]   P.C.J. Clark, H. Radtke, A. Pengpad, A.I. Williamson, B.F. Spencer, S.J.O. Hardman, M.A. Leontiadou, D.C.J. Neo, S.M. Fairclough, A.A.R. Watt, The passivating effect of cadmium in PbS/CdS colloidal quantum dots probed by nm-scale depth profiling, Nanoscale. 9 (2017) 6056–6067. https://doi.org/10.1039/C7NR00672A

[12]   S. Coe-Sullivan, J.S. Steckel, W. Woo, M.G. Bawendi, V. Bulović, Large-area ordered quantum-dot monolayers via phase separation during spin-casting, Adv. Funct. Mater. 15 (2005) 1117–1124. https://doi.org/10.1002/adfm.200400468

[13]   C.P. Collier, T. Vossmeyer, A.J.R. Heath, Nanocrystal superlattices, Annu. Rev. Phys. Chem. 49 (1998) 371–404. https://doi.org/10.1146/annurev.physchem.49.1.371

[14]   A. Das, P.T. Snee, Synthetic developments of nontoxic quantum dots, ChemPhysChem. 17 (2016) 598–617. https://doi.org/10.1002/cphc.201500837

[15]   C. Delerue, M. Lannoo, Nanostructures: theory and modelling. 2004, (n.d.). https://doi.org/10.1007/978-3-662-08903-3_6

[16]   T. Desai, R.H. Daniels, V. Sahi, Medical device applications of nanostructured surfaces, (2010).

[17]   A.I. Ekimov, A.A. Onushchenko, Quantum size effect in three-dimensional microscopic semiconductor crystals, Jetp Lett. 34 (1981) 345–349.

[18]   C. Feldmann, T. Jüstel, C.R. Ronda, P.J. Schmidt, Inorganic luminescent materials: 100 years of research and application, Adv. Funct. Mater. 13 (2003) 511–516. https://doi.org/10.1002/adfm.200301005

[19]   K.A.S. Fernando, S. Sahu, Y. Liu, W.K. Lewis, E.A. Guliants, A. Jafariyan, P. Wang, C.E. Bunker, Y.-P. Sun, Carbon quantum dots and applications in photocatalytic energy conversion, ACS Appl. Mater. Interfaces. 7 (2015) 8363–8376. https://doi.org/10.1021/acsami.5b00448

[20]   A.M. Fox, Fundamentals of Semiconductors: Physics and Materials Properties, 4th Edn., by Peter Y. Yu, Manuel Cardona: Scope: manual. Level: postgraduate, (2012). https://doi.org/10.1080/00107514.2012.661781

[21]   C. Frigerio, D.S.M. Ribeiro, S.S.M. Rodrigues, V.L.R.G. Abreu, J.A.C. Barbosa, J.A. V Prior, K.L. Marques, J.L.M. Santos, Application of quantum dots as analytical tools in automated chemical analysis: a review, Anal. Chim. Acta. 735 (2012) 9–22. https://doi.org/10.1016/j.aca.2012.04.042

[22]   M. Fu, F. Ehrat, Y. Wang, K.Z. Milowska, C. Reckmeier, A.L. Rogach, J.K. Stolarczyk, A.S. Urban, J. Feldmann, Carbon dots: a unique fluorescent cocktail of polycyclic aromatic hydrocarbons, Nano Lett. 15 (2015) 6030–6035. https://doi.org/10.1021/acs.nanolett.5b02215

[23]   B.N.G. Giepmans, S.R. Adams, M.H. Ellisman, R.Y. Tsien, The fluorescent toolbox for assessing protein location and function, Science (80-.). 312 (2006) 217–224. https://doi.org/10.1126/science.1124618

[24]   I.A. Gorbachev, I.Y. Goryacheva, E.G. Glukhovskoy, Investigation of multilayers structures based on the Langmuir-Blodgett films of CdSe/ZnS quantum dots, Bionanoscience. 6 (2016) 153–156. https://doi.org/10.1007/s12668-016-0194-0

[25]   C. Han, N. Zhang, Y.-J. Xu, Structural diversity of graphene materials and their multifarious roles in heterogeneous photocatalysis, Nano Today. 11 (2016) 351–372. https://doi.org/10.1016/j.nantod.2016.05.008

[26]   X. He, Y. Song, Y. Yu, B. Ma, Z. Chen, X. Shang, H. Ni, B. Sun, X. Dou, H. Chen, Quantum light source devices of In (Ga) As semiconductorself-assembled quantum dots, J. Semicond. 40 (2019) 71902.

[27]   C.-Y. Hsieh, P. Hawrylak, Quantum circuits based on coded qubits encoded in chirality of electron spin complexes in triple quantum dots, Phys. Rev. B. 82 (2010) 205311. https://doi.org/10.1103/PhysRevB.82.205311

[28]   C.-Y. Hsieh, A. Rene, P. Hawrylak, Herzberg circuit and Berry's phase in chirality-based coded qubit in a triangular triple quantum dot, Phys. Rev. B. 86 (2012) 115312. https://doi.org/10.1103/PhysRevB.86.115312

[29]   D.L. Huffaker, G. Park, Z. Zou, O.B. Shchekin, D.G. Deppe, 1.3 μm room-temperature GaAs-based quantum-dot laser, Appl. Phys. Lett. 73 (1998) 2564–2566. https://doi.org/10.1063/1.122534

[30]   A.M. Jawaid, S. Chattopadhyay, D.J. Wink, L.E. Page, P.T. Snee, Cluster-seeded synthesis of doped CdSe: Cu4 quantum dots, ACS Nano. 7 (2013) 3190–3197. https://doi.org/10.1021/nn305697q

[31]   X. Jiang, Q. Xu, S.K.W. Dertinger, A.D. Stroock, T. Fu, G.M. Whitesides, A general method for patterning gradients of biomolecules on surfaces using microfluidic networks, Anal. Chem. 77 (2005) 2338–2347. https://doi.org/10.1021/ac048440m

[32]   T. Jüstel, H. Nikol, C. Ronda, New developments in the field of luminescent materials for lighting and displays, Angew. Chemie Int. Ed. 37 (1998) 3084–3103.

[33]   P. Juzenas, W. Chen, Y.-P. Sun, M.A.N. Coelho, R. Generalov, N. Generalova, I.L. Christensen, Quantum dots and nanoparticles for photodynamic and radiation therapies of cancer, Adv. Drug Deliv. Rev. 60 (2008) 1600–1614. https://doi.org/10.1002/(SICI)1521-3773

[34]   C.W. Lee, C.H. Chou, J.H. Huang, C.S. Hsu, T.-P. Nguyen, Investigations of organic light emitting diodes with CdSe (ZnS) quantum dots, Mater. Sci. Eng. B. 147 (2008) 307–311. https://doi.org/10.1016/j.mseb.2007.09.068

[35]   D. Leonard, K. Pond, P.M. Petroff, Critical layer thickness for self-assembled InAs islands on GaAs, Phys. Rev. B. 50 (1994) 11687. https://doi.org/10.1103/PhysRevB.50.11687

[36]   J.W. Lichtman, J.-A. Conchello, Fluorescence microscopy, Nat. Methods. 2 (2005) 910–919. https://doi.org/10.1038/nmeth817

[37] S.A. Lim, M.U. Ahmed, Electrochemical immunosensors and their recent nanomaterial-based signal amplification strategies: a review, RSC Adv. 6 (2016) 24995–25014. https://doi.org/10.1039/C6RA00333H

[38] S.Y. Lim, W. Shen, Z. Gao, Carbon quantum dots and their applications, Chem. Soc. Rev. 44 (2015) 362–381. https://doi.org/10.1039/C4CS00269E

[39] S.S. Lucky, K.C. Soo, Y. Zhang, Nanoparticles in photodynamic therapy, Chem. Rev. 115 (2015) 1990–2042. https://doi.org/10.1021/cr5004198

[40] J.R. Manders, D. Bera, L. Qian, P.H. Holloway, Quantum dots for displays and solid state lighting, in; A. Kitai (Ed.) Materials for Solid State Lighting and Displays. (2017) 31–90.

[41] L. Mangolini, U. Kortshagen, Plasma-assisted synthesis of silicon nanocrystal inks, Adv. Mater. 19 (2007) 2513–2519. https://doi.org/10.1002/adma.200700595

[42] L. Mangolini, E. Thimsen, U. Kortshagen, High-yield plasma synthesis of luminescent silicon nanocrystals, Nano Lett. 5 (2005) 655–659. https://doi.org/10.1021/nl050066y

[43] Y. Masumoto, T. Takagahara, Semiconductor quantum dots: physics, spectroscopy and applications, Springer Science & Business Media, 2013.

[44] Cb. Murray, D.J. Norris, M.G. Bawendi, Synthesis and characterization of nearly monodisperse CdE (E= sulfur, selenium, tellurium) semiconductor nanocrystallites, J. Am. Chem. Soc. 115 (1993) 8706–8715. https://doi.org/10.1021/ja00072a025

[45] K.V.R. Murthy, H.S. Virk, Luminescence phenomena: an introduction, in: Defect Diffus. Forum, Trans Tech Publ, 2014: pp. 1–34. https://doi.org/10.4028/www.scientific.net/DDF.347.1

[46] Z. Ni, X. Pi, M. Ali, S. Zhou, T. Nozaki, D. Yang, Freestanding doped silicon nanocrystals synthesized by plasma, J. Phys. D. Appl. Phys. 48 (2015) 314006. https://doi.org/10.1088/0022-3727/48/31/314006

[47] P.G. Nicholson, F.A. Castro, Organic photovoltaics: principles and techniques for nanometre scale characterization, Nanotechnology. 21 (2010) 492001. https://doi.org/10.1088/0957-4484/21/49/492001

[48] H. Pan, S. Zhu, X. Lou, L. Mao, J. Lin, F. Tian, D. Zhang, Graphene-based photocatalysts for oxygen evolution from water, RSC Adv. 5 (2015) 6543–6552. https://doi.org/10.1039/C4RA09546D

[49] X. Peng, L. Manna, W. Yang, J. Wickham, E. Scher, A. Kadavanich, A.P. Alivisatos, Shape control of CdSe nanocrystals, Nature. 404 (2000) 59–61.

Materials Research Forum LLC
https://doi.org/10.21741/9781644901250-13

[50]   X. Pi, T. Yu, D. Yang, Water-dispersible silicon-quantum-dot-containing micelles self-assembled from an amphiphilic polymer, Part. Part. Syst. Charact. 31 (2014) 751–756. https://doi.org/10.1002/ppsc.201300346

[51]   X.D. Pi, U. Kortshagen, Nonthermal plasma synthesized freestanding silicon–germanium alloy nanocrystals, Nanotechnology. 20 (2009) 295602. https://doi.org/10.1088/0957-4484/20/29/295602

[52]   K.E. Sapsford, T. Pons, I.L. Medintz, H. Mattoussi, Biosensing with luminescent semiconductor quantum dots, Sensors. 6 (2006) 925–953. https://doi.org/10.3390/s6080925

[53]   P. Senellart, G. Solomon, A. White, High-performance semiconductor quantum-dot single-photon sources, Nat. Nanotechnol. 12 (2017) 1026. https://doi.org/10.1038/nnano.2017.218

[54]   N. Sharma, H. Ojha, A. Bharadwaj, D.P. Pathak, R.K. Sharma, Preparation and catalytic applications of nanomaterials: a review, Rsc Adv. 5 (2015) 53381–53403. https://doi.org/10.1039/C5RA06778B

[55]   E. Song, M. Yu, Y. Wang, W. Hu, D. Cheng, M.T. Swihart, Y. Song, Multi-color quantum dot-based fluorescence immunoassay array for simultaneous visual detection of multiple antibiotic residues in milk, Biosens. Bioelectron. 72 (2015) 320–325. https://doi.org/10.1016/j.bios.2015.05.018

[56]   A. Srivastava, M. Sidler, A. V Allain, D.S. Lembke, A. Kis, A. Imamoğlu, Optically active quantum dots in monolayer WSe 2, Nat. Nanotechnol. 10 (2015) 491. https://doi.org/10.1038/nnano.2015.60

[57]   I.N. Stranski, L. Krastanow, Zur Theorie der orientierten Ausscheidung von Ionenkristallen aufeinander, Monatshefte Für Chemie Und Verwandte Teile Anderer Wissenschaften. 71 (1937) 351–364. https://doi.org/10.1007/BF01798103

[58]   D. V Talapin, A.L. Rogach, M. Haase, H. Weller, Evolution of an ensemble of nanoparticles in a colloidal solution: theoretical study, J. Phys. Chem. B. 105 (2001) 12278–12285. https://doi.org/10.1021/jp012229m

[59]   R.R. Turnbull, R.C. Knapp, J.K. Roberts, Illuminator assembly incorporating light emitting diodes, (1998).

[60]   B. Valeur, M.N. Berberan-Santos, A brief history of fluorescence and phosphorescence before the emergence of quantum theory, J. Chem. Educ. 88 (2011) 731–738. https://doi.org/10.1021/ed100182h

[61]   A. Valizadeh, H. Mikaeili, M. Samiei, S.M. Farkhani, N. Zarghami, A. Akbarzadeh, S. Davaran, Quantum dots: synthesis, bioapplications, and toxicity, Nanoscale Res. Lett. 7 (2012) 480. https://doi.org/10.1186/1556-276X-7-480

Materials Research Forum LLC
https://doi.org/10.21741/9781644901250-13

[62]   V. Vatanpour, N. Zoqi, Surface modification of commercial seawater reverse osmosis membranes by grafting of hydrophilic monomer blended with carboxylated multiwalled carbon nanotubes, Appl. Surf. Sci. 396 (2017) 1478–1489. https://doi.org/10.1016/j.apsusc.2016.11.195

[63]   K. Vinothini, M. Rajan, Mechanism for the nano-based drug delivery system, in: Charact. Biol. Nanomater. Drug Deliv., Elsevier, 2019: pp. 219–263.

[64]   F. Wang, V.N. Richards, S.P. Shields, W.E. Buhro, Kinetics and mechanisms of aggregative nanocrystal growth, Chem. Mater. 26 (2014) 5–21. https://doi.org/10.1021/cm402139r

[65]   P.N. Wiecinski, K.M. Metz, T.C. King Heiden, K.M. Louis, A.N. Mangham, R.J. Hamers, W. Heideman, R.E. Peterson, J.A. Pedersen, Toxicity of oxidatively degraded quantum dots to developing zebrafish (Danio rerio), Environ. Sci. Technol. 47 (2013) 9132–9139. https://doi.org/10.1021/es304987r

[66]   Z. Xiao, D. Liu, Z. Tang, H. Li, M. Yuan, Synthesis and characterization of poly (lactic acid)-conjugated CdTe quantum dots, Mater. Lett. 148 (2015) 126–129. https://doi.org/10.1016/j.matlet.2015.01.164

[67]   Y. Xing, J. Rao, Quantum dot bioconjugates for in vitro diagnostics & in vivo imaging, Cancer Biomarkers. 4 (2008) 307–319. https://doi.org/10.3233/CBM-2008-4603

[68]   C. Zhai, H. Zhang, N. Du, B. Chen, H. Huang, Y. Wu, D. Yang, One-pot synthesis of biocompatible CdSe/CdS quantum dots and their applications as fluorescent biological labels, Nanoscale Res Lett. 6 (2011) 31. https://doi.org/10.1007/s11671-010-9774-z

[69]   S. Zhu, Y. Song, J. Shao, X. Zhao, B. Yang, Non-conjugated polymer dots with crosslink-enhanced emission in the absence of fluorophore units, Angew. Chemie Int. Ed. 54 (2015) 14626–14637. https://doi.org/10.1002/anie.201504951

Keyword Index

About the Editors

Dr. Inamuddin is working as Assistant Professor at the Department of Applied Chemistry, Aligarh Muslim University, Aligarh, India. He obtained Master of Science degree in Organic Chemistry from Chaudhary Charan Singh (CCS) University, Meerut, India, in 2002. He received his Master of Philosophy and Doctor of Philosophy degrees in Applied Chemistry from Aligarh Muslim University (AMU), India, in 2004 and 2007, respectively. He has extensive research experience in multidisciplinary fields of Analytical Chemistry, Materials Chemistry, and Electrochemistry and, more specifically, Renewable Energy and Environment. He has worked on different research projects as project fellow and senior research fellow funded by University Grants Commission (UGC), Government of India, and Council of Scientific and Industrial Research (CSIR), Government of India. He has received Fast Track Young Scientist Award from the Department of Science and Technology, India, to work in the area of bending actuators and artificial muscles. He has completed four major research projects sanctioned by University Grant Commission, Department of Science and Technology, Council of Scientific and Industrial Research, and Council of Science and Technology, India. He has published 177 research articles in international journals of repute and nineteen book chapters in knowledge-based book editions published by renowned international publishers. He has published 115 edited books with Springer (U.K.), Elsevier, Nova Science Publishers, Inc. (U.S.A.), CRC Press Taylor & Francis Asia Pacific, Trans Tech Publications Ltd. (Switzerland), IntechOpen Limited (U.K.), Wiley-Scrivener, (U.S.A.) and Materials Research Forum LLC (U.S.A). He is a member of various journals' editorial boards. He is also serving as Associate Editor for journals (Environmental Chemistry Letter, Applied Water Science and Euro-Mediterranean Journal for Environmental Integration, Springer-Nature), Frontiers Section Editor (Current Analytical Chemistry, Bentham Science Publishers), Editorial Board Member (Scientific Reports-Nature), Editor (Eurasian Journal of Analytical Chemistry), and Review Editor (Frontiers in Chemistry, Frontiers, U.K.) He is also guest-editing various special thematic special issues to the journals of Elsevier, Bentham Science Publishers, and John Wiley & Sons, Inc. He has attended as well as chaired sessions in various international and national conferences. He has worked as a Postdoctoral Fellow, leading a research team at the Creative Research Initiative Center for Bio-Artificial Muscle, Hanyang University, South Korea, in the field of renewable energy, especially biofuel cells. He has also worked as a Postdoctoral Fellow at the Center of Research Excellence in Renewable Energy, King Fahd University of Petroleum and Minerals, Saudi Arabia, in the field of polymer electrolyte membrane fuel cells and computational fluid dynamics of polymer electrolyte membrane fuel cells. He is a life member of the Journal of the Indian

Chemical Society. His research interest includes ion exchange materials, a sensor for heavy metal ions, biofuel cells, supercapacitors and bending actuators.

Dr. Tauseef Ahmad Rangreez is working as a postdoctoral fellow at National Institute of Technology, Srinagar, India. He completed his Ph.D in Applied Chemistry, from Aligarh Muslim University, Aligarh, India on the topic "Development of Nanostructure Organic-Inorganic Composite Materials based Sensors for Inorganic Pollutants". He worked as a Project Fellow under the UGC Funded Research Project entitled "Development of Nanostructured Conductive Organic Inorganic Composite Materials based sensors Functionalities for Organic and Inorganic Pollutants". He completed his Masters in Chemistry from Jamia Hamdard, New Delhi. He has published several research articles of international repute. He has edited various books with Springer and Materials Research Forum LLC, U.S.A. His research interest includes ion exchange chromatography, development of nanocomposite sensors for heavy metals and biosensors.

Dr. Mohammad Faraz Ahmer is presently working as Assistant Professor in the Department of Electrical Engineering, Mewat Engineering College, Nuh Haryana, India, since 2012 after working as Guest Faculty in University Polytechnic, Aligarh Muslim University Aligarh, India, during 2009-2011. He completed M.Tech. (2009) and Bachelor of Engineering (2007) degrees in Electrical Engineering from Aligarh Muslim University, Aligarh in the first division. He obtained a Ph.D. degree in 2016 on his thesis entitled "Studies on Electrochemical Capacitor Electrodes". He has published six research papers in reputed scientific journals. He has edited two books with Materials Research Forum, U.S.A. His scientific interests include electrospun nano-composites and supercapacitors. He has presented his work at several conferences. He is actively engaged in searching of new methodologies involving the development of organic composite materials for energy storage systems.

Dr. Rajender Boddula is currently working with Chinese Academy of Sciences-President's International Fellowship Initiative (CAS-PIFI) at National Center for Nanoscience and Technology (NCNST, Beijing). He obtained Master of Science in Organic Chemistry from Kakatiya University, Warangal, India, in 2008. He received his Doctor of Philosophy in Chemistry with the highest honours in 2014 for the work entitled "Synthesis and Characterization of Polyanilines for Supercapacitor and Catalytic Applications" at the CSIR-Indian Institute of Chemical Technology (CSIR-IICT) and Kakatiya University (India). Before joining National Center for Nanoscience and Technology (NCNST) as CAS-PIFI research fellow, China, worked as senior research associate and Postdoc at National Tsing-Hua University (NTHU, Taiwan) respectively in the fields of bio-fuel and CO_2 reduction applications. His academic honors include University Grants Commission National Fellowship and many merit scholarships, study-abroad fellowships from Australian Endeavour Research Fellowship, and CAS-PIFI. He has published many scientific articles in international peer-reviewed journals and has authored around twenty book chapters, and he is also serving as an editorial board member and a referee for reputed international peer-reviewed journals. He has published edited books with Springer (UK), Elsevier, Materials Research Forum LLC (USA), Wiley-Scrivener, (U.S.A.) and CRC Press Taylor & Francis group. His specialized areas of research are energy conversion and storage, which include sustainable nanomaterials, graphene, polymer composites, heterogeneous catalysis for organic transformations, environmental remediation technologies, photoelectrochemical water-splitting devices, biofuel cells, batteries and supercapacitors.

www.ingramcontent.com/pod-product-compliance
Lightning Source LLC
Chambersburg PA
CBHW071320210326
41597CB00015B/1293